This book belongs to

Name: _____

www.math-knots.com

Cover Design by :
Gowri Vemuri

First Edition :
March , 2023

Second Edition:
April,2023

Author :
Gowri Vemuri

Edited by :
Raksha Pothapragada
Ritvik Pothapragada

Questions: mathknots.help@gmail.com

NOTE : CCSSO or NCTM is neither affiliated nor sponsors or endorses this product.

Dedication

This book is dedicated to:

My Mom, who is my best critic, guide and supporter.

To what I am today, and what I am going to become tomorrow,

is all because of your blessings, unconditional affection and support.

This book is dedicated to the

strongest women of my life ,

my dearest mom

and

to all those moms in this universe.

G.V.

         www.math-knots.com

 **Algebra 1**

Math-Knots Grade level enrichment series covers all pre-K to Grade 10 common core math work books.

The Grade 6 math work book is aligned to common core curriculum and more , challenge level questions are included. Content is divided across weeks aligned to school calendar year.
Six assessments are provided based on the topics covered in the previous weeks. The assessments help them to identify learning gaps students can redo the previous weeks content to bridge their learning gaps as needed. End of the year assessment is provided online at the below URL.
www.a4ace.com

For more practice ,you can also subscribe at www.a4ace.com .
All practice sets, and assessments can be taken any number of times within the one-year subscription period.
All our content is written by industry experts with over 30 years of experience.
A4ace.com is part of Math-Knots LLC
Math-Knots is your one stop enrichment place. ~~Learn to think with us.

Note: Video explanations of content in the book is coming up.

Instructions to register on www.a4ace.com
1. Register as a parent and choose a category while registering.
2. Register the student and choose a category while registering.
3. You can add more categories from the dash board as needed.
4. After registering an automatic email will be sent for verification.
5. Verify the account by clicking on the link sent to your email.
6. Login with your credentials and navigate to the dash board.
7. Click on the free test and follow through the instructions.
8. Free test will navigate to a payment page and it will say $0.
   You just have to click Paypal button and you can take the test.
9. Any issues please reach out to mathknots.help@gmail.com

END OF THE YEAR ASSESSMENT : Register for free test online
www.a4ace.com

Advanced Algebra 1 notes

Expressions	Numerical expression	Algebraic/Variable expression
Expressions will not include equal sign	12^2 $8 + 10$ $4^2 - 16 + 9$	$6x + 17$ $5a^2 - 2b$
Equation	Numerical equation	Algebraic/variable equation
Must include an equal sign; One side is equal to the other side	$9 + 8 = 17$ $5^2 - 11 + 2 = 16$	$6x + 5 = 35$ (x=__?__) $9x^2 = 225$ (x=__?__)

Solving Equations :

To solve an equation first add the like terms (if any) and then isolate the variable on to one side of the equation. To solve one - step equations, do the inverse operation to find the value of the variable.

Remember : Always perform the same operation on both sides of the equation to maintain the balance.
Inverse operation for addition is subtraction and vice versa.
Inverse operation for multiplication is division and vice versa.

Polynomial	Highest degree	1st Name by degree	2nd Name by number of terms
8	0	Constant	monomial
$9x + 8$	1	Linear	binomial
$-2x^2 + 7x - 11$	2	Quadratic	trinomial
$8x^3$	3	Cubic	monomial
$2x^4 + x^3$	4	Quartic	binomial
$3x^5 + 5x^3 + 12$	5	Quintic	trinomial
$2x^6 + x^5 - 3x^4 - 21$	6	6th degree	Polynomial with 4 terms
$5x^7 - x^5 + 4x^4 - 7x^2 + x - 3$	7	7th degree	Polynomial with 6 terms

A Polynomial in standard form is always written with terms in sequential order according to their highest degree of exponents (higher exponents to lower exponents).

Parts of the exponent:

7 is the base

4 is the exponent

This is read as "Seven to the fourth power"

Like Terms :

Two or more terms are said to be alike if they have the same variable and the same degree. Coefficients of like terms are not necessarily be same.

An expression is in its simplest form when

1.   All like terms are combined.
2.   All parentheses are opened and simplified.

Like Terms can combined by adding or subtracting their coefficients (pay attention to the positive and negative signs of the coefficient and apply rules of adding integers)

Combining like terms on the opposite side of the equal sign :

When the like terms are on opposite sides, we have to combine like terms by using the inverse operation and by undoing the equation.

Solving equations using the distributive property :

The number in front of the parentheses needs to be multiplied with every term within the parentheses. After the distribution and opening up the parentheses, combine like terms and solve.

Distributing with the negative sign :

Remember to apply the integer rules of positive and negative numbers while distributing.

$$+ \times + = +$$
$$- \times - = +$$
$$- \times + = -$$
$$+ \times - = -$$

www.math-knots.com | www.a4ace.com

Example 1 : $2x + 3 = x + 7$

$2x + 3 = x + 7$

$\underline{-x \;\; - 3 \;\; -x \;\; -3}$

$x + 0 = 0 + 4$

$x = 4$

> Inverse operation for addition is subtraction

Example 2 : $7x + 5 = -3x + 25$

$7x + 5 = -3x + 25$

$\underline{3x \;\; -5 \;\;\;\; 3x \;\;\;\; -5}$

$10x + 0 = 0 + 20$

$10x = 20$

$\dfrac{\cancel{10}x}{\cancel{10}} = \dfrac{\cancel{20}^2}{\cancel{10}}$

$\boxed{x = 2}$

> Inverse operation for addition is subtraction and vice versa

> Inverse operation for multiplication is division

Example 3 : $\dfrac{2x}{5} + 5 = 15$

$\dfrac{2x}{5} + 5 = 15$

$\underline{\qquad\qquad -5 \quad -5}$

$\dfrac{2x}{5} + 0 = 10$

$\dfrac{2x}{5} = 10$

$\cancel{5} \cdot \dfrac{2x}{\cancel{5}} = 5 \cdot 10$

$\dfrac{\cancel{2}x}{\cancel{2}} = \dfrac{\cancel{50}^{25}}{\cancel{2}}$

$\boxed{x = 25}$

> Inverse operation for addition is subtraction and vice versa

> Inverse operation for division is multiplication

> Inverse operation for multiplication is division

 **Algebra 1**

Inequality :

An inequality is a relation between two expressions that are not equal. As a mathematical statement an inequality states one side of the equation is less than, less than or equal to or greater than or greater than equal to the other side.

If the inequality has **less than** or **greater than** symbol,
1.   The graph starts with the open circle.
2.   For less than the graphing line goes toward the left.
3.   For greater than the graphing line goes toward the right.

If the inequality has **less than or equal to** or **greater than or equal** to symbol,
1.   The graph starts with the closed circle.
2.   For less than or equal to the graphing line goes toward the left.
3.   For greater than or equal to the graphing line goes toward the right.

Inequality statement	Inequality verbal expression	Inequality graph
x > -3 or -3 < x	x is greater than -3	
x < 3 or 3 > x	x is less than 3	
x >= -1 or -1 <= x	x is greater than or equal to -1	
x <= 1 or 1 <= x	x is less than or equal to 1	

Basic inequalities :

Solving inequalities is same as solving for an equation except for one special rule.

14

Compound Inequalities :

x < 0 or x ≥ 5 means all values less than 0 or 5 and more. In other words we are excluding the values 0,1,2,3,4.

Absolute Value :

Absolute value of a number is its distance from 0. Since the distance cannot be negative absolute value is always positive.

$$|7| = 7 \qquad\qquad |-11| = 11$$

$$|2.3| = 2.3 \qquad\qquad |-0.75| = 0.75$$

Equations involving absolute values can be solved similar to regular algebraic equations solving. Absolute value should be treated as parentheses when applying PEDMAS rules.

Steps to solve absolute value equations.

Step 1 : Solve the expression within the absolute value.
       (As applicable with PEDMAS rules)

Step 2 : Isolate the absolute value to one side of the equation.

Step 3 : Verify the value on the other side of the equation.
       If the value is positive move to step 4.
       If the value is negative there is no solution.

Step 4 : The expression inside the absolute value equals to positive and negative values of the other side of the equation.

Step 5 : Make the expression equal to positive value and solve for the variable.

Step 6 : Make the expression equal to negative value and solve for the variable.

Step 7 : The value obtained in step 5 and step 6 or the solutions to the absolute value equation.

Note : Absolute value equations can have two solutions. Since the absolute value of a number and its opposite are the same.
       Absolute value can never be negative.

www.math-knots.com | www.a4ace.com

Absolute value Inequalities :

Absolute value inequalities are similar to absolute value equations.
Absolute value inequalities can have the below solutions
1. Two solutions
2. No solution
3. All real numbers

Steps to solve absolute value Inequalities are similar to solving the absolute value equations.

$|x| < a$ can be rewritten as $-a < x < a$ (where a is positive)
can also be written as $x < a$ **and** $x > -a$

$|x| \leq a$ can be rewritten as $-a \leq x \leq a$ (where a is positive)
can also be written as $x \leq a$ **and** $x \geq -a$

$|x| > a$ can be rewritten as $x > a$ **or** $x < -a$ (where a is positive)

Note : $<$ or $\leq$ are represented by the word **and**
$>$ or $\geq$ are represented by the word **or**

SLOPE :

Slope of a straight line is how far the line is away from the horizontal line in other words how slanted or angular the straight line is.

Slope of a straight line is the rate of change for a given set of points.
Example : $(x_1, y_1), (x_2, y_2)$

$$\text{Slope (m)} = \frac{\text{Rise}}{\text{Run}} = \frac{\text{difference in the y coordinates}}{\text{difference in the x coordinates}} = \frac{y_2 - y_1}{x_2 - x_1} = \frac{y_1 - y_2}{x_1 - x_2}$$

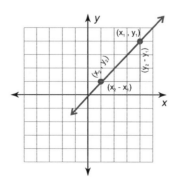

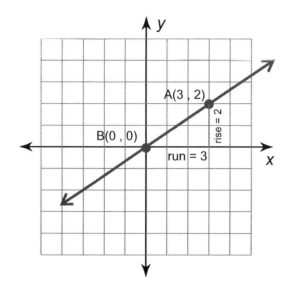

A straight line with a **positive** slope always **rises** from left to right

A(3 , 2)   B(0 , 0)

(x_2 , y_2) (x_1 , y_1)

$$\text{Slope} = \frac{y_2 - y_1}{x_2 - x_1} = \frac{2 - 0}{3 - 0} = \frac{2}{3}$$

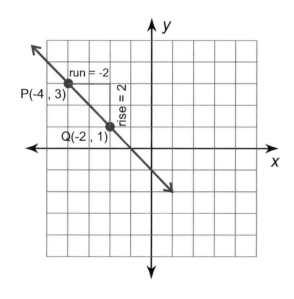

A straight line with a **negative** slope always **falls** from left to right

P(-4 , 3)   Q(-2 , 1)

(x_2 , y_2) (x_1 , y_1)

$$\text{Slope} = \frac{y_2 - y_1}{x_2 - x_1} = \frac{3 - 1}{-4 - (-2)} = \frac{2}{-4 + 2}$$

$$= \frac{2}{-2} = -1$$

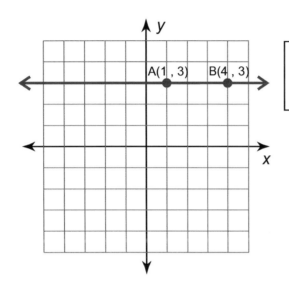

A straight line parallel to x axis always has a slope = 0

A(1 , 3)   B(4 , 3)
(x_2 , y_2) (x_1 , y_1)

$$\text{Slope} = \frac{y_2 - y_1}{x_2 - x_1} = \frac{3 - 3}{1 - 4} = \frac{0}{-3} = 0$$

Note : A straight line of the form y = k, where k is a constant always has a zero slope.

Tip : When x coordinates are different and y coordinates are same slope is always zero.

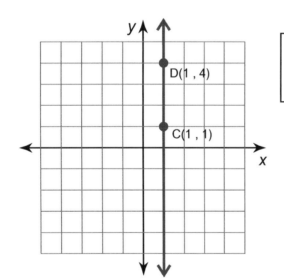

A straight line parallel to y axis always has an undefined slope

C(1 , 1)   D(1 , 4)
(x_2 , y_2) (x_1 , y_1)

$$\text{Slope} = \frac{y_2 - y_1}{x_2 - x_1} = \frac{1 - 4}{1 - 1} = \frac{-3}{0} = \text{undefined}$$

Any number when you divide by 0 the value is undefined.

Note : A straight line of the form x = k, where k is a constant always has an undefined slope.

Tip : When x coordinates are same slope is undefined.

 **Algebra 1**

Slope intercept form of the straight line :

y = mx + b is the slope intercept form of the straight line where m is the slope of the straight line and b is the y intercept.

Examples : y = 7x - 1 ; Slope = 7 , y intercept = -1

y = -11x + 8 ; Slope = -11 , y intercept = 8

Standard form of the straight line :

ax + by = c is the Standard form of the straight line.

Example : 2x + 3y = 10

Point slope form of the straight line :

$y - y_1 = m(x - x_1)$ is the point slope form of the straight line where m is the slope and (x1 , y1) is any given point on the straight line.

Example : y - 2 = 3(x - 5) then slope of the straight line is 3 and (5 , 2) is a given point on the straight line.

Example : Given two points A(2 , 3) and B(5 , 8). Find the equation of a straight line in point slope form. Also express it in slope intercept form and standard form.

Step 1 : Find the slope

$$m = \frac{y_2 - y_1}{x_2 - x_1} = \frac{8 - 3}{5 - 2} = \frac{5}{3}$$

Step 2 : Substitute the slope value obtained in step 1 and any one point A or B in the equation $y - y_1 = m(x - x_1)$

$$y - 3 = \frac{5}{3} (x - 2)$$ Equation of the straight line in point slope form

Step 3 : Simplify the equation obtained in step 2 to rewrite in the form of y = mx + b

$$3 (y - 3) = \frac{5}{\cancel{3}} \cancel{3}.(x - 2)$$

$3y - 9 = 5x - 10$

$3y - 9 + 9 = 5x - 10 + 9$

$$\frac{3y}{3} = \frac{5x - 1}{3}$$

$$\boxed{y = \frac{5}{3}x - \frac{1}{3}}$$ Equation of the straight line in slope intercept form

Step 4 : Ax + By = C is the standard form of the straight line

Lets use the point slope of the straight line obtained in step 2.

$$y - 3 = \frac{5}{3}(x - 2)$$

$$3(y - 3) = \frac{5}{\cancel{3}}\cancel{3}.(x - 2)$$

$3y - 9 = 5x - 10$

$3y - 9 + 10 = 5x - 10 + 10$

$3y + 1 = 5x$

$3y + 1 - 3y = 5x - 3y$

$$\boxed{5x - 3y = 1}$$ Standard form of the straight line.

Important :
To find the x intercept of a given straight line substitute y = 0 and solve the equation. The value obtained is the x intercept.

Important :
To find the y intercept of a given straight line substitute x = 0 and solve the equation. The value obtained is the y intercept.

---

                   www.math-knots.com | www.a4ace.com

Parallel lines :

Two straight lines are parallel if and only if they are of the same slope but the y intercepts are different.

Example : y = 2x + 3 and y = 2x - 7 have the same slope = 3

Perpendicular lines :

Two straight lines are perpendicular if and only if their slopes are negative reciprocals of each other. Their y intercepts can be same or different.

Example : y = 5x + 9 and y = - $\frac{1}{5}$ x + 2 are perpendicular to each other as their slopes

5 and - $\frac{1}{5}$ are negative reciprocals of each other.

Important :
When the product of slopes of two straight lines equals to -1 then the lines are perpendicular to each other

---

Graphing the linear inequalities :

Graphing a linear inequality is similar to graphing any linear equation except ,

   1. If the inequality has ≤ or ≥ use a solid line to graph.

   2. If the inequality has < or > then use a dotted line to graph.

After plotting the solid or dotted line

   1. Test a point outside of the line to check if it is a solution of the given inequality.

   2. If the point tested is a solution then shade that side of the graph otherwise shade the opposite side.

   3. The shaded region and the solid line represents all the possible solutions of the given linear inequality. If the graph has the dotted line then only the shaded region represents all the solutions satisfying the linear inequality.

System of Linear equations :

Two or more linear equations can be solved by using elimination or substitution methods to find the common point or the points that satisfy both the equations.
The common points obtained are the solutions for the system of equations given.

Graphing System of Linear equations :

Two or more linear equations are graphed in the same way as graphing a linear equation.
The point at which the two lines intersect is the solution of the given pair of equations.
The solution can be cross verified by solving the equations as explained above.

Graphing System of Linear Inequalities :

Graphing two or more linear inequalities is similar to graphing a linear inequality and extending to more. The overlapping area of the inequalities graphed is the set of solutions that work for both system of linear inequalities graphed.

Graphing Absolute value of equations :

Graphing an absolute value of equation is similar to graphing an equation for y = absolute value of x consider all the possible order pair solutions to graph.
The point at which the graph turns to another direction is called the **vertex**.
The vertical line of symmetry for any graph is at its vertex.

System of Linear equations :

Two or more linear equations can be solved by using elimination or substitution methods to find the common point or the points that satisfy both the equations.
The common points obtained are the solutions for the system of equations given.

Graphing System of Linear equations :

Two or more linear equations are graphed in the same way as graphing a linear equation.
The point at which the two lines intersect is the solution of the given pair of equations.
The solution can be cross verified by solving the equations as explained above.

Graphing System of Linear Inequalities :

Graphing two or more linear inequalities is similar to graphing a linear inequality and extending to more. The overlapping area of the inequalities graphed is the set of solutions that work for both system of linear inequalities graphed.

Graphing Absolute value of equations :

Graphing an absolute value of equation is similar to graphing an equation for y = absolute value of x consider all the possible order pair solutions to graph.
The point at which the graph turns to another direction is called the **vertex**.
The vertical line of symmetry for any graph is at its vertex.

Polynomials :

A polynomial is an expression or an equation with more than one term containing variables. A polynomial is named

1. Its highest power of a variable
Example : Linear (x), Quadratic (a^2), Cubic (b^3)

2. By the number of terms of the polynomial
Example : Monomial (x), Binomial (a^2 + b), Trinomial (a^3 + b^2 + c)

Polynomial is identified its two parts as described above. Polynomials must be simplified to the lowest possible terms before they are named.

Simplifying Standard form of polynomials :

Step 1 : Regroup the like terms and keep them in parentheses

Step 2 : Add or subtract the like terms as defined by signs preceeding the coefficients

Step 3 : Open the parentheses and make sure all the terms are simplified to the lowest possible.

Step 4 : Write the term with the highest followed by next lowest degree and so on.

This is the standard form of writing a polynomial

Example : $5x^3 + 3x - 6x^2 + 2x - 7x + x^3 + 9$

$$= (5x^3 + x^3) - 6x^2 + (3x + 2x - 7x) + 9$$

$$= 6x^3 - 6x^2 - 2x + 9$$

23   www.math-knots.com | www.a4ace.com

Simplifying expressions :

Simplifying polynomial expressions is similar to simplifying any expressions.

1. Like terms can only be combined.
2. Distribute the numerical value or the variable as necessary or as given
3. Remember the unlike terms can never be combined together.
Example : We cannot add an x to an x^2 , $2a^3$ to a^4 , xy^2 to x^2y
4. Make sure all parentheses are simplified and opened.
5. Make sure everything that can be combined is combined.
6. Remember the sign before the term goes along with the term.
7. While using algebraic method when terms move from one side of the equation to the other side they change the sign before them to their opposite.

Solving proportions using equations :

Step 1 : Cross multiply the numerator of one side of the equation with the denominator in the opposite side also called as butterfly method.

Example : $\dfrac{x+2}{3} \diagdown \diagup \dfrac{3x+5}{5}$

$$5(x + 2) = 3(3x + 5)$$

Step 2 : Distribute where necessary

$$5(x + 10) = 3(3x + 5)$$

$$5x + 5(10) = 3(3x) + 3(5)$$
$$5x + 50 = 9x + 15$$

Step 3 : Bring all the like terms to one side of the equations, combine the like terms and solve using algebraic expressions method.

$$5x + 50 = 9x + 15$$
$$-15 -15 \longrightarrow \text{Solving by algebraic method}$$

$$5x + 35 = 9x$$
$$-5x -5x$$

$$35 = 9x - 5x \longrightarrow \text{Combining like terms}$$
$$35 = 4x$$

$$\dfrac{4x}{4} = \dfrac{35}{4} = \boxed{x = \dfrac{35}{4}}$$

www.math-knots.com | www.a4ace.com

Multiplying binomials :

F.O.I.L acronym stands for "**Front, Outer, Inner, Last**"

F.O.I.L method is used to multiply binomials.
It can also be called as the **extended distributive property**

Lets multiply the binomials (ax + b) and (cx + d)

(ax + b) (cx + d)

Front : Multiplying the first terms of both binomials

(ax + b) (cx + d)

$ax.cx = acx^2$

Outer : Multiplying the first term of the first binomial with the second term (last term) of the second binomial

(ax + b) (cx + d)

ax.d = adx

Inner : Multiplying the second term of the first binomial with the first terms of the second binomial

(ax + b) (cx + d)

b.cx = bcx

Last : Multiplying the second terms (last terms) of both binomials

(ax + b) (cx + d)

b.d = bd

Write down all the terms obtained by using F.O.I.L method (Remember to consider the signs before the terms)
$acx^2 + adx + bcx + bd$

Combining the like terms, it can be simplified as $acx^2 + (ad + bc)x + bd$

 **Algebra 1**

Multiplying a binomial and a trinomial and more :

To multiply a binomial with a trinomial or a trinomial with a trinomial we use the extended distributive property.

Step 1 : Distribute the first term of the binomial or the trinomial into the trinomial

Step 2 : Distribute the second term of the binomial into the trinomial.

Step 3 : Repeat the above process for all the terms in the first trinomial or more.

Step 4 : Add all the products together obtained above. Remember to consider the signs in front of the terms while adding.

Step 5 : Combine the like terms. Remember to consider the signs in front of the terms while combining.

Step 6 : Make sure all the terms that can be combined are combined.

Step 7 : Write the polynomial in the standard simplified form.

Example : $(2x + 1)(x - 5)$

$(2x + 1)(x - 5)$ FRONT

$2x \cdot x = 2x^2$

$(2x + 1)(x - 5)$ OUTER
$2x \cdot -5 = -10x$

$(2x + 1)(x - 5)$ INNER

$1 \cdot x = x$

$(2x + 1)(x - 5)$ LAST

$1 \cdot -5 = -5$

Combining all the terms
$2x^2 + (-10x) + x + (-5)$
Combining like terms
$2x^2 -10x + x + -5$
$2x^2 -9x -5$

$(3x^2 + x - 2)(5x^2 - 3x + 7)$

$(3x^2)(5x^2 - 3x + 7) + (x)(5x^2 - 3x + 7) + (-2)(5x^2 - 3x + 7)$

$= 3x^2(5x^2 - 3x + 7) + x(5x^2 - 3x + 7) -2(5x^2 - 3x + 7)$

$= 3x^2(5x^2) + 3x^2(- 3x) + 3x^2(7) + x(5x^2)$
$- x(3x) + x(7) -2(5x^2) - (-2)(3x) + (-2)(7)$

$= 15x^4 - 9x^3 + 21x^2 + 5x^3 - 3x^2 + 7x -10x^2 + 6x - 14$

Regroup the like terms

$= 15x^4 - 9x^3 + 5x^3 + 21x^2 - 3x^2 -10x^2 + 7x + 6x - 14$

$= 15x^4 - 4x^3 + 21x^2 - 13x^2 + 13x - 14$

$= 15x^4 - 4x^3 + 8x^2 + 13x - 14$

Factoring a quadratic equation of the form $x^2 + bx + c$:

The coefficient of x^2 is 1 in this special case.

$(x + 5)(x - 3) = x^2 - 3x + 5x - 15 = x^2 + 2x - 15$

A quadratic equation $x^2 + 2x - 15$ can be factored by finding the factors for -15

Factors for -15 are : 1 & -15, -1 & 15, 3 & -5 and -3 & 5

Chose a pair of factors so that their sum is equal to +2

-3 and 5 are the right pair of factors as -3 + 5 = 2 and the remaining three pairs does not satisfy this criteria.

Rewrite $x^2 + 2x - 15 = x^2 - 3x + 5x - 15$

Take the greatest common factor x from the first two terms and 5 from the last two terms.

$x(x - 3) + 5(x - 3)$

Now x - 3 is the greatest common factor in both the terms

$(x - 3)(x + 5)$

$x^2 + 2x - 15 = (x - 3)(x + 5)$

Important : Remember to consider the sign before the terms when finding the common factors to factorize

---

Factoring a quadratic equation of the form $ax^2 + bx + c$:

Step 1 : Multiply the coefficient "a" and "c". Find the product.

Step 2 : Find all the factor products of "ac"

Step 3 : Identify a pair such that the product of two factors equals to "ac" and when the factors are added the sum equals to "b". Remember to consider the sign before the terms.

Step 4 : Rewrite the middle term as the sum of the identified factors.

Step 5 : Take the greatest common factor out of the first two terms and then from the last two terms.

Step 6 : If step 1 to 5 is done accurately we should see the matching binomial in the two terms.

Step 7 : Take the matching binomial out as the common factor.

Step 8 : After taking the matching binomial as a factor the coefficient of terms in step 6 become the second binomial.

Example : $2x^2 - 9x - 5$

$2 \times -5 = -10$. Factors of -10 are 1 & -10, 10 & -1, 5 & -2, 2 & -5

Out of the factors 1 & -10 fit the criteria
$1 + (-10) = -9$ (Sum of the factors = -9)

10 & -1 does not fit the criteria.
$10 + (-1) = 10 - 1 = 9$ (We need the sum value as -9 not +9

$2x^2 - 9x - 5 = 2x^2 - 10x + x - 5$
$= 2x(x - 5) + 1(x - 5)$
$= (x - 5)(2x + 1)$

$2x^2 - 9x - 5 = (x - 5)(2x + 1)$

Solving a quadratic equation :

The quadratic equation $ax^2 + bx + c = 0$ can be solved by finding the factors and making teach of the factors equal to zero to find the values of x for which the quadratic equation becomes zero.

Example : $2x^2 - 9x - 5 = 0$

$(x - 5)(2x + 1) = 0$

$x - 5 = 0$ or $2x + 1 = 0$
$ + 5 +5$ $- 1 - 1$

$x = 5$ or $\dfrac{2x}{2} = \dfrac{-1}{2}$

$x = 5$ or $x = \dfrac{-1}{2}$

The quadratic function $2x^2 - 9x - 5 = 0$ when

$x = 5$ or $x = \dfrac{-1}{2}$

The solutions of the quadratic equation are

$5 , \dfrac{-1}{2}$

Graphing quadratic functions :

An equation of the standard form $y = ax^2 + bx + c$ is a quadratic function that has an x^2

The graph of a quadratic function is called as a parabola. The parabola is always symmetrical at its vertex. The vertex is a point where the graph changes its direction.

1. If a is positive in the quadratic function $y = ax^2 + bx + c$ then the parabola opens up.

2. If a is negative in the quadratic function $y = ax^2 + bx + c$ the parabola opens down.

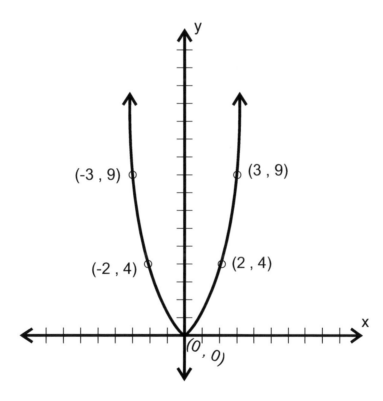

The graph of $y = x^2$ opens up as the coefficient of x^2 is positive and the vertex is (0,0) which is the lowest point on the graph. The graph is symmetrical at the vertex.

www.math-knots.com | www.a4ace.com

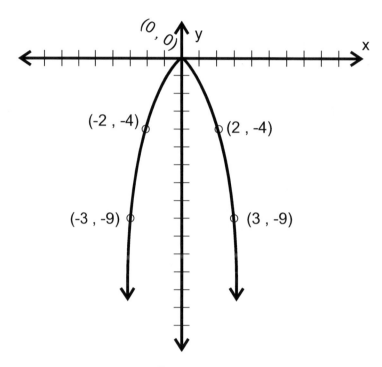

The graph of $y = -x^2$ opens down as the coefficient of x^2 is negative and the vertex is (0,0) which is the highest point on the graph. The graph is symmetrical at the vertex.

<u>Finding x intercept :</u>

The x intercept of any function are the values of x when y = 0.

1. For linear equations we can set y = 0 and find the value of x which is the x intercept.

2. For quadratic equations, factor the quadratic equation as explained in the previous sections and make it equal to 0 to find the values of x, which are x intercepts.

> x intercept is the point where the function value becomes 0.

Example : $y = 2x^2 - 9x - 5 = (x - 5)(2x + 1)$

To find x intercept make y = 0
so, $2x^2 - 9x - 5 = 0 = (x - 5)(2x + 1) = 0$

$(x - 5) = 0$, $(2x + 1) = 0$

$x = 5$ and $x = \dfrac{-1}{2}$

So the x intercepts are 5 and $\dfrac{-1}{2}$

If we plot a graph, the graph should touch at (5,0) and ($\dfrac{-1}{2}$, 0)

Finding y intercept :

The y intercept of any function are the values of y when x = 0.

1. For linear equations we can set x = 0 and find the value of y which is the y - intercept.

2. For quadratic equations, substitute x = 0 to find the values of y - intercept.

     y intercept is the point when x = 0.

Example : $y = 2x^2 - 9x - 5$

To find y intercept substitute x = 0
so, $2x^2 - 9x - 5 = y$
$y = 2(0^2) - 9(0) - 5$
$y = -5$

y - intercept is -5 for the quadratic equation $2x^2 - 9x - 5$.
The graph of the quadratic function $y = 2x^2 - 9x - 5$ touches the y axis at the point (0,-5)

Finding x intercept using quadratic formula :

The quadratic equation $y = ax^2 + bx + c = 0$ can also be factored by using the quadratic formula $\dfrac{-b \pm \sqrt{b^2 - 4ac}}{2a}$

Step 1 : Make sure the quadratic equation is in the standard form $ax^2 + bx + c$ (convert it into standard form if it is not given)

Step 2 : If y is given then set y = 0.

Step 3 : Note down the values of "a" , "b" , "c". Make sure to consider the signs before each term.

Step 4 : Substitute the values "a" , "b" , "c" into the quadratic formula.

Step 5 : Since y = 0 the values found using the quadratic formula are the x - intercepts.

> It is not always possible to find the factors of the quadratic equation. In such a scenario the quadratic formula plays an important role. However we should get the same x - intercept values with both the methods.

The discriminant (b^2 - 4ac) :

The discriminant value (b^2 - 4ac) can be used to identified the number of x - intercepts for a given quadratic function.

1. If b^2 - 4ac = 0, the quadratic function will have one x - intercept which is $\dfrac{-b}{2a}$

2. If b^2 - 4ac is negative value and $\sqrt{b^2 - 4ac}$ is undefined (or imaginary), there are no x - intercepts for the given quadratic function.
In other words the graph will never touch the x axis.

3. If b^2 - 4ac is positive value, then the quadratic function will have two x - intercepts.

One intercept will be $\dfrac{-b + \sqrt{b^2 - 4ac}}{2a}$ and the other will be $\dfrac{-b - \sqrt{b^2 - 4ac}}{2a}$

Graphing a quadratic function :

The line of symmetry of any quadratic function of the standard form $y = ax^2 + bx + c = 0$ is at x = $\dfrac{-b}{2a}$

Step 1 : Point the line of symmetry by finding the value of $\dfrac{-b}{2a}$

Step 2 : Substitute the above value obtained in step 1 for x and find the y value of the vertex.

Step 3 : Find at least two other points to plot the graph.

Tip : You can substitute x = 0 to find the y - intercept value.

Step 4 : Plot the points and join them. You should see the graph as a parabola with the vertex and line of symmetry at $\dfrac{-b}{2a}$

Graphing a pair of quadratic functions :

Follow the same steps as explained above and graph each quadratic function.

Graphing a quadratic inequalities :

To graph a quadratic inequality perform the same steps as given for graphing the quadratic function and follow the below steps after plotting.

Step 5 : Draw a solid line for the quadratic function from step 4 if the inequality is $\geq$ or $\leq$ and move to step 7 otherwise move to step 6.

Step 6 : Draw a broken line for the quadratic function from step 4 if the inequality is $>$ or $<$ otherwise move to step 7.

Step 7 : Test one or two points outside the graph. If the test points satisfy the inequality meaning there are one of many possible solutions. Shade that side of the graph.

If the test points does not satisfy the inequality meaning there are not one of many possible solutions to the inequality. Shade the other side of the graph.

Graphing a pair of inequalities :

Follow the steps as explained in graphing the inequality and graph both the inequalities. Find the common area of the solutions and shade only that area. Any point lying in this common area is one of many solutions that satisfy the pair of quadratic inequalities.

Transformations

1. Rotation of 90 (clockwise) Rotation of 270 (counter clockwise) $(x, y) \longrightarrow (y, -x)$

2. Rotation of 90 (counter clockwise) Rotation of 270 (clockwise) $(x, y) \longrightarrow (-y, x)$

3. Rotation of 180 (clockwise) & (counter clockwise) $(x, y) \longrightarrow (-x, -y)$

4. Reflection over x-axis : $(x, y) \longrightarrow (x, -y)$

5. Reflection over y-axis : $(x, y) \longrightarrow (-x, y)$

6. Reflection over x = y : $(x, y) \longrightarrow (y, x)$

7. Reflection over y = -x : $(x, y) \longrightarrow (-y, -x)$

www.math-knots.com | www.a4ace.com

1) Write the equation of the straight line passing through the point A (-1 , 4) and parallel to $y = 9x + 1$.

    A) $y = -9x + 13$ B) $y = 13x - 9$

    C) $y = -13x - 9$ D) $y = 9x + 13$

2) Write the equation of the straight line passing through the point A (4 , − 4) and parallel to $y = -\dfrac{9}{4}x + 3$.

    A) $y = -x + 5$ B) $y = x + 5$

    C) $y = 5x - 1$ D) $y = -\dfrac{9}{4}x + 5$

3) Write the equation of the straight line passing through the point A (-1 , − 2) and parallel to $y = x - 5$.

    A) $y = -x - 5$ B) $y = x - 1$

    C) $y = -5x - 1$ D) $y = -x - 1$

4) Write the equation of the straight line passing through the point A (− 5 , − 2) and parallel to $y = \dfrac{1}{3}x + 2$.

    A) $y = -x - \dfrac{1}{3}$ B) $y = -\dfrac{1}{3}x - \dfrac{1}{3}$

    C) $y = x - \dfrac{1}{3}$ D) $y = \dfrac{1}{3}x - \dfrac{1}{3}$

5) Write the equation of the straight line passing through the point A (− 4 , − 2) and parallel to $y = -\dfrac{3}{4}x$.

    A) $y = -\dfrac{3}{4}x - 5$ B) $y = 5x - \dfrac{3}{4}$

    C) $y = -\dfrac{3}{4}x + 5$ D) $y = -5x - \dfrac{3}{4}$

6) Write the equation of the straight line passing through the point A (− 4 , − 1) and parallel to $y = x - 5$.

    A) $y = x + 3$ B) $y = -3x + 3$

    C) $y = 2x + 3$ D) $y = 3x + 3$

7) Write the equation of the straight line passing through the point A (1 , − 5) and parallel to $y = x + 2$.

    A) $y = x - 6$ B) $y = 6x + 1$

    C) $y = -2x + 1$ D) $y = -6x + 1$

8) Write the equation of the straight line passing through the point A (− 2 , − 5) and parallel to $y = \dfrac{5}{2}x + 1$

    A) $y = \dfrac{3}{2}x$ B) $y = \dfrac{3}{2}$

    C) $y = \dfrac{5}{2}x$ D) $x = 3$

9) Write the equation of the straight line passing through the point A (− 3 , − 3) and parallel to y = − 5.

A) $y = -3$

B) $y = -\dfrac{1}{4}x - \dfrac{3}{4}$

C) $y = \dfrac{1}{2}x - \dfrac{3}{4}$

D) $y = -\dfrac{5}{4}x - \dfrac{3}{4}$

10) Write the equation of the straight line passing through the point A (− 3 , − 3) and parallel to $y = \dfrac{5}{3}x - 4$.

A) $y = 2x + \dfrac{5}{3}$

B) $y = \dfrac{5}{3}x + 2$

C) $y = -2x + \dfrac{5}{3}$

D) $y = \dfrac{5}{3}x - 2$

11) Write the equation of the straight line passing through the point A (− 5 , 5) and parallel to $y = -\dfrac{6}{5}x - 2$.

A) $y = -x - \dfrac{3}{5}$

B) $y = \dfrac{3}{5}x - 1$

C) $y = -\dfrac{6}{5}x - 1$

D) $y = -\dfrac{3}{5}x - 1$

12) Write the equation of the straight line passing through the point A (2 , 2) and parallel to $y = \dfrac{5}{2}x + 3$.

A) $y = \dfrac{5}{2}x - 3$

B) $y = -\dfrac{5}{2}x - 3$

C) $y = -3x + \dfrac{5}{2}$

D) $y = x + \dfrac{5}{2}$

13) Write the equation of the straight line passing through the point A (4 , − 2) and parallel to $y = -\dfrac{3}{4}x + 4$.

A) $y = -\dfrac{3}{4}x + 1$

B) $y = x + \dfrac{5}{4}$

C) $y = -x + \dfrac{5}{4}$

D) $y = \dfrac{5}{4}x + 1$

14) Write the equation of the straight line passing through the point A (− 4 , 5) and parallel to $y = -\dfrac{2}{3}x + 2$.

A) $y = -\dfrac{5}{3}x - \dfrac{2}{3}$

B) $y = -\dfrac{1}{3}x - \dfrac{2}{3}$

C) $y = -\dfrac{2}{3}x + \dfrac{7}{3}$

D) $y = \dfrac{7}{3}x - \dfrac{2}{3}$

15) Write the equation of the straight line passing through the point A (5 , − 3) and parallel to $y = -\dfrac{1}{10}x + 2$.

A) $y = -\dfrac{5}{2}x + \dfrac{2}{5}$

B) $y = -\dfrac{1}{10}x - \dfrac{5}{2}$

C) $y = \dfrac{2}{5}x - \dfrac{5}{2}$

D) $y = \dfrac{1}{10}x - \dfrac{5}{2}$

16) Write the equation of the straight line passing through the point A (− 3 , − 1) and parallel to $y = -\dfrac{4}{3}x - 3$.

A) $y = -\dfrac{1}{3}x - \dfrac{4}{3}$

B) $y = -\dfrac{4}{3}x - 5$

C) $y = 5x - \dfrac{4}{3}$

D) $y = -5x - \dfrac{4}{3}$

17) Write the equation of the straight line passing through the point A (3 , 5) and parallel to $y = x - 3$.

    A) $y = -x + 2$ B) $y = -5x + 2$

    C) $y = -3x + 2$ D) $y = x + 2$

18) Write the equation of the straight line passing through the point A (− 2 , − 1) and parallel to $y = 2$.

    A) $x = 1$ B) $x = -1$

    C) $y = -1$ D) $y = 1$

19) Write the equation of the straight line passing through the point A (− 3 , − 3) and parallel to $y = \dfrac{4}{3} x + 4$.

    A) $y = x - \dfrac{4}{3}$ B) $y = \dfrac{4}{3} x + 1$

    C) $y = -\dfrac{4}{3} x + 1$ D) $y = \dfrac{2}{3} x + 1$

20) Write the equation of the straight line passing through the point A (1 , − 2) and parallel to $y = \dfrac{3}{2} x + 4$.

    A) $y = 2x - \dfrac{7}{2}$ B) $y = -2x - \dfrac{7}{2}$

    C) $y = -\dfrac{5}{2} x - \dfrac{7}{2}$ D) $y = \dfrac{3}{2} x - \dfrac{7}{2}$

21) Write the equation of the straight line passing through the point A (1 , 4) and parallel to $y = 5 x + 5$.

    A) $y + 1 = 5(x + 4)$

    B) $y + 4 = -3(x - 1)$

    C) $y - 4 = 5(x - 1)$

    D) $y - 4 = 0$

22) Write the equation of the straight line passing through the point A (− 1 , − 4) and parallel to $y = - x - 4$.

    A) $y - 4 = x + 1$

    B) $y + 1 = -\dfrac{1}{3}(x + 4)$

    C) $y + 1 = \dfrac{4}{3}(x - 4)$

    D) $y + 4 = -(x + 1)$

www.math-knots.com | www.a4ace.com

23) Write the equation of the straight line passing through the point A (− 4 , 4) and parallel to y = − 2 x.

A) $y - 4 = -2(x + 4)$

B) $y - 4 = -\frac{1}{5}(x + 4)$

C) $y - 4 = -\frac{1}{5}(x - 4)$

D) $y + 4 = \frac{1}{5}(x + 4)$

24) Write the equation of the straight line passing through the point A (5 , − 5) and parallel to $y = -\frac{4}{5}x - 5$.

A) $y - 5 = \frac{1}{5}(x + 5)$

B) $y + 5 = -\frac{4}{5}(x - 5)$

C) $y - 5 = \frac{4}{5}(x + 5)$

D) $y - 5 = -\frac{4}{5}(x - 5)$

25) Write the equation of the straight line passing through the point A (− 3 , − 4) and parallel to y = x + 1.

A) y + 4 = 3 (x + 3)

B) y + 4 = x + 3

C) y + 3 = − 5 (x − 4)

D) y + 3 = 5 (x + 4)

26) Write the equation of the straight line passing through the point A (− 5 , − 5) and parallel to y = x − 5.

A) y + 5 = x + 5

B) y − 5 = − (x + 5)

C) y − 5 = x + 5

D) y − 5 = 5(x − 5)

27) Write the equation of the straight line passing through the point A (− 3 , 3) and parallel to $y = -\frac{5}{3}x + 4$.

A) $y - 3 = -\frac{5}{3}(x + 3)$

B) $y - 3 = -\frac{3}{5}(x + 3)$

C) $y - 3 = -\frac{3}{5}(x - 3)$

D) $y + 3 = \frac{3}{5}(x + 3)$

28) Write the equation of the straight line passing through the point A (1 , − 1) and parallel to y = − 5 x + 1.

A) $y + 1 = \frac{1}{5}(x + 1)$

B) $y - 1 = -\frac{1}{5}(x + 1)$

C) y + 1 = −5 (x −1)

D) $y + 1 = -\frac{5}{3}(x + 1)$

29) Write the equation of the straight line passing through the point A (0 , 0) and parallel to $y = \dfrac{3}{2}x + 3$.

A) $y = \dfrac{3}{2}x$ B) $y = 2x$

C) $y = \dfrac{1}{2}x$ D) $y = -\dfrac{3}{2}x$

30) Write the equation of the straight line passing through the point A (4 , − 3) and parallel to $y = -\dfrac{5}{4}x + 1$.

A) $y + 4 = -\dfrac{5}{12}(x - 3)$

B) $y + 3 = -\dfrac{5}{4}(x - 4)$

C) $y + 4 = -\dfrac{5}{12}(x + 3)$

D) $y - 4 = -\dfrac{36}{5}(x - 3)$

31) Write the equation of the straight line passing through the point A (− 2 , 2) and parallel to y = 7 x + 2.

A) $y - 2 = 7(x + 2)$

B) $y + 2 = -\dfrac{1}{4}(x - 2)$

C) $y + 2 = \dfrac{1}{4}(x + 2)$

D) $y + 2 = \dfrac{7}{4}(x + 2)$

32) Write the equation of the straight line passing through the point A (− 3 , − 5) and parallel to y = 2 x + 4.

A) $y + 5 = -(x - 3)$

B) $y + 3 = 3(x - 5)$

C) $y + 5 = 2(x - 3)$

D) $y + 5 = 2(x + 3)$

33) Write the equation of the straight line passing through the point A (5 , 4) and parallel to $y = \dfrac{1}{3}x - 2$.

A) $y + 4 = \dfrac{2}{3}(x - 5)$

B) $y - 4 = \dfrac{1}{3}(x - 5)$

C) $y + 4 = x - 5$

D) $y - 5 = -\dfrac{1}{3}(x + 4)$

34) Write the equation of the straight line passing through the point A (− 1 , − 4) and parallel to y = 9 x + 4.

A) $y - 1 = 9(x - 4)$

B) $y - 4 = 9(x + 1)$

C) $y - 1 = 9(x + 4)$

D) $y + 4 = 9(x + 1)$

35) Write the equation of the straight line passing through the point A (5 , 1) and parallel to $y = \dfrac{6}{5} x - 3$.

   A) $y - 5 = -4 (x + 1)$

   B) $y + 1 = -(x - 5)$

   C) $y - 5 = -\dfrac{6}{5} (x - 1)$

   D) $y - 1 = \dfrac{6}{5} (x - 5)$

36) Write the equation of the straight line passing through the point A (1 , 5) and parallel to $y = 3 x + 3$.

   A) $y - 5 = 3 (x - 1)$

   B) $y - 1 = -\dfrac{1}{3} (x + 5)$

   C) $y + 5 = 3 (x - 1)$

   D) $y + 5 = -3 (x - 1)$

37) Write the equation of the straight line passing through the point A (4 , - 1) and parallel to $y = x - 4$.

   A) $y - 1 = -(x - 4)$

   B) $y + 1 = -(x - 4)$

   C) $y + 1 = x - 4$

   D) $y - 1 = 3 (x - 4)$

38) Write the equation of the straight line passing through the point A (0 , 3) and parallel to $y = \dfrac{5}{2} x - 2$.

   A) $y = \dfrac{5}{3} (x - 3)$ B) $y = -\dfrac{2}{5} (x - 3)$

   C) $y = \dfrac{2}{5} (x + 3)$ D) $y - 3 = \dfrac{5}{2} x$

39) Write the standard form of the equation of the straight line passing through the point A (- 5 , - 5) and parallel to $y = \dfrac{6}{5} x - 1$.

   A) $5 x - 6 y = - 5$

   B) $x + 3 y = 15$

   C) $5 x + 6 y = 30$

   D) $6 x - 5 y = - 5$

40) Write the standard form of the equation of the straight line passing through the point A (5 , - 2) and parallel to $y = -\dfrac{2}{9} x - 4$.

   A) $2x + 9 y = -36$ B) $2x - 9 y = -36$

   C) $2x + 9 y = -8$ D) $36x + 2 y = -9$

41) Write the standard form of the equation of the straight line passing through the point A (2 , 5) and parallel to y = 4 x − 1.

A) 4 x − y = 3

B) 4 x + y = 3

C) 3 x + y = −3

D) 3 x + y = 4

44) Write the standard form of the equation of the straight line passing through the point P (-3 , − 2)

and parallel to $y = -\dfrac{1}{3}x - 1$.

A) x − y = 0

B) x + 3 y = 0

C) 2 x + y = 0

D) x + 3 y = − 9

42) Write the standard form of the equation of the straight line passing through the point A (4 , 4) and parallel to y = 4 x − 1.

A) 2 x + 4y = − 1

B) 4 x − y = 12

C) 4 x + y = − 12

D) 4 x − y = − 12

45) Write the standard form of the equation of the straight line passing through the point A (− 5 , − 5)

and parallel to $y = \dfrac{1}{5}x + 4$.

A) 3 x + 5 y = 20

B) x + y = − 4

C) x − 5 y = 20

D) 3 x − 5 y = 20

43) Write the standard form of the equation of the straight line passing through the point A (2 , 4) and parallel to

$y = \dfrac{4}{3}x - 3$.

A) 4 x − 3 y = 4

B) 3 x + 4 y = − 4

C) 4 x − 3 y = − 4

D) 4 x + 3 y = − 12

46) Write the standard form of the equation of the straight line passing through the point A (5 , − 5) and parallel to y = − 2 x − 2.

A) 2 x − y = 5

B) 2 x + y = − 5

C) 2 x + 5 y = 1

D) 2 x + y = 5

47) Write the standard form of the equation of the straight line passing through the point A (− 4, 5) and parallel to

$$y = -\frac{7}{5}x + 1.$$

A) 7 x + 5 y = − 3   B) 10 x + 5 y = 4

C) 2 x − 5 y = 1   D) 4 x + 5 y = 10

48) Write the standard form of the equation of the straight line passing through the point A (− 2 , − 5) and parallel to y = 5 x + 3.

A) 2 x + y = − 4   B) 5 x + y = 4

C) 5 x − y = − 4   D) 5 x − y = − 5

49) Write the standard form of the equation of the straight line passing through the point A (4 , 2) and parallel to y = 5.

A) 3 x − y = 4   B) 4 x + y = 3

C) y = 2   D) 4 x − y = 1

50) Write the standard form of the equation of the straight line passing through the point A (− 1 , − 3) and parallel to y = − x − 2.

A) x + 5 y = − 1   B) x + y = 4

C) x + y = − 4   D) x + y = 5

51) Write the standard form of the equation of the straight line passing through the point A (− 1 , − 1) and parallel to y = − 4 x.

A) 4 x − y = 5

B) 4 x − 5 y = 1

C) 4 x + 5 y = 1

D) 4 x + y = − 5

52) Write the standard form of the equation of the straight line passing through the point A (− 2 , 1) and parallel to y = x − 4.

A) x + 4 y = − 1   B) x − y = − 3

C) x − 4 y = − 1   D) x + 4 y = 1

53) Write the standard form of the equation of the straight line passing through the point A (− 5, 4) and parallel to

$$y = -\frac{3}{5}x + 3.$$

A) $3x - 5y = 5$ B) $x + y = -5$

C) $3x + 5y = 5$ D) $x - y = 1$

54) Write the standard form of the equation of the straight line passing through the point A (− 1 , − 5) and parallel to $y = 4x$.

A) $2x - 4y = 1$ B) $x - 4y = -1$

C) $x + y = -4$ D) $4x - y = 1$

55) Write the standard form of the equation of the straight line passing through the point A (5 , 5) and parallel to $y = \frac{5}{8}x$.

A) $3x - 8y = -24$ B) $3x - 8y = -15$

C) $5x - 8y = -15$ D) $3x + 8y = -24$

56) Write the standard form of the equation of the straight line passing through the point A (− 3 , − 3) and parallel to

$$y = -\frac{5}{2}x + 3.$$

A) $5x - 2y = 21$ B) $2x + 2y = 21$

C) $3x + 2y = -21$ D) $5x + 2y = -21$

57) Write the equation of the straight line passing through the point A (− 2 , − 1) and perpendicular to $y = \frac{2}{3}x + 5$.

A) $y = -4x - \frac{3}{2}$ B) $y = \frac{5}{2}x - \frac{3}{2}$

C) $y = -\frac{3}{2}x - 4$ D) $y = -\frac{5}{2}x - \frac{3}{2}$

58) Write the equation of the straight line passing through the point A (2 , 4) and perpendicular to $y = -x - 5$.

A) $y = -2x - 2$ B) $y = -2x + 2$

C) $y = 2x - 2$ D) $y = x + 2$

59) Write the equation of the straight line passing through the point A (− 3 , − 4) and perpendicular to x = 0.

    A) y = − 4   B) x = − 4

    C) x = 4   D) x = 3

60) Write the equation of the straight line passing through the point A (4 , − 5) and perpendicular to y = 4 x − 2.

    A) $y = -\dfrac{1}{2}x - 4$ B) $y = -\dfrac{3}{2}x - 4$

    C) $y = -\dfrac{1}{4}x - 4$ D) $y = \dfrac{1}{2}x - 4$

61) Write the equation of the straight line passing through the point A (2 , 3) and perpendicular to $y = \dfrac{1}{3}x - 4$.

    A) y = − x + 9   B) y = 3 x + 9

    C) y = − 3 x + 9   D) y = 9 x + 3

62) Write the equation of the straight line passing through the point A (5 , 2) and perpendicular to y = 4 x − 2.

    A) $y = -\dfrac{1}{4}x + \dfrac{13}{4}$

    B) $y = \dfrac{1}{4}x - \dfrac{1}{4}$

    C) $y = \dfrac{1}{2}x - \dfrac{1}{4}$

    D) $y = \dfrac{13}{4}x - \dfrac{1}{4}$

63) Write the equation of the straight line passing through the point A (− 1 , 4) and perpendicular to $y = \dfrac{1}{4}x - 4$.

    A) y = 4x   B) y = −4 x

    C) y = 4   D) y = −2 x

64) Write the equation of the straight line passing through the point A (1 , − 3) and perpendicular to $y = -\dfrac{1}{2}x - 4$.

    A) y = − 5 x + 4   B) y = − 2 x − 5

    C) y = 2 x − 5   D) y = 4 x − 5

www.math-knots.com | www.a4ace.com

65) Write the equation of the straight line passing through the point A (− 1, − 4) and perpendicular to $y = -\dfrac{1}{4}x - 4$.

A) x = − 4　　　　B) y = 4 x

C) y = 4　　　　D) x = 4

66) Write the equation of the straight line passing through the point A (5 , − 2) and perpendicular to y = x + 4.

A) y = 2 x + 3　　　　B) y = 3 x − 2

C) y = − x + 3　　　　D) y = − 2 x + 3

67) Write the equation of the straight line passing through the point A (− 3 , 0) and perpendicular to y = 5.

A) y = − 1　　　　B) x = 1

C) $y = \dfrac{1}{3}$　　　　D) x = − 3

68) Write the equation of the straight line passing through the point Q(-4 , -1) and perpendicular to $y = -\dfrac{1}{5}x - 2$.

A) y = 5 x + 19　　　　B) y = − 5 x + 19

C) y = 19 x + 5　　　　D) y = − 2 x + 5

69) Write the equation of the straight line passing through the point A (5 , 2) and perpendicular to $y = \dfrac{5}{3}x - 5$.

A) $y = -\dfrac{3}{5}x + 5$

B) $y = -5x - \dfrac{4}{5}$

C) $y = 5x - \dfrac{4}{5}$

D) $y = -\dfrac{4}{5}x + 5$

70) Write the equation of the straight line passing through the point A (5 , 5) and perpendicular to $y = -\dfrac{10}{7}x - 1$.

A) $y = -\dfrac{3}{2}x - \dfrac{7}{10}$

B) $y = \dfrac{7}{10}x + \dfrac{3}{2}$

C) $y = \dfrac{3}{2}x - \dfrac{7}{10}$

D) $y = -\dfrac{7}{10}x + \dfrac{3}{2}$

Algebra 1

71) Write the equation of the straight line passing through the point A (5 , − 3) and perpendicular to $y = \dfrac{5}{8}x - 3$.

A) $y = x + 5$

B) $y = 5x + \dfrac{8}{5}$

C) $y = \dfrac{8}{5}x + 5$

D) $y = -\dfrac{8}{5}x + 5$

72) Write the equation of the straight line passing through the point A (1 , 3) and perpendicular to $y = \dfrac{1}{2}x + 5$.

A) $y = 5x - 5$

B) $y = 5x + 5$

C) $y = -2x + 5$

D) $y = -5x + 5$

73) Write the equation of the straight line passing through the point A (− 5 , 4) and perpendicular to $y = \dfrac{10}{3}x - 3$.

A) $y = -\dfrac{3}{10}x + \dfrac{3}{10}$

B) $y = \dfrac{3}{10}x + \dfrac{5}{2}$

C) $y = \dfrac{5}{2}x + \dfrac{3}{10}$

D) $y = -\dfrac{3}{10}x + \dfrac{5}{2}$

74) Write the equation of the straight line passing through the point A (− 5 , 1) and perpendicular to y = 0.

A) $y = -5x$

B) $x = -5$

C) $y = 5x$

D) $y = x$

75) Write the equation of the straight line passing through the point A (0 , − 2) and perpendicular to y = 0.

A) $y = \dfrac{1}{4}$

B) $x = 0$

C) $y = 0$

D) $y = \dfrac{1}{4}x$

76) Write the point slope of the equation of the straight line passing through the point A (3 , 2) and perpendicular to y = 3 x + 4.

A) $y + 3 = -3(x + 2)$

B) $y - 2 = -\dfrac{1}{3}(x - 3)$

C) $y - 3 = -3(x + 2)$

D) $y + 2 = \dfrac{1}{3}(x - 3)$

77) Write the point slope of the equation of the straight line passing through the point A (2 , − 4) and perpendicular to y = x − 5.

A) $y + 4 = \dfrac{1}{2}(x + 2)$

B) $y + 4 = -\dfrac{1}{2}(x + 2)$

C) $y + 4 = -2(x - 2)$

D) $y + 4 = -(x - 2)$

78) Write the point slope of the equation of the straight line passing through the point A (4 , − 1) and perpendicular to $y = \dfrac{4}{5}x + 4.$

A) $y + 4 = -\dfrac{4}{5}(x + 1)$

B) $y - 4 = -\dfrac{4}{5}(x - 1)$

C) $y + 4 = -\dfrac{4}{5}(x - 1)$

D) $y + 1 = -\dfrac{5}{4}(x - 4)$

79) Write the point slope of the equation of the straight line passing through the point A (2 , 1) and perpendicular to y = − 2 x + 2.

A) $y + 1 = -5(x - 2)$

B) $y - 1 = -\dfrac{1}{2}(x + 2)$

C) $y + 2 = x + 1$

D) $y - 1 = \dfrac{1}{2}(x - 2)$

80) Write the point slope of the equation of the straight line passing through the point A (− 4 , 1) and perpendicular to $y = \dfrac{4}{3}x + 1.$

A) $y - 1 = -5(x - 4)$

B) $y - 4 = 2(x + 1)$

C) $y - 1 = -\dfrac{3}{4}(x + 4)$

D) $y + 1 = -\dfrac{1}{2}(x + 4)$

81) Write the point slope of the equation of the straight line passing through the point A (3 , − 4) and perpendicular to $y = \dfrac{3}{8}x + 4.$

A) $y + 3 = -\dfrac{8}{3}(x + 4)$

B) $y + 4 = \dfrac{8}{3}(x - 3)$

C) $y + 4 = x + 3$

D) $y + 4 = -\dfrac{8}{3}(x - 3)$

82) Write the point slope of the equation of the straight line passing through the point A (1 , 0) and perpendicular to y = − x − 1.

A) $y = -(x + 1)$ B) $y = \dfrac{1}{5}(x - 1)$

C) $y - 1 = -5x$ D) $y = x - 1$

83) Write the point slope of the equation of the straight line passing through the point A (5 , 3) and perpendicular to

$$y = -\frac{5}{7}x + 4.$$

  A) $y + 3 = -(x - 5)$

  B) $y - 3 = \frac{7}{5}(x - 5)$

  C) $y - 3 = -(x - 5)$

  D) $y - 5 = -(x - 3)$

84) Write the point slope of the equation of the straight line passing through the point A (0 , 1) and perpendicular to y = 2 x − 5.

  A) $y - 1 = -\frac{1}{2}x$

  B) $y + 1 = \frac{1}{10}x$

  C) $y - 1 = -\frac{1}{10}x$

  D) $y = -10(x - 1)$

85) Write the point slope of the equation of the straight line passing through the point A (1 , − 3) and perpendicular to

$$y = -\frac{1}{2}x + 2.$$

  A) $y + 1 = x + 3$

  B) $y + 3 = -5(x - 1)$

  C) $y + 3 = -(x - 1)$

  D) $y + 3 = 2(x - 1)$

86) Write the point slope of the equation of the straight line passing through the point A (− 3 , 4) and perpendicular to

$$y = \frac{1}{3}x.$$

  A) $y + 3 = 3(x - 4)$

  B) $y + 4 = x - 3$

  C) $y - 4 = -3(x + 3)$

  D) $y + 4 = -(x - 3)$

87) Write the point slope of the equation of the straight line passing through the point A (− 2 , 1) and perpendicular to

$$y = \frac{2}{3}x - 3.$$

  A) $y - 1 = 3(x + 2)$

  B) $y - 1 = -\frac{3}{2}(x + 2)$

  C) $y + 1 = 2(x - 2)$

  D) $y + 2 = \frac{3}{2}(x + 1)$

88) Write the point slope of the equation of the straight line passing through the point A (4 , − 5) and perpendicular to

$$y = \frac{2}{3}x + 2.$$

  A) $y - 5 = -\frac{3}{2}(x - 4)$

  B) $y - 4 = \frac{2}{3}(x - 5)$

  C) $y - 4 = \frac{3}{2}(x + 5)$

  D) $y + 5 = -\frac{3}{2}(x - 4)$

89) Write the point slope of the equation of the straight line passing through the point A (− 4 , − 3) and perpendicular to y = − x − 2.

   A) $y + 4 = 5 (x - 3)$

   B) $y - 4 = - 5 (x - 3)$

   C) $y + 3 = x + 4$

   D) $y - 3 = - (x + 4)$

90) Write the point slope of the equation of the straight line passing through the point A (2 , 3) and perpendicular to y = x.

   A) $y + 2 = - (x - 3)$

   B) $y - 3 = - (x - 2)$

   C) $y - 2 = 0$

   D) $y - 2 = x + 3$

91) Write the point slope of the equation of the straight line passing through the point A (4 , 2) and perpendicular to

$y = \dfrac{1}{3} x - 1.$

   A) $y - 2 = - 3 (x - 4)$

   B) $y - 4 = x - 2$

   C) $y - 4 = - (x - 2)$

   D) $y - 2 = - 5 (x + 4)$

92) Write the point slope of the equation of the straight line passing through the point A (3 , 4) and perpendicular to y = − 2 x + 2.

   A) $y - 4 = \dfrac{1}{2} (x - 3)$

   B) $y + 4 = - \dfrac{5}{2} (x + 3)$

   C) $y - 4 = - \dfrac{6}{5} (x + 3)$

   D) $y + 4 = - 2 (x - 3)$

93) Write the point slope of the equation of the straight line passing through the point A (− 3 , 0) and perpendicular to y = − x.

   A) $y = \dfrac{1}{2} (x + 3)$

   B) $y = - (x - 3)$

   C) $y = x + 3$

   D) $y = - 2 (x - 3)$

94) Write the point slope of the equation of the straight line passing through the point A (− 5 , 5) and perpendicular to y = x.

   A) $y + 5 = 3 (x - 5)$

   B) $y - 5 = 2 (x - 5)$

   C) $y - 5 = - (x - 5)$

   D) $y - 5 = - (x + 5)$

95) Write the standard form of the equation of the straight line passing through the point A (2 , – 4) and perpendicular to x = 0.

A) x – y = – 4

B) 2 x + y = – 4

C) x + 4 y = 1

D) y = – 4

96) Write the standard form of the equation of the straight line passing through the point A (– 5 , – 4) and perpendicular to

$y = -\dfrac{5}{7} x - 4$

A) 5 x + 10 y = – 7

B) 5 x – 10 y = 7

C) 7 x – 5 y = – 15

D) 10 x – 4 y = – 7

97) Write the standard form of the equation of the straight line passing through the point A (– 4 , – 5) and perpendicular to

$y = -\dfrac{1}{8} x - 2$

A) 4 x – y = 27

B) 8 x – y = – 27

C) 4 x + y = 27

D) 4 x – y = – 27

98) Write the standard form of the equation of the straight line passing through the point A (– 3 , 3) and perpendicular to y = – 3 x + 3.

A) x + 3 y = – 12

B) x + 3 y = – 9

C) x – 3 y = – 12

D) x – 3 y = 0

99) Write the standard form of the equation of the straight line passing through the point A (– 2 , 3) and perpendicular to

$y = \dfrac{5}{6}x + 4.$

A) 5 x – 6 y = 3

B) 6 x – y = – 3

C) x + 6 y = – 3

D) 6 x + 5 y = 3

100) Write the standard form of the equation of the straight line passing through the point A (1 , 4) and perpendicular to y = x – 4.

A) x – y = – 3

B) 3 x – y = – 1

C) x + y = – 5

D) x + y = 5

101) Write the standard form of the equation of the straight line passing through the point A (5 , − 3) and perpendicular to y = x − 3.

A) x + y = 2

B) 5 x − y = − 2

C) 5 x + y = 2

D) x − 5 y = − 2

102) Write the standard form of the equation of the straight line passing through the point A (0 , − 3) and perpendicular to y = 2.

A) x = 0

B) y = 0

C) 5 x = 0

D) 3 x = 0

103) Write the standard form of the equation of the straight line passing through the point A (0 , − 4) and perpendicular to $y = -\dfrac{1}{6}x - 5$.

A) x − 6 y = − 3

B) 6 x − y = 4

C) x − 6 y = 3

D) 6 x − y = − 5

104) Write the standard form of the equation of the straight line passing through the point A (− 4 , 0) and perpendicular to $y = -\dfrac{4}{3}x$.

A) 4 x − 3 y = − 12

B) 4 x − 12 y = − 3

C) 3 x + 4 y = − 12

D) 3 x − 4 y = − 12

105) Write the standard form of the equation of the straight line passing through the point A (2 , − 3) and perpendicular to $y = \dfrac{2}{3}x + 3$.

A) 4 x + y = 2

B) 3 x − y = − 2

C) 3 x + 2 y = 0

D) 3 x − y = − 1

106) Write the standard form of the equation of the straight line passing through the point A (− 4 , 3) and perpendicular to y = − 7 x + 4.

A) 7 x − 25 y = − 1

B) 25 x − y = − 7

C) 25 x + 7 y = 1

D) x − 7 y = − 25

107) Write the standard form of the equation of the straight line passing through the point A (− 2 , − 4) and perpendicular to $y = -\dfrac{1}{2}x + 5.$

A) − y = 2

B) 2 x − y = 0

C) − y = − 2

D) 4 x − y = 2

108) Write the standard form of the equation of the straight line passing through the point A (5, − 3) and perpendicular to $y = \dfrac{5}{2}x - 3.$

A) x + 2 y = 0

B) 2 x + 5 y = − 5

C) 5 x − 2 y = − 5

D) 5 x − 2 y = 0

109) Write the standard form of the equation of the straight line passing through the point A (− 5 , − 2) and perpendicular to $y = -\dfrac{5}{4}x + 4.$

A) 2 x + 10 y = 5

B) 3 x − 10 y = − 5

C) 5 x − 2 y = 10

D) 4 x − 5 y = − 10

110) Write the standard form of the equation of the straight line passing through the point A (− 1 , 5) and perpendicular to $y = -\dfrac{3}{10}x - 3.$

A) x + 3 y = 15

B) 10 x − 3 y = − 25

C) x − 3 y = − 15

D) 2 x + 3 y = − 15

111) Which graph below represents as x-intercept as 4 and y-intercept as 3.

A)
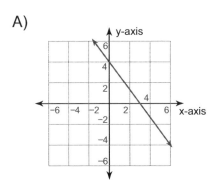

B)
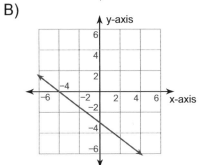

C)
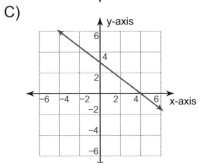

D)
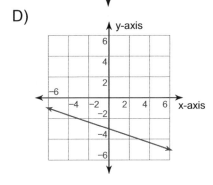

112) Which graph below represents as x-intercept as 3 and y-intercept as 2.

A)
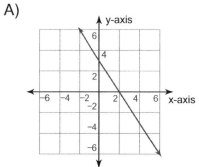

B)
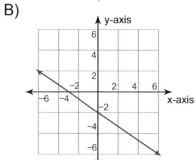

C)
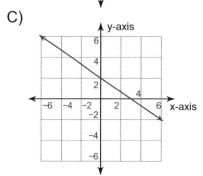

D)
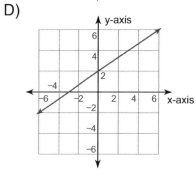

www.math-knots.com | www.a4ace.com

113) Which graph below represents as x-intercept as 1 and y-intercept as − 5.

A)
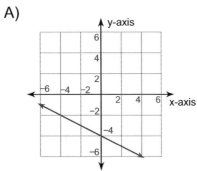

B)
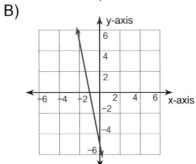

C)
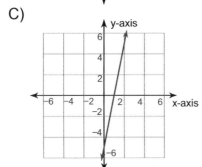

D)
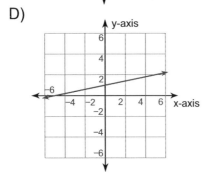

114) Which graph below represents as x-intercept as − 3 and y-intercept as − 1.

A)
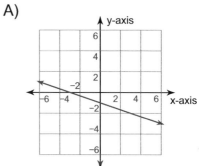

B)
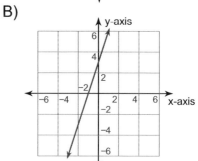

C)
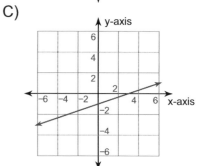

D)
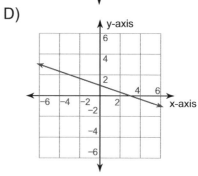

115) Which graph below represents as x-intercept as − 3 and y-intercept as − 2.

A)
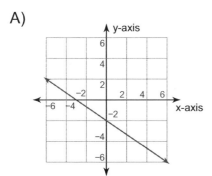

B)
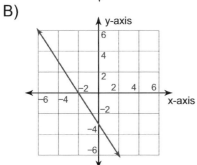

C)
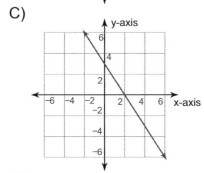

D)
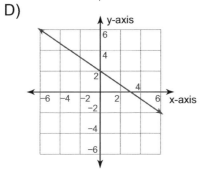

116) Which graph below represents as x-intercept as − 1 and y-intercept as − 1.

A)
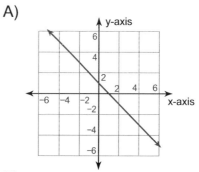

B)
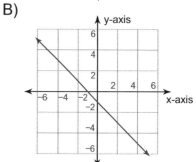

C)
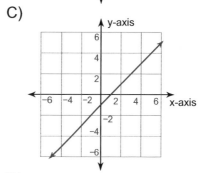

D)
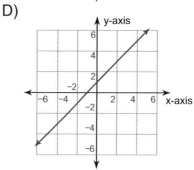

www.math-knots.com | www.a4ace.com

117) Which graph below represents as x-intercept as − 2 and y-intercept as − 3.

A)
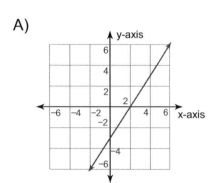

B)
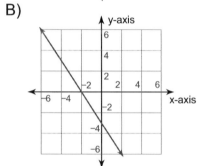

C)
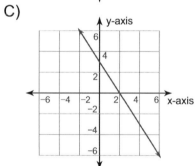

D)
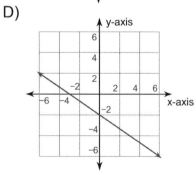

118) Which graph below represents as x-intercept as − 5 and y-intercept as − 5.

A)
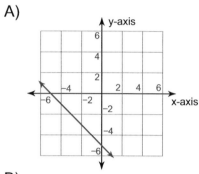

B)
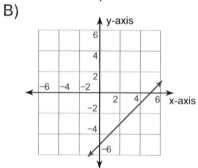

C)
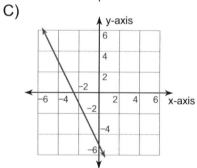

D)
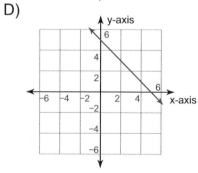

www.math-knots.com | www.a4ace.com

119) Which graph below represents as x-intercept as 1 and y-intercept as − 2.

A)

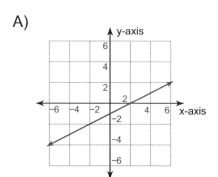

B)

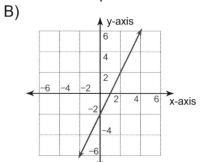

C)

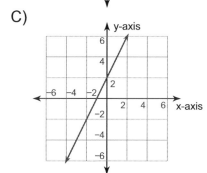

D)

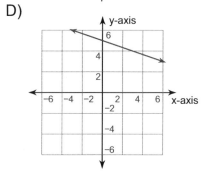

120) Which graph below represents as x-intercept as 1 and y-intercept as − 4.

A)

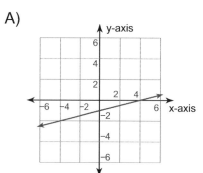

B)

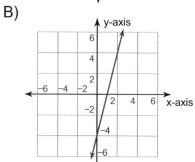

C)

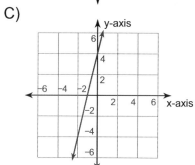

D)

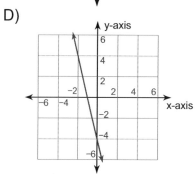

www.math-knots.com | www.a4ace.com

Algebra 1

121) Which graph below represents as x-intercept as − 2 and y-intercept as 4.

A)
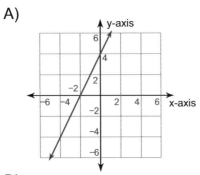

B)
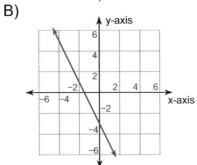

C)
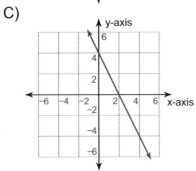

D)
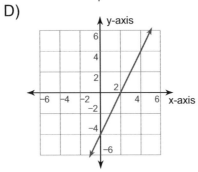

122) Which graph below represents as x-intercept as − 1 and y-intercept as − 5.

A)
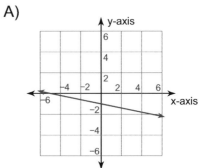

B)
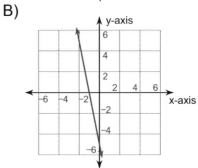

C)
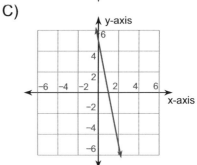

D)
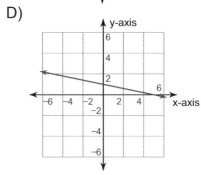

www.math-knots.com | www.a4ace.com

 **Algebra 1**

123) Which graph below represents as x-intercept as − 1 and y-intercept as 2.

A)
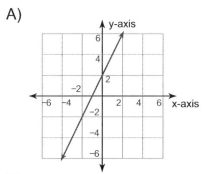

B)
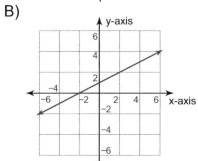

C)
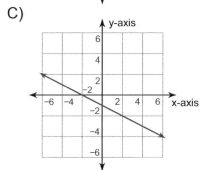

D)
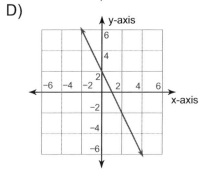

124) Which graph below represents as x-intercept as 4 and y-intercept as − 3.

A)
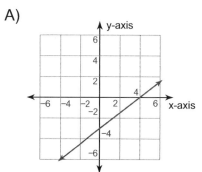

B)
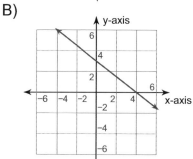

C)
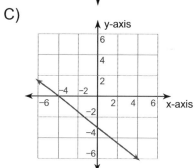

D)
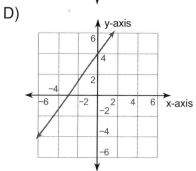

125) Which graph below represents as x-intercept as − 2 and y-intercept as − 4.

A)
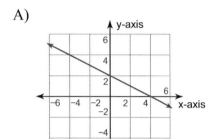

B)
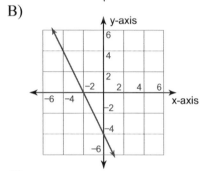

C)
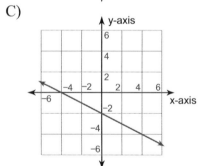

D)
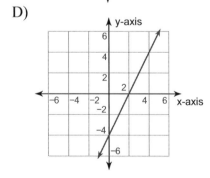

126) Which graph below represents as x-intercept as − 4 and y-intercept as 3.

A)
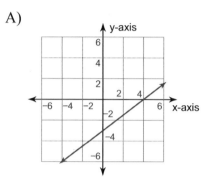

B)
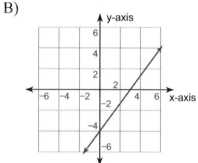

C)
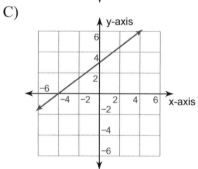

D)
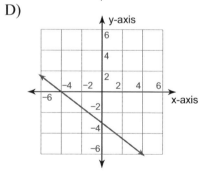

www.math-knots.com | www.a4ace.com

127) Which graph below represents as x-intercept as 3 and y-intercept as − 3.

A)

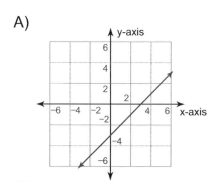

B)

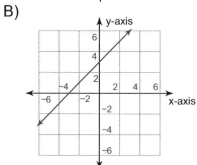

C)

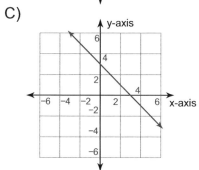

D)

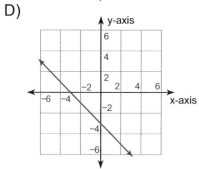

128) Which graph below represents as x-intercept as − 3 and y-intercept as − 3.

A)

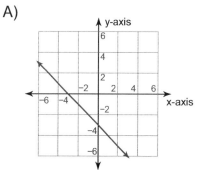

B)

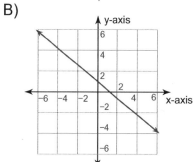

C)

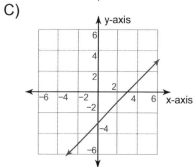

D)

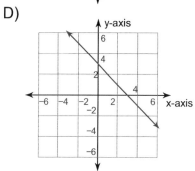

www.math-knots.com | www.a4ace.com

129) Plot the straight-line $y = \dfrac{3}{2}x + 2$

130) Plot the straight-line $y = -2x + 5$

A)

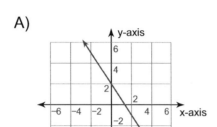

B)

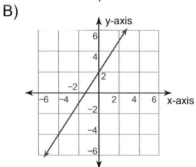

C)

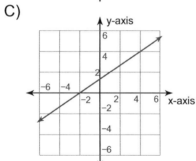

D)

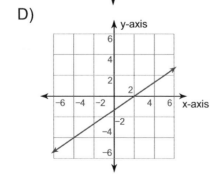

A)

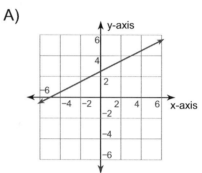

B)

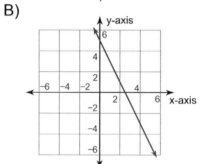

C)

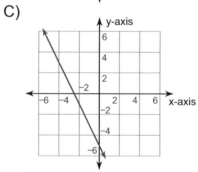

D)

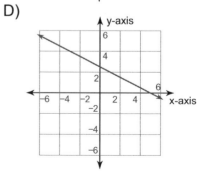

Algebra 1

131) Plot the straight-line $y = -\dfrac{4}{5}x + 3$

A)

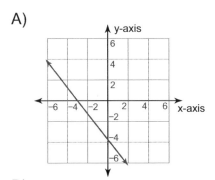

B)

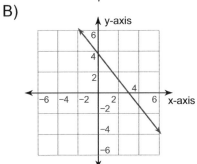

C)

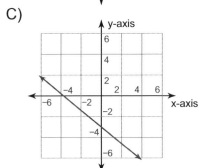

D)

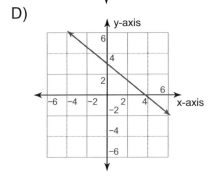

132) Plot the straight-line $y = \dfrac{6}{5}x - 3$

A)

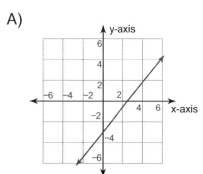

B)

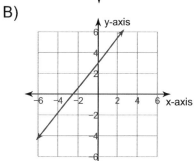

C)

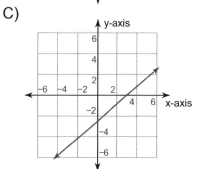

D)

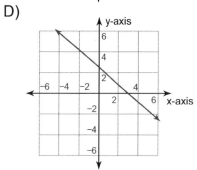

www.math-knots.com | www.a4ace.com

 **Algebra 1**

133) Plot the straight-line y = − 6 x − 4

134) Plot the straight-line y = − 2 x + 2

A)
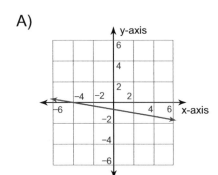

B)
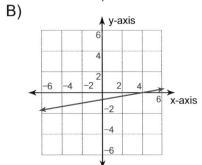

C)
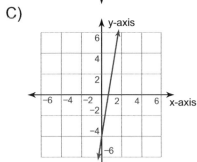

D)
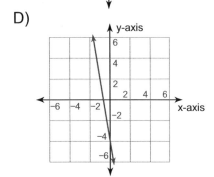

A)
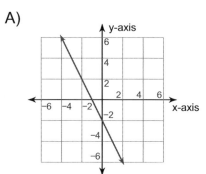

B)
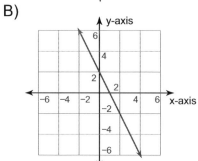

C)
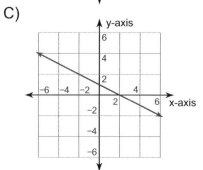

D)
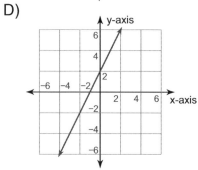

Algebra 1

135) Plot the straight-line $y = -\dfrac{3}{2}x + 5$

A)

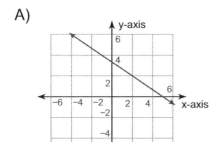

B)

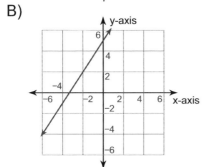

C)

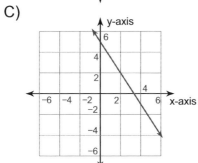

D)

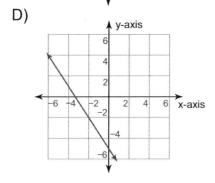

136) Plot the straight-line $y = -\dfrac{2}{5}x - 1$

A)

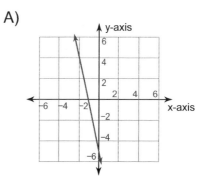

B)

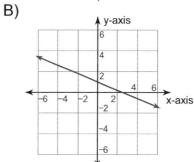

C)

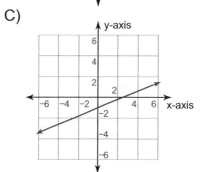

D)

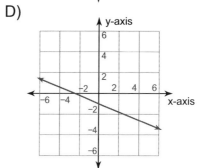

www.math-knots.com | www.a4ace.com

137) Plot the straight-line
$y = x$

138) Plot the straight-line $y = -\dfrac{5}{3}x + 4$

A)

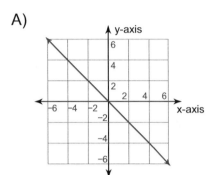

A)

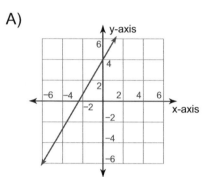

B)

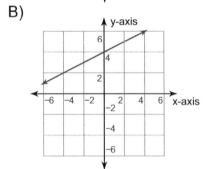

B)

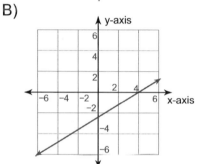

C)

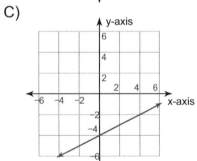

C)

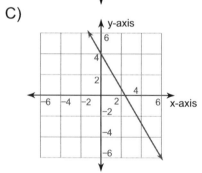

D)

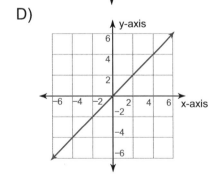

D)

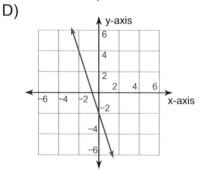

139) Plot the straight-line
$y = -x - 1$

A)
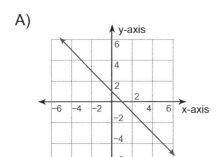

B)
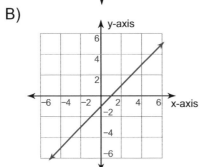

C)
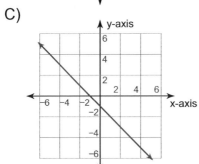

D)
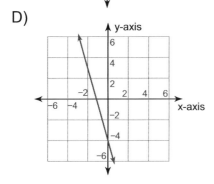

140) Plot the straight-line $y = -1$ in the

A)
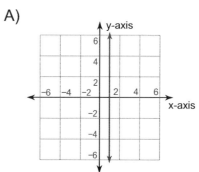

B)
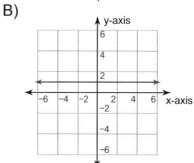

C)
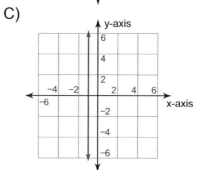

D)
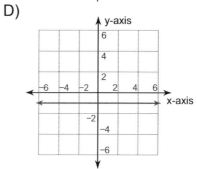

www.math-knots.com | www.a4ace.com

141) Plot the straight-line $y = -\dfrac{2}{3}x + 2$

142) Plot the straight-line $y = \dfrac{3}{2}x + 4$

A)

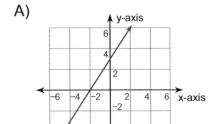

B)

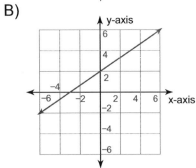

C)

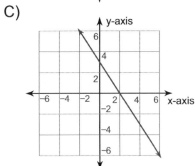

D)

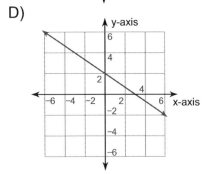

A)

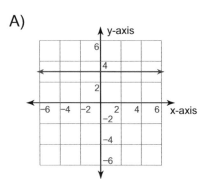

B)

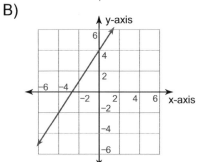

C)

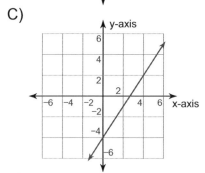

D)

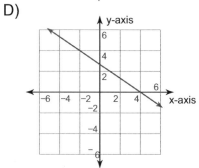

www.math-knots.com | www.a4ace.com

Algebra 1

143) Plot the straight-line $y = \frac{3}{5}x + 2$

A)

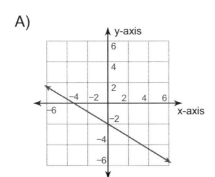

B)

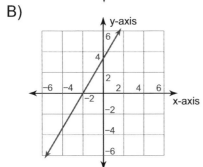

C)

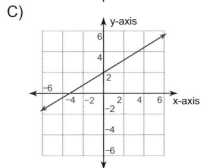

D)

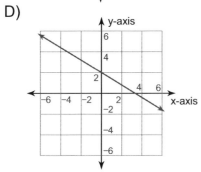

144) Plot the straight-line $y = \frac{1}{5}x + 3$

A)

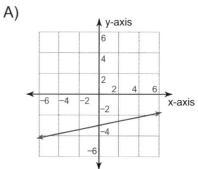

B)

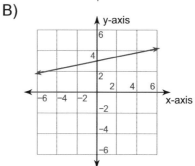

C)

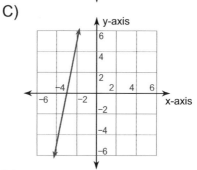

D)

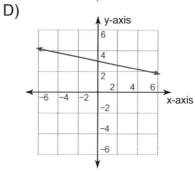

www.math-knots.com | www.a4ace.com

145) Plot the straight-line $y = -\dfrac{7}{3}x + 5$

146) Plot the straight-line $y = -x + 3$

A)
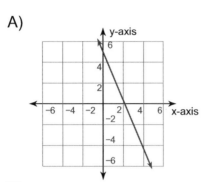

B)
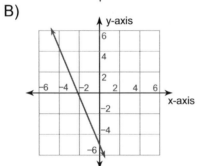

C)
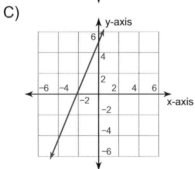

D)
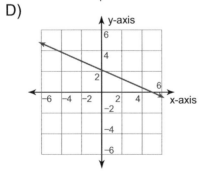

A)
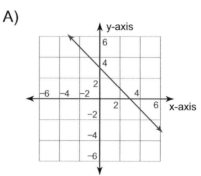

B)
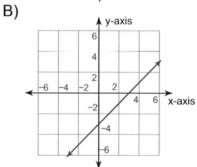

C)
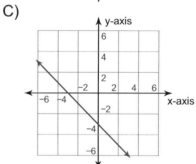

D)
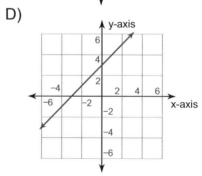

www.math-knots.com | www.a4ace.com

Algebra 1

147) Plot the straight-line $1 + \frac{1}{5}y + \frac{1}{10}x = 0$

148) Plot the straight-line $-21x - 6y = 30$

A)

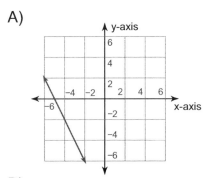

A)

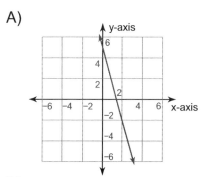

B)

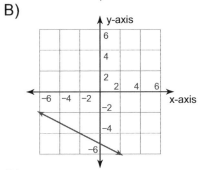

B)

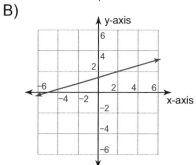

C)

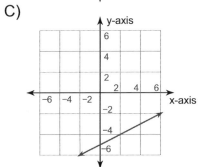

C)

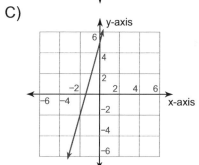

D)

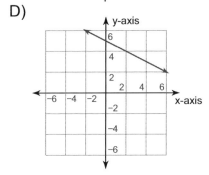

D)

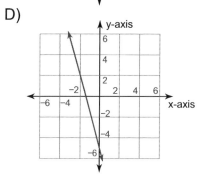

www.math-knots.com | www.a4ace.com

149) Plot the straight-line $3x = -y + 3$

150) Plot the straight-line $5y - 7x = 10$

A)
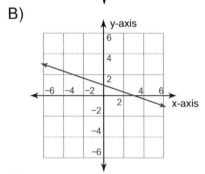

150) A)
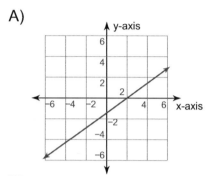

B)
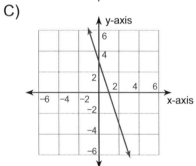

B)
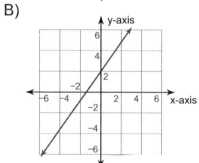

C)

C)
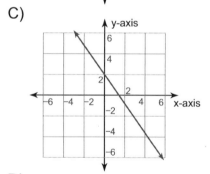

D)
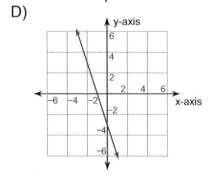

D)
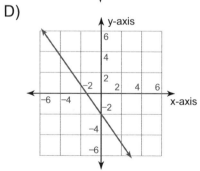

www.math-knots.com | www.a4ace.com

151) Plot the straight-line $-3 + 2x = -y$

152) Plot the straight-line $-2 = -3x - 2y$

A)

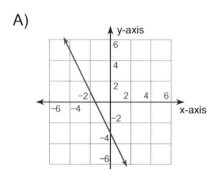

B)

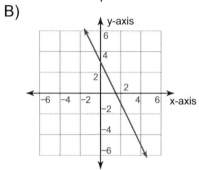

C)

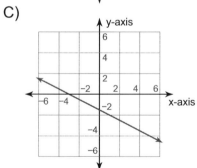

D)

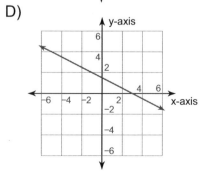

A)

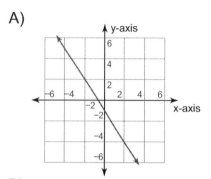

B)

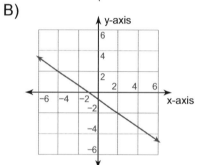

C)

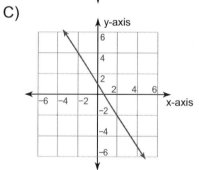

D)

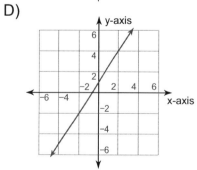

Algebra 1

153) Plot the straight-line $-\dfrac{3}{7}y = -x + \dfrac{15}{7}$

154) Plot the straight-line $-x - y = -1$

A)

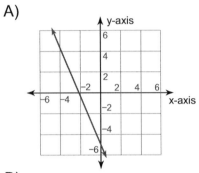

B)

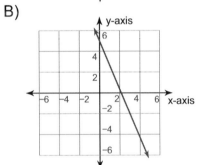

C)

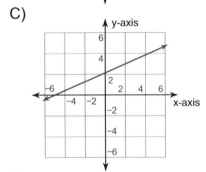

D)

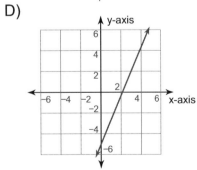

A)

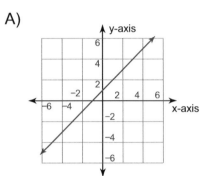

B)

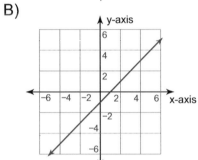

C)

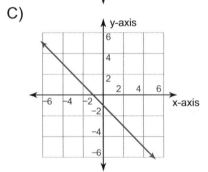

D)

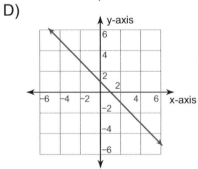

www.math-knots.com | www.a4ace.com

Algebra 1

155) Plot the straight-line $0 = -x + 2 + 2y$

156) Plot the straight-line $-2y = -4 + 3x$

A)

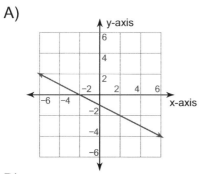

A)

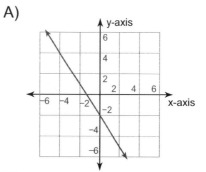

B)

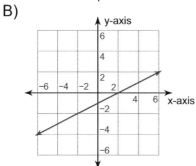

B)

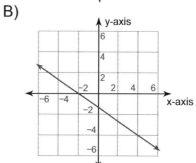

C)

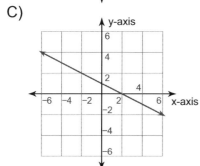

C)

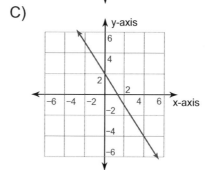

D)

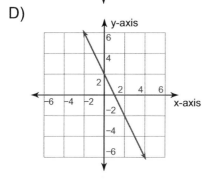

D)

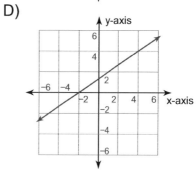

www.math-knots.com | www.a4ace.com

157) Plot the straight-line $-3 - y = 2x$

158) Plot the straight-line $5 + x = -y$

A)
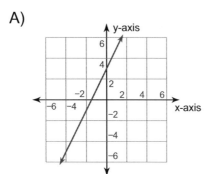

B)
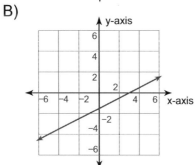

C)
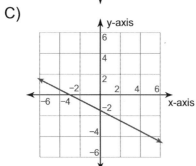

D)
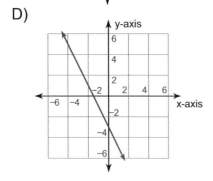

A)
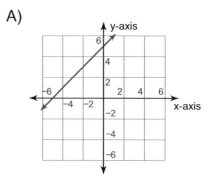

B)
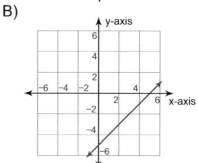

C)
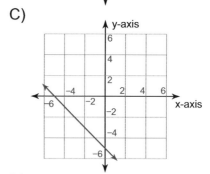

D)
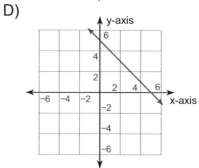

www.math-knots.com | www.a4ace.com

159) Plot the straight-line y + 3 x = 2

160) Plot the straight-line − 2 y − 8 − x = 0

A)
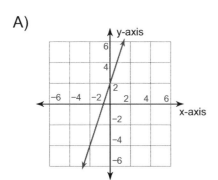

B)
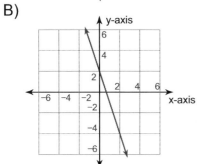

C)
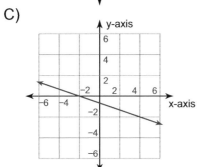

D)
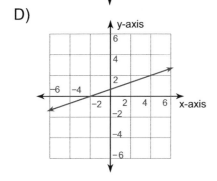

A)
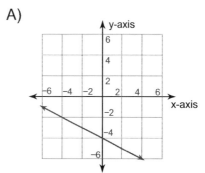

B)
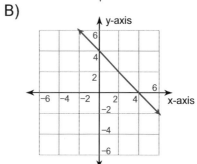

C)
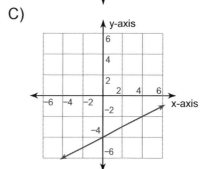

D)
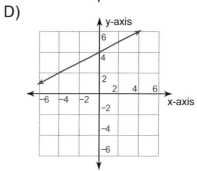

Algebra 1

161) Plot the straight-line $-y + 1 = -4x$

A)

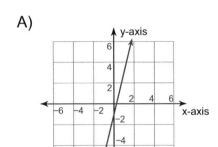

B)

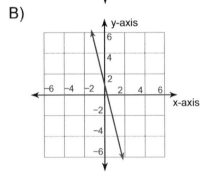

C)

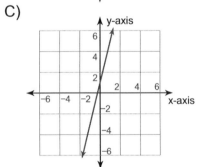

D)

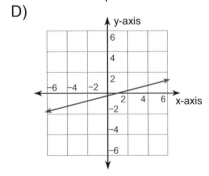

162) Plot the straight-line $-4x = -3 + 3y$

A)

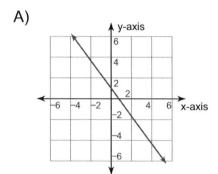

B)

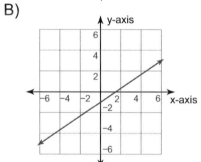

C)

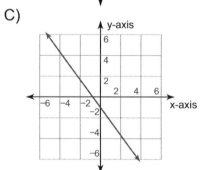

D)

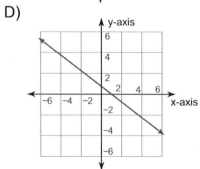

www.math-knots.com | www.a4ace.com

Algebra 1

163) Plot the straight-line $15 - 8x = 5y$

A)

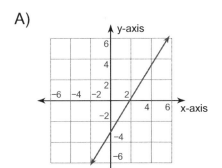

B)

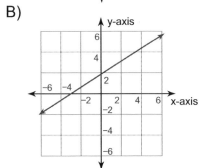

C)

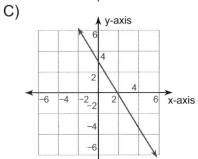

D)

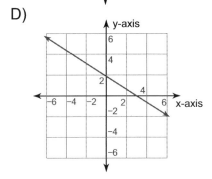

164) Plot the straight-line $-5y - 10 = 6$

A)

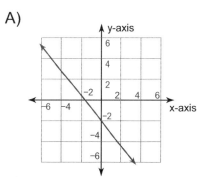

B)

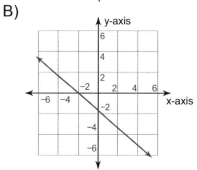

C)

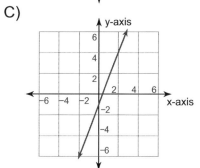

D)

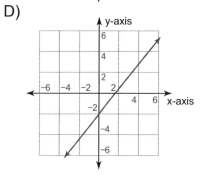

www.math-knots.com | www.a4ace.com

1) Below points F (− 3 , − 1) , G (− 2 , 3) , H (− 1 , 3) , I (0 , 1) are translated to 2 units right and 4 units down. Find the new coordinates.

   A) F' (− 3 , − 3) , G' (− 2 , 1) , H' (− 1 , 1) , I' (0 , − 1)

   B) F' (0 , 1) , G' (1 , 5) , H' (2 , 5) , I' (3 , 3)

   C) F' (− 1 , − 5) , G' (0 , − 1) , H' (1 , − 1) , I' (2 , − 3)

   D) F' (1 , 1) , G' (2 , 5) , H' (3 , 5) , I' (4 , 3)

2) Below points L (1 , 1) , K (2 , 5) , J (5 , 1) are translated to 2 units down. Find the new coordinates.

   A) L' (− 3 , − 3) , K' (− 2 , 1) , J' (1 , − 3)

   B) L' (− 4 , − 2) , K' (− 3 , 2) , J' (0 , − 2)

   C) L' (1 , − 1) , K' (2 , 3) , J' (5 , − 1)

   D) L' (− 5 , − 1) , K' (− 4 , 3) , J' (− 1 , − 1)

3) Below points E (− 4 , − 1) , D (− 4 , 3) , C (− 3 , 3) , B (0 , 1) are translated to 4 units right and 1 unit down. Find the new coordinates.

   A) E' (0 , − 2) , D' (0 , 2) , C' (1 , 2) , B' (4 , 0)

   B) E' (− 4 , 1) , D' (− 3 , 5) , C' (− 3 , 5) , B' (0 , 3)

   C) E' (− 2 , − 1) , D' (− 2 , 3) , C' (− 1 , 3) , B' (2 , 1)

   D) E' (− 5 , − 1) , D' (− 5 , 3) , C' (− 4 , 3) , B' (− 1 , 1)

4) Below points U (− 4 , − 5) , V (− 3 , − 2) , W (0 , − 2) , X (1 , − 4) are translated to 3 units right and 2 units up. Find the new coordinates.

   A) U' (− 5 , − 5) , V' (− 4 , − 2) , W' (− 1 , − 2) , X' (0 , − 4)

   B) U' (− 1 , − 3) , V' (0 , 0) , W' (3 , 0) , X' (4 , − 2)

   C) U' (− 5 , − 3) , V' (− 4 , 0) , W' (− 1 , 0) , X' (0 , − 2)

   D) U' (− 1 , − 1) , V' (0 , 2) , W' (3 , 2) , X' (4 , 0)

5) Below points D (− 1 , − 2) , C (0 , 0) , B (4 , − 4) are translated to 1 unit left and 2 units up. Find the new coordinates.

   A) D' (− 2 , 0) , C' (− 1 , 2) , B' (3 , − 2)

   B) D' (− 3 , − 1) , C' (− 2 , 1) , B' (2 , − 3)

   C) D' (− 3 , 0) , C' (− 2 , 2) , B' (2 , − 2)

   D) D' (− 3 , − 3) , C' (− 2 , − 1) , B' (2 , − 5)

6) Below points C (1 , 1) , D (1 , 5) , E (3 , 5) , F (2 , 1) are translated to 4 units left and 1 unit down. Find the new coordinates.

   A) C' (2 , − 2) , D' (2 , 2) , E' (4 , 2) , F' (3 , − 2)

   B) C' (− 5 , 0) , D' (− 5 , 4) , E' (− 3 , 4) , F' (− 4 , 0)

   C) C' (0 , − 5) , D' (0 , − 1) , E' (2 , − 1) , F' (1 , − 5)

   D) C' (− 3 , 0) , D' (− 3 , 4) , E' (− 1 , 4) , F' (− 2 , 0)

7) Below points I (3 , 1) , H (4 , 4) , G (5 , 4) , F (3 , − 1) are translated to 4 units left. Find the new coordinates.

A) I' (1 , − 1) , H' (2 , 2) , G' (3 , 2) , F' (1 , − 3)

B) I' (− 2 , 2) , H' (− 1 , 5) , G' (0 , 5) , F' (− 2 , 0)

C) I' (0 , 1) , H' (1 , 4) , G' (2 , 4) , F' (0 , − 1)

D) I' (− 1 , 1) , H' (0 , 4) , G' (1 , 4) , F' (− 1 , − 1)

8) Below points I (− 1 , − 2) , H (− 3 , 3) , G (− 2 , 3) , F (0 , − 2) are translated to 2 units left and 1 unit down. Find the new coordinates.

A) I' (− 3 , − 3) , H' (− 5 , 2) , G' (− 4 , 2) , F' (− 2 , − 3)

B) I' (1 , 0) , H' (− 1 , 5) , G' (0 , 5) , F' (2 , 0)

C) I' (1 , − 2) , H' (− 1 , 3) , G' (0 , 3) , F' (2 , − 2)

D) I' (1 , − 1) , H' (− 1 , 4) , G' (0 , 4) , F' (2 , − 1)

9) Below points J (− 5 , − 2) , K (− 3 , 0) , L (− 2 , − 3) are translated to 7 units right and 4 units up. Find the new coordinates.

A) J' (− 4 , − 2) , K' (− 2 , 0) , L' (− 1 , − 3)

B) J' (− 5 , − 3) , K' (− 3 , − 1) , L' (− 2 , − 4)

C) J' (2 , 2) , K' (4 , 4) , L' (5 , 1)

D) J' (− 1 , − 3) , K' (1 , − 1) , L' (2 , − 4)

10) Below points V (3 , 1) , U (2 , 5) , T (5 , 2) are translated to 3 units left and 4 units down. Find the new coordinates.

A) V' (− 1 , − 1) , U' (− 2 , 3) , T' (1 , 0)

B) V' (0 , − 3) , U' (− 1 , 1) , T' (2 , − 2)

C) V' (− 2 , − 1) , U' (− 3 , 3) , T' (0 , 0)

D) V' (3 , − 1) , U' (2 , 3) , T' (5 , 0)

11) Below points V (− 2 , − 4) , U (− 3 , 0) , T (− 1 , − 1) , S (− 1 , − 4) are translated to 2 units left and 5 units up. Find the new coordinates.

A) V' (1 , − 3) , U' (0 , 1) , T' (2 , 0) , S' (2 , − 3)

B) V' (0 , − 5) , U' (− 1 , − 1) , T' (1 , − 2) , S' (1 , − 5)

C) V' (3 , 1) , U' (2 , 5) , T' (4 , 4) , S' (4 , 1)

D) V' (− 4 , 1) , U' (− 5 , 5) , T' (− 3 , 4) , S' (− 3 , 1)

12) Below points J (− 4 , − 3) , K (− 3 , 0) , L (− 1 , − 1) are translated to 2 units right. Find the new coordinates.

A) J' (− 4 , − 5) , K' (− 3 , − 2) , L' (− 1 , − 3)

B) J' (− 3 , − 1) , K' (− 2 , 2) , L' (0 , 1)

C) J' (− 2 , − 3) , K' (− 1 , 0) , L' (1 , − 1)

D) J' (2 , 2) , K' (3 , 5) , L' (5 , 4)

Algebra 1

13) Below points H (− 3 , − 4) , G (− 2 , − 2) , F (1 , − 5) are translated to 3 units right and 3 units up. Find the new coordinates.

A) H' (− 4 , 3) , G' (− 3 , 5) , F' (0 , 2)

B) H' (0 , − 1) , G' (1 , 1) , F' (4 , − 2)

C) H' (− 3 , 1) , G' (− 2 , 3) , F' (1 , 0)

D) H' (0 , 1) , G' (1 , 3) , F' (4 , 0)

14) Below points B (3 , − 3) , C (5 , − 2) , D (5 , − 4) are translated to 2 units left and 4 units up. Find the new coordinates.

A) B' (0 , − 1) , C' (2 , 0) , D' (2 , − 2)

B) B' (1 , 1) , C' (3 , 2) , D' (3 , 0)

C) B' (2 , 1) , C' (4 , 2) , D' (4 , 0)

D) B' (− 1 , 1) , C' (1 , 2) , D' (1 , 0)

15) Below points I (− 4 , 3) , J (− 2 , 4) , K (1 , 2) are translated to 1 unit left and 5 units down. Find the new coordinates.

A) I' (− 1 , 2) , J' (1 , 3) , K' (4 , 1)

B) I' (− 1 , − 3) , J' (1 , − 2) , K' (4 , − 4)

C) I' (0 , − 2) , J' (2 , − 1) , K' (5 , − 3)

D) I' (− 5 , − 2) , J' (− 3 , − 1) , K' (0 , − 3)

16) Below points S (0 , 1) , R (− 1 , 3) , Q (2 , 4) , P (3 , 2) are translated to 1 unit left and 6 units down. Find the new coordinates.

A) S' (− 4 , 2) , R' (− 5 , 4) , Q' (− 2 , 5) , P' (− 1 , 3)

B) S' (1 , 0) , R' (0 , 2) , Q' (3 , 3) , P' (4 , 1)

C) S' (− 1 , − 5) , R' (− 2 , − 3) , Q' (1 , − 2) , P' (2 , − 4)

D) S' (1 , − 1) , R' (0 , 1) , Q' (3 , 2) , P' (4 , 0)

17) Below points I (− 1 , − 3) , H (2 , 1) , G (4 , − 4) are translated to 3 units up. Find the new coordinates.

A) I' (− 1 , 0) , H' (2 , 4) , G' (4 , − 1)

B) I' (0 , − 3) , H' (3 , 1) , G' (5 , − 4)

C) I' (− 5 , 1) , H' (− 2 , 5) , G' (0 , 0)

D) I' (− 1 , 1) , H' (2 , 5) , G' (4 , 0)

18) Below points K (1 , 4) , L (4 , 5) , M (1 , 3) are translated to 1 unit left and 8 units down. Find the new coordinates.

A) K' (− 4 , − 3) , L' (− 1 , − 2) , M' (− 4 , − 4)

B) K' (1 , − 2) , L' (4 , − 1) , M' (1 , − 3)

C) K' (0 , − 4) , L' (3 , − 3) , M' (0 , − 5)

D) K' (0 , 3) , L' (3 , 4) , M' (0 , 2)

19) Below points H (− 2 , 1) , I (− 3 , 4) , J (0 , 1) are translated to 1 unit left and 6 units down. Find the new coordinates.

   A) H' (0 , 1) , I' (− 1 , 4) , J' (2 , 1)

   B) H' (− 3 , 1) , I' (− 4 , 4) , J' (− 1 , 1)

   C) H' (2 , − 5) , I' (1 , − 2) , J' (4 , − 5)

   D) H' (− 3 , − 5) , I' (− 4 , − 2) , J' (− 1 , − 5)

20) Below points S (− 5 , − 3) , T (− 5 , − 1) , U (− 2 , 0) , V (− 4 , − 3) are translated to 4 units right and 1 unit down. Find the new coordinates.

   A) S' (− 4 , − 4) , T' (− 4 , − 2) , U' (− 1 , − 1) , V' (− 3 , − 4)

   B) S' (− 1 , − 4) , T' (− 1 , − 2) , U' (2 , − 1) , V' (0 , − 4)

   C) S' (− 1 , − 5) , T' (− 1 , − 3) , U' (2 , − 2) , V' (0 , − 5)

   D) S' (− 3 , − 2) , T' (− 3 , 0) , U' (0 , 1) , V' (− 2 , − 2)

21) Below points J (0 , − 3) , K (3 , 1) , L (2 , − 4) are translated to 4 units left and 4 units up. Find the new coordinates.

   A) J' (2 , 0) , K' (5 , 4) , L' (4 , − 1)

   B) J' (− 4 , 0) , K' (− 1 , 4) , L' (− 2 , − 1)

   C) J' (− 4 , 1) , K' (− 1 , 5) , L' (− 2 , 0)

   D) J' (− 5 , − 3) , K' (− 2 , 1) , L' (− 3 , − 4)

22) Below points B (1 , − 1) , C (4 , 0) , D (2 , − 4) are translated to 1 unit left and 5 units up. Find the new coordinates.

   A) B' (0 , 4) , C' (3 , 5) , D' (1 , 1)

   B) B' (− 2 , 3) , C' (1 , 4) , D' (− 1 , 0)

   C) B' (1 , − 2) , C' (4 , − 1) , D' (2 , − 5)

   D) B' (− 1 , 3) , C' (2 , 4) , D' (0 , 0)

23) Below points U (− 4 , − 4) , V (− 2 , 0) , W (0 , − 3) are translated to 2 units right and 5 units up. Find the new coordinates.

   A) U' (− 1 , − 5) , V' (1 , − 1) , W' (3 , − 4)

   B) U' (− 1 , − 4) , V' (1 , 0) , W' (3 , − 3)

   C) U' (− 2 , 1) , V' (0 , 5) , W' (2 , 2)

   D) U' (− 4 , 0) , V' (− 2 , 4) , W' (0 , 1)

24) Below points E (2 , − 3) , D (4 , − 1) , C (3 , − 5) are translated to 5 units left and 5 units up. Find the new coordinates.

   A) E' (− 4 , 2) , D' (− 2 , 4) , C' (− 3 , 0)

   B) E' (1 , 1) , D' (3 , 3) , C' (2 , − 1)

   C) E' (2 , − 1) , D' (4 , 1) , C' (3 , − 3)

   D) E' (− 3 , 2) , D' (− 1 , 4) , C' (− 2 , 0)

25) Below points H (– 3 , 0) , G (– 2 , 3) , F (0 , – 1) are translated to 1 unit left and 2 units down. Find the new coordinates.

A) H' (– 4 , 2) , G' (– 3 , 5) , F' (– 1 , 1)

B) H' (– 4 , – 2) , G' (– 3 , 1) , F' (– 1 , – 3)

C) H' (0 , – 1) , G' (1 , 2) , F' (3 , – 2)

D) H' (– 1 , – 3) , G' (0 , 0) , F' (2 , – 4)

26) Below points E (2 , – 4) , D (3 , – 2) , C (4 , – 5) are translated to 1 unit right . Find the new coordinates.

A) E' (3 , – 4) , D' (4 , – 2) , C' (5 , – 5)

B) E' (3 , 1) , D' (4 , 3) , C' (5 , – 5)

C) E' (– 3 , – 4) , D' (– 2 , – 2) , C' (– 1 , – 5)

D) E' (– 1 , – 3) , D' (0 , – 1) , C' (1 , – 4)

27) Below points F (– 3 , – 3) , G (– 2 , 0) , H (– 1 , – 5) are translated to 6 units right and 5 units up. Find the new coordinates.

A) F' (– 4 ,– 1) , G' (– 3 , 2) , H' (– 2 , – 3)

B) F' (– 3 , – 2) , G' (– 2 , 1) , H' (– 1 , – 4)

C) F' (– 1 , – 3) , G' (0 , 0) , H' (1 , – 5)

D) F' (3 , 2) , G' (4 , 5) , H' (5 , 0)

28) Below points I (2 , – 3) , H (4 , – 1) , G (4 , – 4) are translated to 4 units left and 1 unit up. Find the new coordinates.

A) I' (– 2 , – 2) , H' (0 , 0) , G' (0 , – 3)

B) I' (– 3 , 1) , H' (– 1 , 3) , G' (– 1 , 0)

C) I' (– 4 , 2) , H' (– 2 , 4) , G' (– 2 , 1)

D) I' (– 3 , 3) , H' (– 1 , 5) , G' (– 1 , 2)

29) Find the rule of translation from U (1 , 3) , V (3 , 5) , W (5 , 4) , X (2 , 1) to U' (– 4 , – 3) , V'(– 2 , – 1) , W' (0 , – 2) , X' (– 3 , – 5)

A) translation: 2 units left
B) translation: 2 units left and 3 units down
C) translation: 5 units left and 6 units down
D) translation: 1 unit left and 6 units down

30) Find the rule of translation from V (1 , – 3) , W (0 , 0) , X (4 , – 1) , Y (5 , – 3) to V' (– 1 , – 4) , W'(– 2 , – 1) , X' (2 ,– 2) , Y' (3 , – 4)

A) translation: 1 unit left and 2 units up
B) translation: 1 unit left and 2 units down
C) translation: 2 units left and 1 unit down
D) translation: 2 units left and 2 units down

 **Algebra 1**

31) Find the rule of translation from
I (− 5 , − 1) , J (− 5 , 0) , K (− 1 , 3) ,
L (− 2 , 0)
to
I' (0 , − 3) , J' (0 , − 2) , K' (4 , 1) ,
L' (3 , − 2)

A) translation: 5 units right and 2 units down

B) translation: 1 unit up

C) translation: 4 units right and 4 units down

D) translation: 6 units right and 1 unit down

32) Find the rule of translation from
J (0 , − 3) , I (1 , − 1) , H (5 , − 2)
to
J' (− 2 , − 5) , I' (− 1 , − 3) , H' (3 , − 4)

A) translation: 4 units left and 1 unit down

B) translation: 5 units left and 1 unit down

C) translation: 3 units up

D) translation: 2 units left and 2 units down

33) Find the rule of translation from'

E (− 2 , − 3) , F (− 3 , − 1) , G (− 1 , 1) ,
H (2 , − 1)
to
E' (− 4 , − 5) , F' (− 5 , − 3) , G' (− 3 , − 1)
H' (0 , − 3)

A) translation: 4 units up

B) translation: 2 units left and 2 units down

C) translation: 1 unit right and 1 unit up

D) translation: 1 unit right and 3 units up

34) Find the rule of translation from
G (− 3 , − 4) , H (− 1 , 1) ,
I (0 , 1) , J (− 1 , − 2)
to
G' (− 1 , − 5) , H' (1 , 0) ,
I' (2 , 0) , J' (1 , − 3)

A) translation: 3 units up

B) translation: 2 units right
and 1 unit down

C) translation: 2 units right
and 4 units up

D) translation: 1 unit down

35) Find the rule of translation from
G (2 , − 2) , F (1 , 1) , E (5 , 1) ,
D (3 , − 3)
to
G' (− 4 , − 3) , F' (− 5 , 0) ,
E' (− 1 , 0) , D' (− 3 , − 4)

A) translation: 2 units up

B) translation: 2 units down

C) translation: 4 units left and
2 units up

D) translation: 6 units left and
1 unit down

36) Find the rule of translation from
F (− 3 , 1) , G (− 5 , 4) ,
H (− 3 , 5) , I (1 , 2)
to
F' (0 , − 1) , G' (− 2 , 2) ,
H' (0 ,3) , I' (4 , 0)

A) translation: 6 units down

B) translation: 1 unit right and
1 unit down

C) translation: 3 units right and
2 units down

D) translation: 1 unit right and
4 units down

 www.math-knots.com | www.a4ace.com

37) Find the rule of translation from

U (− 5 , − 2) , V (− 3 , − 1) , W (− 5 , − 5)
to
U' (2 , 0) , V' (4 , 1) , W' (2 , − 3)

A) translation: 4 units up

B) translation: 7 units right and 2 units up

C) translation: 8 units right

D) translation: 1 unit right and 5 units up

38) Find the rule of translation from

J (− 2 , − 4) , K (− 3 , − 2) , L (1 , − 1) ,
M (0 , − 4)
to
J' (− 4 , 0) , K' (− 5 , 2) , L' (− 1 , 3) ,
M' (− 2 , 0)

A) translation: 3 units right and 6 units up

B) translation: 2 units left and 4 units up

C) translation: 3 units right and 1 unit down

D) translation: 4 units up

39) Find the rule of translation from

K (− 3 , − 4) , L (− 5 , 0) , M (− 5 , 1) ,
N (0 , − 1)
to
K' (− 2 , − 4) , L' (− 4 , 0) , M' (− 4 , 1) ,
N' (1 , − 1)

A) translation: 1 unit right and 2 units up

B) translation: 5 units right and 3 units up

C) translation: 4 units right and 2 units up

D) translation: 1 unit right

40) Find the rule of translation from
F (− 4 , − 3) , E (− 2 , 0) ,
D (0 , − 3) ,
to
F' (− 2 , 0) , E' (0 , 3) ,
D' (2 , 0)

A) translation: 1 unit right
and 1 unit up

B) translation: 2 units right

C) translation: 1 unit left and
5 units up

D) translation: 2 units right
and 3 units up

41) Find the rule of translation from
R (1 , 4) , S (4 , 5) , T (1 , 2)
to
R' (− 3 , − 1) , S' (0 , 0) ,
T' (− 3 , − 3)

A) translation: 4 units left
and 5 units down

B) translation: 3 units left
and 6 units down

C) translation: 3 units left
and 4 units down

D) translation: 6 units left
and 3 units down

42) Find the rule of translation from
E (− 1 , − 2) , D (0 , 1) ,
C (4 , − 1) , B (3 , − 3)
to
E' (− 2 , − 3) , D' (− 1 , 0) ,
C' (3 , − 2) , B' (2 , − 4)

A) translation: 4 units left
and 4 units up

B) translation: 1 unit left
and 4 units up

C) translation: 1 unit left
and 1 unit down

D) translation: 1 unit right
and 1 unit up

www.math-knots.com | www.a4ace.com

43) Find the rule of translation from

I (− 5 , 4) , J (− 4 , 5) , K (− 2 , 3)
to
I' (− 2 , − 3) , J' (− 1 , − 2) , K' (1 , − 4)

A) translation: 7 units right

B) translation: 5 units right and 5 units down

C) translation: 3 units right and 6 units down

D) translation: 3 units right and 7 units down

44) Find the rule of translation from

D (− 1 , − 3) , E (− 1 , − 2) , F (3 , − 1) ,
G (1 , − 5)
to
D' (1 , 2) , E' (1 , 3) , F' (5 , 4) ,
G' (3 , 0)

A) translation: 4 units up

B) translation: 2 units right and 5 units up

C) translation: 4 units left and 6 units up

D) translation: 2 units right and 1 unit up

45) Find the rule of translation from

H (3 , − 4) , I (2 , − 1) , J (4 , − 4)
to
H' (4 , 1) , I' (3 , 4) , J' (5 , 1)

A) translation: 1 unit left and 1 unit down

B) translation: 6 units left

C) translation: 1 unit right and 5 units up

D) translation: 3 units left and 3 units up

46) Find the rule of translation from

J (− 3 , 0) , K (− 3 , 5) ,
L (− 1 , 5) , M (0 , 0)
to
J' (− 1 , − 5) , K' (− 1 , 0) ,
L' (1 , 0) ,M' (2 , − 5)

A) translation: 1 unit down

B) translation: 1 unit right and
4 units down

C) translation: 5 units right and
3 units down

D) translation: 2 units right and
5 units down

47) Find the rule of translation from

F (− 3 , − 2) , G (− 3 , 1) ,
H (1 ,1) ,I (− 1 , − 3)
to
F' (− 4 , − 2) , G' (− 4 , 1) ,
H' (0 , 1) ,I' (− 2 , − 3)

A) translation: 1 unit left

B) translation: 1 unit left
and 2 units up

C) translation: 3 units up

D) translation: 4 units right

48) Find the rule of translation from

H (− 5 , 0) , I (− 4 , 3) ,
J (− 3 , 2)
to
H' (− 3 , 1) , I' (− 2 , 4) ,
J' (− 1 , 3)

A) translation: 3 units down

B) translation: 2 units right
and 1 unit up

C) translation: 2 units up

D) translation: 1 unit right

49) Find the rule of translation from

K (2 , − 3) , L (4 , 2) , M (5 , 0)
to
K' (− 1 , − 1) , L' (1 , 4) , M' (2 , 2)

A) translation: 3 units left and 2 units up

B) translation: 5 units left and 1 unit down

C) translation: 6 units left and 3 units up

D) translation: 3 units left

50) Find the rule of translation from

Y (3 , 0) , X (4 , 3) , W (5 , 1)
to
Y' (− 4 , − 2) , X' (− 3 , 1) , W' (− 2 , − 1)

A) translation: 2 units left and 5 units down

B) translation: 4 units left and 2 units up

C) translation: 7 units left and 2 units down

D) translation: 4 units left and 3 units down

51) Find the rule of translation from

H (2 , − 3) , G (4 , − 1) , F (5 , − 2)
to
H' (− 1 , 2) , G' (1 , 4) , F' (2 , 3)

A) translation: 2 units left and 4 units up

B) translation: 5 units left and 5 units up

C) translation: 3 units left and 5 units up

D) translation: 2 units up

52) Find the rule of translation from
C (− 4 , − 1) , D (− 2 , 3) ,
E (− 1 , 2) , F (1 , − 3)
to
C' (− 3 , 1) , D' (− 1 , 5) ,
E' (0 , 4) ,F' (2 , − 1)

A) translation: 3 units right and 2 units down

B) translation: 3 units right and 1 unit down

C) translation: 1 unit right and 2 units up

D) translation: 4 units right and 2 units up

53) Find the rule of translation from
J (− 4 , 2) , I (− 1 , 5) , H (1 ,1)
to
J' (0 , − 2) , I' (3 , 1) , H' (5 , − 3)

A) translation: 4 units right and 1 unit down

B) translation: 3 units right and 5 units down

C) translation: 2 units right

D) translation: 4 units right and 4 units down

54) Find the rule of translation from
Z (0 , − 4) , Y (− 2 , − 1) ,
X (1 , 0) ,W (4 , − 1)
to
Z' (− 2 , − 4) , Y' (− 4 , − 1) ,
X' (− 1 , 0) ,W' (2 , − 1)

A) translation: 1 unit right and 2 units up

B) translation: 2 units left and 4 units up

C) translation: 2 units left

D) translation: 1 unit left and 1 unit down

Algebra 1

55) Find the rule of translation from

$$K(-3,-1), L(-3,4), M(0,2)$$
to
$$K'(-1,0), L'(-1,5), M'(2,3)$$

A) translation: 1 unit right and 1 unit down

B) translation: 2 units right and 1 unit up

C) translation: 5 units right and 1 unit up

D) translation: 4 units right and 3 units down

56) Find the rule of translation from

$$W(-5,-2), V(-3,-1), U(-4,-3)$$
to
$$W'(2,2), V'(4,3), U'(3,1)$$

A) translation: 7 units right and 4 units up

B) translation: 4 units right and 1 unit up

C) translation: 4 units right and 1 unit down

D) translation: 1 unit right and 2 units up

57) Graph the transformation of the figure plotted below to 2 units right and 1 unit up.

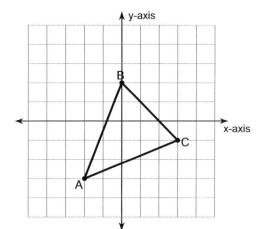

58) Graph the transformation of the figure plotted below to 4 units right and 3 units down.

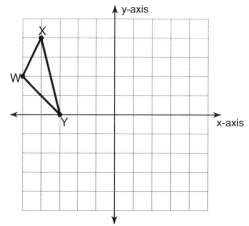

59) Graph the transformation of the figure plotted below to 3 units right and 7 units down.

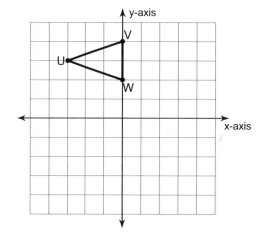

60) Graph the transformation of the figure plotted below to 1 unit left

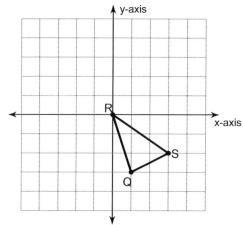

Algebra 1

61) Graph the transformation of the figure plotted below to 6 units left and 4 units down.

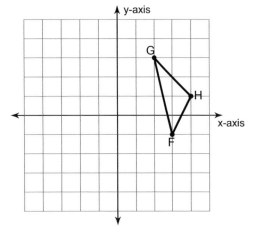

62) Graph the transformation of the figure plotted below to 5 units left and 1 unit down.

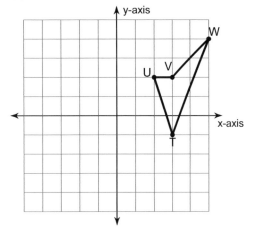

63) Graph the transformation of the figure plotted below to 7 units left and 1 unit down.

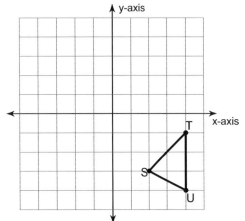

64) Graph the transformation of the figure plotted below to 5 units left and 3 units up.

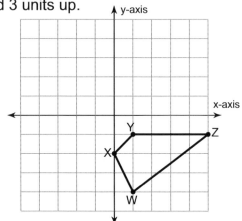

65) Graph the transformation of the figure plotted below to 2 units left and 1 unit up.

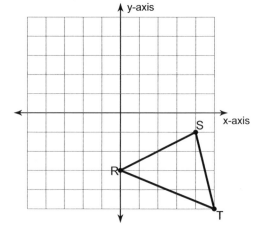

66) Graph the transformation of the figure plotted below to 4 units right and 1 unit down.

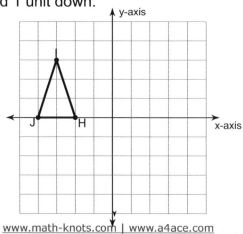

www.math-knots.com | www.a4ace.com

67) Graph the transformation of the figure plotted below to 3 units right and 5 units down.

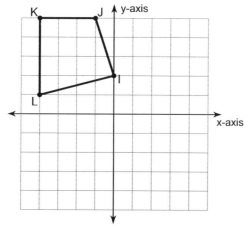

68) Graph the transformation of the figure plotted below to 6 units left and 1 unit up.

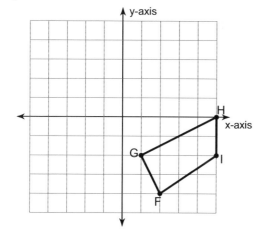

69) Graph the transformation of the figure plotted below to 4 units right and 3 units down.

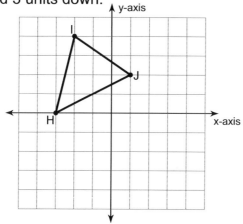

70) Graph the transformation of the figure plotted below to 4 units right.

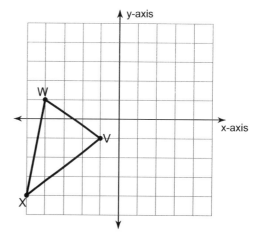

71) Graph the transformation of the figure plotted below to 2 units right and 6 units up.

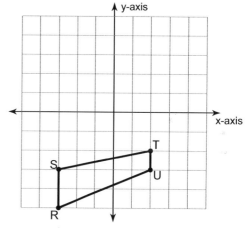

72) Graph the transformation of the figure plotted below to 6 units up.

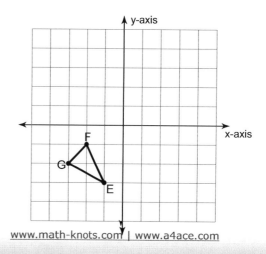

Algebra 1

73) Graph the transformation of the figure plotted below to 5 units left and 2 units down.

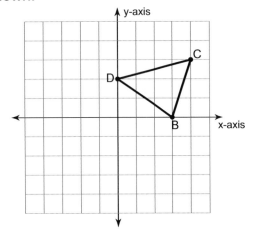

74) Graph the transformation of the figure plotted below to 3 units left and 2 units down.

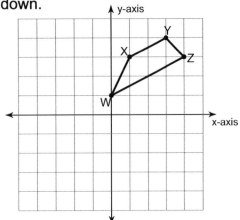

75) Graph the transformation of the figure plotted below to 2 units right and 2 units down.

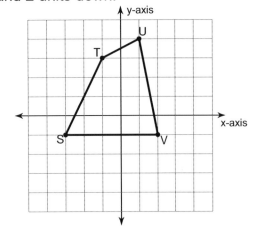

76) Graph the transformation of the figure plotted below to 1 unit right and 4 units up.

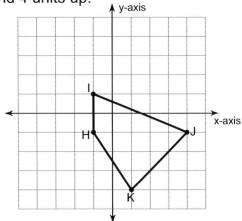

77) Graph the transformation of the figure plotted below to 5 units right and 2 units down.

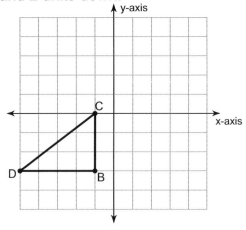

78) Graph the transformation of the figure plotted below to 5 units right.

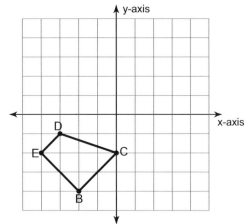

79) Graph the transformation of the figure plotted below to 4 units right and 2 units up.

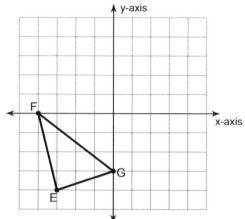

80) Graph the transformation of the figure plotted below to 1 unit right and 1 unit up.

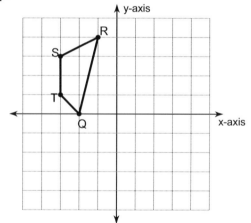

81) Graph the transformation of the figure plotted below to 2 units right and 2 units up.

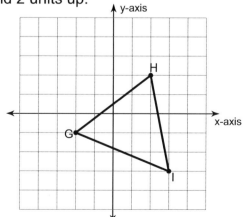

82) Graph the transformation of the figure plotted below to 5 units left and 3 units up.

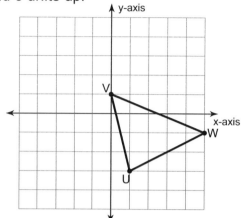

83) Graph the transformation of the figure plotted below to 6 units left and 2 units down.

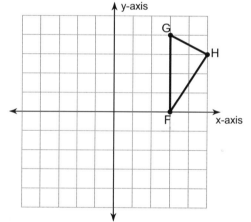

84) Graph the transformation of the figure plotted below to 2 units left and 2 units up.

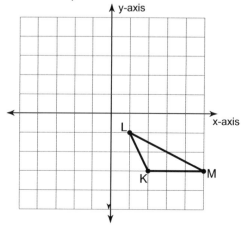

85) Below points A (1 , 3) , B (5 , 4) , C (4 , 1) are reflected across x = 2. Find the new coordinates.

A) B' (1 , 4) , C' (2 , 1) , A' (5 , 3)

B) B' (5 , – 4) , C' (4 , 1) , A' (1 , – 1)

C) B' (– 1 , 4) , C' (0 , 1) , A' (3 , 3)

D) B' (5 , 0) , C' (4 , 3) , A' (1 , 1)

86) Below points A (– 3 , 2) , B (1 , 4) , C (2 , 2) are reflected across x = – 1. Find the new coordinates.

A) B' (– 1 , 4) , C' (– 2 , 2) , A' (3 , 2)

B) B' (– 3 , 4) , C' (– 4 , 2) , A' (1 , 2)

C) B' (1 , – 4) , C' (2 , – 2) , A' (– 3 , – 2)

D) B' (1 , 2) , C' (2 , 4) , A' (– 3 , 4)

87) Below points P (– 4 , – 5) , Q (– 4 , – 1) , R (– 2 , – 5) are reflected across y-axis. Find the new coordinates.

A) Q' (– 4 , – 3) , R' (– 2 , 1) , P' (– 4 , 1)

B) Q' (– 4 , 1) , R' (– 2 , 5) , P' (– 4 , 5)

C) Q' (4 , – 1) , R' (2 , – 5) , P' (4 , – 5)

D) Q' (– 2 , – 1) , R' (– 4 , – 5) , P' (– 2 , – 5)

88) Below points G(2 , -3), H(4 , -2), I(4 , -3) are reflected across x = 1. Find the new coordinates.

A) H' (0 , – 2) , I' (0 , – 3) , G' (2 , – 3)

B) H' (– 2 , – 2) , I' (– 2 , – 3) , G' (0 , – 3)

C) H' (2 , – 2) , I' (2 , – 3) , G' (4 , – 3)

D) H' (4 , 2) , I' (4 , 3) , G' (2 , 3)

89) Below points P(2 , -4), Q(2 , -1), R(5 , -1) , S(4 , -4) are reflected across y = -2. Find the new coordinates.

A) Q' (4 , – 1) , R' (1 , – 1) , S' (2 , – 4) , P' (4 , – 4)

B) Q' (2 , – 3) , R' (5 , – 3) , S' (4 , 0) , P' (2 , 0)

C) Q' (0 , – 1) , R' (– 3 , – 1) , S' (– 2 , – 4) , P' (0 , – 4)

D) Q' (2 , – 1) , R' (– 1 , – 1) , S' (0 , – 4) , P' (2 , – 4)

90) Below points U(-5 , -4), V(-5 , -2), W(0 , -1) , X(-1 , -3) are reflected across y = -1. Find the new coordinates.

A) V' (3 , – 2) , W' (– 2 , – 1) , X' (– 1 , – 3) , U' (3 , – 4)

B) V' (– 5 , 0) , W' (0 , – 1) , X' (– 1 , 1) , U' (– 5 , 2)

C) V' (– 5 , 2) , W' (0 , 1) , X' (– 1 , 3) , U' (– 5 , 4)

D) V' (5 , – 2) , W' (0 , – 1) , X' (1 , – 3) , U' (5 , – 4)

91) Below points C (− 5 , − 1) , D (− 1 , 2) ,
E (− 3 , − 3) are reflected across x = − 1.
Find the new coordinates.

A) D' (1 , 2) , E' (3 , − 3) , C' (5 , − 1)

B) D' (− 1 , 0) , E' (− 3 , 5) , C' (− 5 , 3)

C) D' (− 1 , 2) , E' (1 , − 3) , C' (3 , − 1)

D) D' (− 1 , − 4) , E' (− 3 , 1) ,
C' (− 5 , − 1)

92) Below points G (− 5 , 3) , F (− 3 , 4) ,
E (− 3 , 1) are reflected across x = − 2.
Find the new coordinates.

A) F' (− 3 , − 2) , E' (− 3 , 1) , G' (− 5 , − 1)

B) F' (− 3 , 4) , E' (− 3 , 1) , G' (− 1 , 3)

C) F' (− 3 , 2) , E' (− 3 , 5) , G' (− 5 , 3)

D) F' (− 1 , 4) , E' (− 1 , 1) , G' (1 , 3)

93) Below points D (− 5 , 3) , E (− 5 , 4) ,
F (0 , 2) , G (− 3 , 2) are reflected across
x-axis. Find the new coordinates.

A) E' (− 5 , − 4) , F' (0 , − 2) ,
G' (− 3 , − 2) , D' (− 5 , − 3)

B) E' (3 , 4) , F' (− 2 , 2) ,
G' (1 , 2) , D' (3 , 3)

C) E' (− 5 , − 2) , F' (0 , 0) ,
G' (− 3 , 0) , D' (− 5 , − 1)

D) E' (− 5 , 2) , F' (0 , 4) ,
G' (− 3 , 4) , D' (− 5 , 3)

94) Below points U(-5 , -4), V(-3 , -2),
W(-3 , -4) are reflected across x = -4.
Find the new coordinates.

A) V' (− 3 , 0) , W' (− 3 , 2) ,
U' (− 5 , 2)

B) V' (− 5 , − 2) , W' (− 5 , − 4) ,
U' (− 3 , − 4)

C) V' (− 3 , − 4) , W' (− 3 , − 2) ,
U' (− 5 , − 2)

D) V' (− 3 , 2) , W' (− 3 , 4) ,
U' (− 5 , 4)

95) Below points B(-4 , -3), C(-2 , -3),
D(-1 , -5) are reflected across x = -2.
Find the new coordinates.

A) C' (− 2 , − 5) , D' (− 1 , − 3) ,
B' (− 4 , − 5)

B) C' (− 2 , − 3) , D' (− 3 , − 5) ,
B' (0 , − 3)

C) C' (− 2 , − 1) , D' (− 1 , 1) ,
B' (− 4 , − 1)

D) C' (− 2 , − 3) , D' (− 1 , − 1) ,
B' (− 4 , − 3)

96) Below points D(-3 , 1), E(0 , 5),
F(1 , 3) are reflected across y-axis.
Find the new coordinates.

A) E' (− 2 , 5) , F' (− 3 , 3) ,
D' (1 , 1)

B) E' (2 , 5) , F' (1 , 3) ,
D' (5 , 1)

C) E' (0 , − 1) , F' (1 , 1) ,
D' (− 3 , 3)

D) E' (0 , 5) , F' (− 1 , 3) ,
D' (3 , 1)

97) Below points K (3 , − 4) , J (3 , − 1) ,
I (5 , − 1) are reflected across y = − 3.
Find the new coordinates.

A) J' (1 , − 1) , I' (− 1 , − 1) ,
   K' (1 , − 4)

B) J' (3 , − 5) , I' (5 , − 5) ,
   K' (3 , − 2)

C) J' (3 , − 1) , I' (5 , − 1) ,
   K' (3 , 2)

D) J' (5 , − 1) , I' (3 , − 1) ,
   K' (5 , − 4)

98) Below points D (− 3 , − 3) , E (− 2 , − 1) ,
F (2 , − 3) are reflected across y = − 3.
Find the new coordinates.

A) E' (− 2 , − 3) , F' (2 , − 1) ,
   D' (− 3 , − 1)

B) E' (0 , − 1) , F' (− 4 , − 3) ,
   D' (1 , − 3)

C) E' (2 , − 1) , F' (− 2 , − 3) ,
   D' (3 , − 3)

D) E' (− 2 , − 5) , F' (2 , − 3) ,
   D' (− 3 , − 3)

99) Below points J (4 , − 4) , I (3 , − 2) ,
H (4 , 0) , G (5 , − 4) are reflected across
y = − 1. Find the new coordinates.

A) I' (3 , 2) , H' (4 , 0) , G' (5 , 4) ,
   J' (4 , 4)

B) I' (5 , − 2) , H' (4 , 0) ,
   G' (3 , − 4) ,J' (4 , − 4)

C) I' (− 1 , − 2) , H' (− 2 , 0) ,
   G' (− 3 , − 4) , J' (− 2 , − 4)

D) I' (3 , 0) , H' (4 , − 2) ,
   G' (5 , 2) , J' (4 , 2)

100) Below points B(-1 , -4), C(2 , -1),
D(4 , -3) are reflected across x-axis.
Find the new coordinates.

A) C' (2 , − 1) , D' (4 , 1) ,
   B' (− 1 , 2)

B) C' (− 2 , − 1) , D' (− 4 , − 3) ,
   B' (1 , − 4)

C) C' (2 , − 3) , D' (4 , − 1) ,
   B' (− 1 , 0)

D) C' (2 , 1) , D' (4 , 3) ,
   B' (− 1 , 4)

101) Below points U(-2 , -4), V(-3 , 1),
W(1 , -3) are reflected across x = -1.
Find the new coordinates.

A) V' (1 , 1) , W' (− 3 , − 3) ,
   U' (0 , − 4)

B) V' (− 3 , − 5) , W' (1 , − 1) ,
   U' (− 2 , 0)

C) V' (5 , 1) , W' (1 , − 3) ,
   U' (4 , − 4)

D) V' (− 3 , − 1) , W' (1 , 3) ,
   U' (− 2 , 4)

102) Below points H(-3 , -4), I(-2 , -1),
J(1 , -1) , K(0 , -3) are reflected
across y-axis.
Find the new coordinates.

A) I' (− 2 , − 5) , J' (1 , − 5) ,
   K' (0 , − 3) , H' (− 3 , − 2)

B) I' (− 2 , − 3) , J' (1 , − 3) ,
   K' (0 , − 1) ,H' (− 3 , 0)

C) I' (4 , − 1) , J' (1 , − 1) ,
   K' (2 , − 3) , H' (5 , − 4)

D) I' (2 , − 1) , J' (− 1 , − 1) ,
   K' (0 , − 3) , H' (3 , − 4)

103) Below points E (3 , – 1) , D (3 , 3) , C (5 , 2) are reflected across x-axis. Find the new coordinates.

A) D' (3 , 1) , C' (5 , 2) , E' (3 , 5)

B) D' (3 , – 3) , C' (5 , – 2) , E' (3 , 1)

C) D' (– 1 , 3) , C' (– 3 , 2) , E' (– 1 , – 1)

D) D' (1 , 3) , C' (– 1 , 2) , E' (1 , – 1)

104) Below points G(-1 , -4), H(-1, -2), I(4 , -2) are reflected across x-axis. Find the new coordinates.

A) H' (– 1 , 0) , I' (4 , 0) , G' (– 1 , 2)

B) H' (– 1 , 2) , I' (4 , 2) , G' (– 1 , 4)

C) H' (– 1 , – 2) , I' (4 , – 2) , G' (– 1 , 0)

D) H' (3 , – 2) , I' (– 2 , – 2) , G' (3 , – 4)

105) Below points K (3 , – 4) , J (2 , – 1) , I (5 , – 2) are reflected across x = 3. Find the new coordinates.

A) J' (2 , – 1) , I' (5 , 0) , K' (3 , 2)

B) J' (2 , – 5) , I' (5 , – 4) , K' (3 , – 2)

C) J' (– 2 , – 1) , I' (– 5 , – 2) , K' (– 3 , – 4)

D) J' (4 , – 1) , I' (1 , – 2) , K' (3 , – 4)

106) Below points K (0 , – 3) , J (3 , 1) , I (3 , – 2) are reflected across y = 1. Find the new coordinates.

A) J' (3 , – 3) , I' (3 , 0) , K' (0 , 1)

B) J' (– 3 , 1) , I' (– 3 , – 2) , K' (0 , – 3)

C) J' (3 , – 1) , I' (3 , 2) , K' (0 , 3)

D) J' (3 , 1) , I' (3 , 4) , K' (0 , 5)

107) Below points G(-1 , -3), H(4 , -2), I(4 , -2) are reflected across y = -2. Find the new coordinates.

A) H' (– 2 , – 2) , I' (0 , – 5) , G' (1 , – 3)

B) H' (4 , – 2) , I' (2 , 1) , G' (1 , – 1)

C) H' (4 , – 2) , I' (2 , – 1) , G' (1 , – 3)

D) H' (4 , 0) , I' (2 , 3) , G' (1 , 1)

108) Below points F(-3 , -3), E(-3 , -2), D(-1 , -2) , C(0 , -4) are reflected across x = 1. Find the new coordinates.

A) E' (5 , – 2) , D' (3 , – 2) , C' (2 , – 4) , F' (5 , – 3)

B) E' (3 , – 2) , D' (1 , – 2) , C' (0 , – 4) , F' (3 , – 3)

C) E' (– 1 , – 2) , D' (– 3 , – 2) , C' (– 4 , – 4) , F' (– 1 , – 3)

D) E' (– 3 , – 2) , D' (– 1 , – 2) , C' (0 , 0) , F' (– 3 , – 1)

 **Algebra 1**

Vol 2
Week 20
Reflections

109) Below points S(-4 , -1), T(-3 , 3), U(0 , 2) , V(1 , -3) are reflected across y = 1.
Find the new coordinates.

A) T' (− 3 , − 3) , U' (0 , − 2) , V' (1 , 3) , S' (− 4 , 1)

B) T' (− 3 , − 1) , U' (0 , 0) , V' (1 , 5) , S' (− 4 , 3)

C) T' (− 1 , 3) , U' (− 4 , 2) , V' (− 5 , − 3) , S' (0 , − 1)

D) T' (3 , 2) , U' (0 , 2) , V' (− 1 , − 3) , S' (4 , − 1)

110) Below points C(2 , -1), D(1 , 2), E(4 , 4) , F(5 , 2) are reflected across x-axis.
Find the new coordinates.

A) D' (5 , 2) , E' (2 , 4) , F' (1 , 2) , C' (4 , − 1)

B) D' (− 1 , 2) , E' (− 4 , 4) , F' (− 5 , 2) , C' (− 2 , − 1)

C) D' (3 , 2) , E' (0 , 4) , F' (− 1 , 2) , C' (2 , − 1)

D) D' (1 , − 2) , E' (4 , − 4) , F' (5 , − 2) , C' (2 , 1)

111) Find the rule of translation from

H (1 , 1) , I (3 , 5) , J (5 , 3)
to
I' (3 , − 1) , J' (5 , 1) , H' (1 , 3)

A) reflection across the x-axis

B) reflection across the y-axis

C) reflection across x = 2

D) reflection across y = 2

112) Find the rule of translation from

E (0 , − 4) , F (1 , − 1) , G (3 , − 4)
to
F' (− 1 , − 1) , G' (− 3 , − 4) , E' (0 , − 4)

A) reflection across y = −1

B) reflection across x = 1

C) reflection across the x-axis

D) reflection across the y-axis

113) Find the rule of translation from

Z (− 2 , − 3) , Y (− 2 , − 1) , X (2 , − 2) , W (3 , − 3)
to
Y' (− 2 , 3) , X' (2 , 4) , W' (3 , 5) , Z' (− 2 , 5)

A) reflection across y = 1

B) reflection across x = 1

C) reflection across y = −3

D) reflection across the y-axis

114) Find the rule of translation from

K (2 , − 4) , J (2 , − 1) , I (4 , − 3)
to
J' (2 , − 1) , I' (0 , − 3) , K' (2 , − 4)

A) reflection across x = 3

B) reflection across x = 1

C) reflection across y = − 2

D) reflection across x = 2

115) Find the rule of translation from

B (1 , − 2) , C (1 , 0) , D (3 , − 1) ,
E (2 , − 3)
to
C' (1 , − 4) , D' (3 , − 3) , E' (2 , − 1) ,
B' (1 , − 2)

A) reflection across y = −2

B) reflection across y = −1

C) reflection across y = 1

D) reflection across x = 2

116) Find the rule of translation from

S (− 3 , − 2) , T (− 1 , 3) , U (1 , 3) ,
V (− 1 , − 2)
to
T' (− 3 , 3) , U' (− 5 , 3) , V' (− 3 , − 2) ,
S' (− 1 , − 2)

A) reflection across x = − 2

B) reflection across x = 1

C) reflection across y = 1

D) reflection across y = − 1

117) Find the rule of translation from

W (− 4 , − 3) , V (− 5 , 0) , U (0 , 3) ,
T (0 , 0)
to
V' (1 , 0) , U' (− 4 , 3) , T' (− 4 , 0) ,
W' (0 , − 3)

A) reflection across y = − 1

B) reflection across the x-axis

C) reflection across x = − 2

D) reflection across y = 1

118) Find the rule of translation from

P (− 5 , 1) , Q (− 5 , 3) ,
R (− 1 , 4)
to
Q' (− 5 , − 3) , R' (− 1 , − 4) ,
P'(− 5 , − 1)

A) reflection across y = 2

B) reflection across y = 3

C) reflection across the x-axis

D) reflection across the y-axis

119) Find the rule of translation from

N (− 5 , 3) , M (− 1 , 5) , L (0 , 1)
to
M' (1 , 5) , L' (0 , 1) , N' (5 , 3)

A) reflection across y = 1

B) reflection across y = 3

C) reflection across x = − 1

D) reflection across the y-axis

120) Find the rule of translation from

R (− 3 , − 2) , S (− 4 , 3) ,
T (0 , − 1)
to
S' (− 4 , − 5) , T' (0 , − 1) ,
R' (− 3 , 0)

A) reflection across y = − 1

B) reflection across x = − 2

C) reflection across y = 1

D) reflection across x = − 1

www.math-knots.com | www.a4ace.com

121) Find the rule of translation from

$$P(-4,1), Q(-2,4), R(-1,3)$$
to
$$Q'(2,4), R'(1,3), P'(4,1)$$

A) reflection across the y-axis

B) reflection across y = 1

C) reflection across the x-axis

D) reflection across x = −1

122) Find the rule of translation from

$$K(-4,2), J(-4,5), I(-1,5),$$
$$H(-1,1)$$
to
$$J'(0,5), I'(-3,5), H'(-3,1),$$
$$K'(0,2)$$

A) reflection across x = −2

B) reflection across x = −1

C) reflection across y = 3

D) reflection across y = 2

123) Find the rule of translation from

$$E(-3,-5), D(-3,-1),$$
$$C(-1,0), B(-1,-1)$$
to
$$D'(-3,-1), C'(-1,-2),$$
$$B'(-1,-1), E'(-3,3)$$

A) reflection across x = 1

B) reflection across y = −1

C) reflection across x = −3

D) reflection across y = −2

124) Find the rule of translation from

$$T(-3,-3), S(-3,-1),$$
$$R(0,-3)$$
to
$$S'(-3,-3), R'(0,-1),$$
$$T'(-3,-1)$$

A) reflection across x = 1

B) reflection across y = −3

C) reflection across the y-axis

D) reflection across y = −2

125) Find the rule of translation from

$$U(-4,2), V(-3,4),$$
$$W(-2,3), X(-2,2)$$
to
$$V'(3,4), W'(2,3),$$
$$X'(2,2), U'(4,2)$$

A) reflection across x = −2

B) reflection across y = 1

C) reflection across x = −1

D) reflection across the y-axis

126) Find the rule of translation from

$$K(2,-3), L(2,-1),$$
$$M(5,-3)$$
to
$$L'(2,-1), M'(-1,-3),$$
$$K'(2,-3)$$

A) reflection across y = −2

B) reflection across the y-axis

C) reflection across x = 2

D) reflection across y = −1

www.math-knots.com | www.a4ace.com

127) Find the rule of translation from

K (− 5 , 2) , L (− 4 , 5) ,
M (− 4 , 2)
to
L' (− 4 , 1) , M' (− 4 , 4) ,
K' (− 5 , 4)

A) reflection across y = 3

B) reflection across the y-axis

C) reflection across y = 2

D) reflection across x = −3

128) Find the rule of translation from

D (− 3 , − 2) , C (− 4 , 0) ,
B (1 , − 1) , A (− 1 , − 3)
to
C' (− 4 , − 4) , B' (1 , − 3) ,
A' (− 1 , − 1) , D' (− 3 , − 2)

A) reflection across y = − 2

B) reflection across x = − 1

C) reflection across x = − 2

D) reflection across the x-axis

129) Find the rule of translation from

V (1 , 1) , W (0 , 5) , X (3 , 1)
to
W' (0 , − 3) , X' (3 , 1) , V' (1 , 1)

A) reflection across y = 2

B) reflection across y = 1

C) reflection across the y-axis

D) reflection across y = 3

130) Find the rule of translation from

Y (4 , − 5) , X (2 , 0) , W (4 , 2) ,
V (5 , − 3)
to
X' (2 , 0) , W' (0 , 2) ,
V' (− 1 , − 3) , Y' (0 , − 5)

A) reflection across x = 1

B) reflection across y = −1

C) reflection across x = 2

D) reflection across the x-axis

131) Find the rule of translation from

X (− 4 , 2) , W (− 2 , 4) ,
V (− 1 , 0)
to
W' (− 2 , 0) , V' (− 1 , 4) ,
X' (− 4 , 2)

A) reflection across y = 2

B) reflection across the x-axis

C) reflection across x = − 3

D) reflection across x = − 2

132) Find the rule of translation from

W (− 5 , − 4) , V (− 4 , 0) ,
U (− 2 , 0)
to
V' (− 4 , 0) , U' (− 2 , 0) ,
W' (− 5 , 4)

A) reflection across x = − 3

B) reflection across the x-axis

C) reflection across y = − 2

D) reflection across y = − 1

133) Find the rule of translation from

H (− 5 , − 5) , I (− 2 , − 1) ,
J (− 1 , − 3)
to
I' (− 2 , − 3) , J' (− 1 , − 1) ,
H' (− 5 , 1)

A) reflection across x = − 3

B) reflection across the x-axis

C) reflection across y = − 2

D) reflection across x = − 2

134) Find the rule of translation from

I (0 , − 1) , J (− 3 , 3) , K (1 , 5) ,
L (3 , 0)
to
J' (− 3 , − 3) , K' (1 , − 5) , L' (3 , 0) ,
I' (0 , 1)

A) reflection across y = 2

B) reflection across x = 1

C) reflection across y = 1

D) reflection across the x-axis

135) Find the rule of translation from

R (− 1 , − 5) , S (0 , − 1) ,
T (3 , − 2)
to
S' (0 , − 1) , T' (− 3 , − 2) ,
R' (1 , − 5)

A) reflection across the x-axis

B) reflection across the y-axis

C) reflection across x = 2

D) reflection across y = − 1

136) Find the rule of translation from

R (0 , 1) , S (2 , 3) , T (3 , 3) ,
U (3 , 1)
to
S' (0 , 3) , T' (− 1 , 3) , U'(− 1 , 1) ,
R' (2 , 1)

A) reflection across x = 1

B) reflection across y = − 1

C) reflection across the x-axis

D) reflection across y = 2

137) Graph the transformation of the figure reflected across y = − 1.

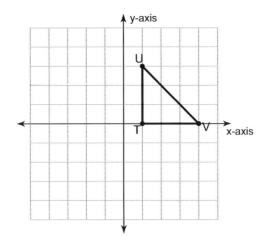

138) Graph the transformation of the figure reflected across y = 1.

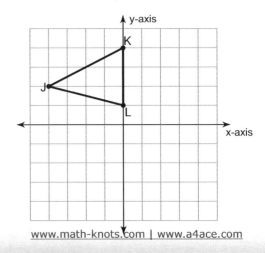

www.math-knots.com | www.a4ace.com

139) Graph the transformation of the figure reflected across y-axis.

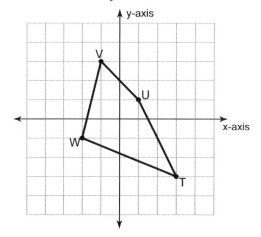

140) Graph the transformation of the figure reflected across x-axis.

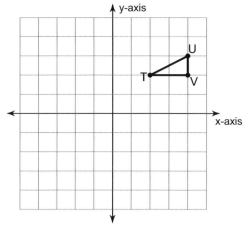

141) Graph the transformation of the figure reflected across y = 1.

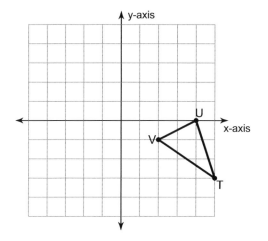

142) Graph the transformation of the figure reflected across x = 1.

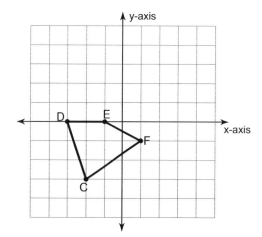

143) Graph the transformation of the figure reflected across y = − 1.

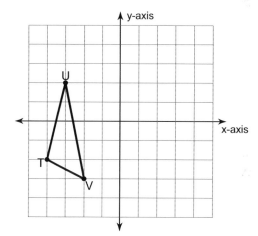

144) Graph the transformation of the figure reflected across y = 1.

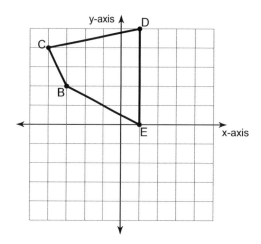

www.math-knots.com | www.a4ace.com

Algebra 1

145) Graph the transformation of the figure reflected across x-axis.

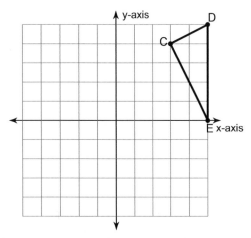

146) Graph the transformation of the figure reflected across y = − 1.

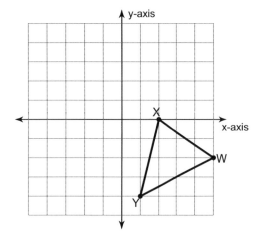

147) Graph the transformation of the figure reflected across y = − 1.

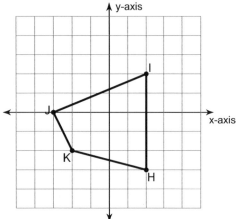

148) Graph the transformation of the figure reflected across x = − 3.

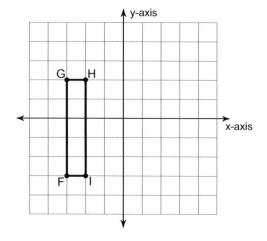

149) Graph the transformation of the figure reflected across x = 1.

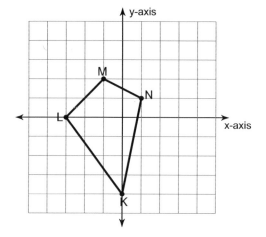

150) Graph the transformation of the figure reflected across x = − 2.

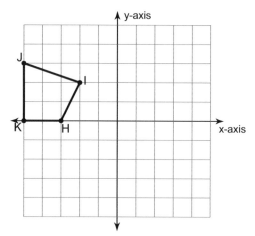

www.math-knots.com | www.a4ace.com

151) Graph the transformation of the figure reflected across y = 2.

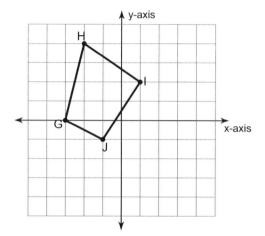

154) Graph the transformation of the figure reflected across x = − 1.

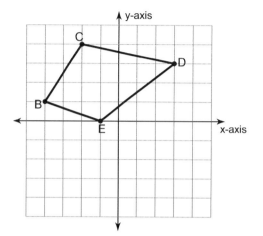

152) Graph the transformation of the figure reflected across y-axis.

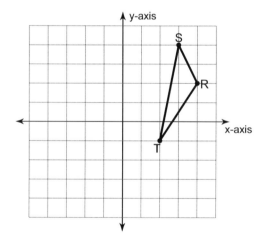

155) Graph the transformation of the figure reflected across y = − 1.

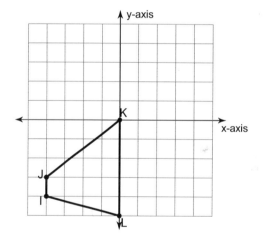

153) Graph the transformation of the figure reflected across x = − 2.

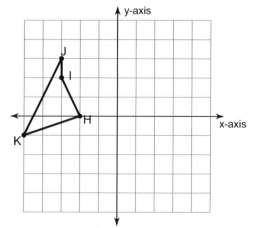

156) Graph the transformation of the figure reflected across y = − 2.

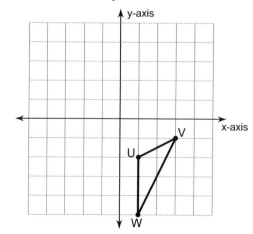

www.math-knots.com | www.a4ace.com

157) Graph the transformation of the figure reflected across x = 3.

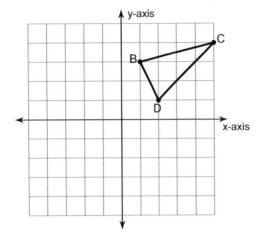

158) Graph the transformation of the figure reflected across y = − 1.

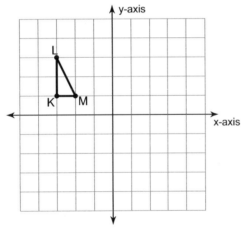

159) Graph the transformation of the figure reflected across x = − 2.

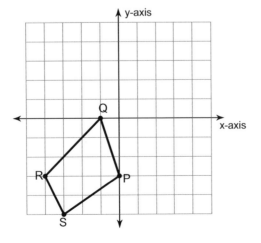

160) Graph the transformation of the figure reflected across x-axis.

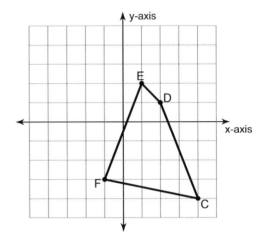

161) Graph the transformation of the figure reflected across x = − 1.

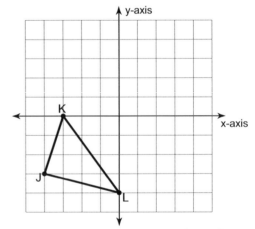

162) Graph the transformation of the figure reflected across x-axis.

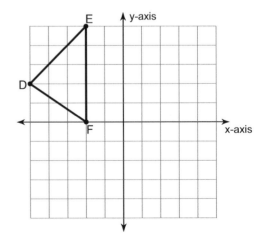

163) Graph the transformation of the figure reflected across y = − 1.

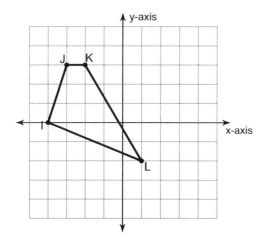

164) Graph the transformation of the figure reflected across x = − 3.

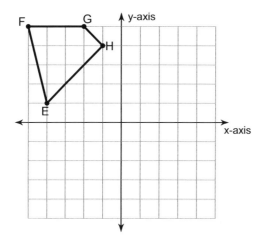

1) Below points J (1 , − 4) , I (1 , − 3) , H (4 , − 4) , G (3 , − 5) are rotated to 90° counterclockwise about the origin. Find the new co-ordinates.

A) J' (− 4 , − 1) , I' (− 3 , − 1) , H' (− 4 , − 4) , G' (− 5 , − 3)

B) J' (− 1 , 4) , I' (− 1 , 3) , H' (− 4 , 4) , G' (− 3, 5)

C) J' (4 , 1) , I' (3 , 1) , H' (4 , 4) , G' (5 , 3)

D) J' (− 4 , 1) , I' (− 4 , 2) , H' (− 1 , 1) , G' (− 2 , 0)

2) Below points C (0 , − 5) , D (− 1 , − 2) , E (3 , − 3) , F (5 , − 5) are rotated to 180° about the origin. Find the new co-ordinates.

A) C' (− 3 , − 5) , D' (− 4 , − 2) , E' (0 , − 3) , F' (2 , − 5)

B) C' (0 , 5) , D' (1 , 2) , E' (− 3 , 3) , F' (− 5 , 5)

C) C' (− 5 , 0) , D' (− 2 , 1) , E' (− 3 , − 3) , F' (− 5 , − 5)

D) C' (5 , 0) , D' (2 , − 1) , E' (3 , 3) , F' (5 , 5)

3) Below points K (− 3 , 1) , L (− 3 , 3) , M (0 , 1) are rotated to 180° about the origin. Find the new co-ordinates.

A) K' (3 , − 1) , L' (3 , − 3) , M' (0 , − 1)

B) K' (1 , 3) , L' (3 , 3) , M' (1 , 0)

C) K' (− 1 , − 3) , L' (− 3 , − 3) , M' (− 1 , 0)

D) K' (− 2 , 2) , L' (− 2 , 4) , M' (1 , 2)

4) Below points C (3 , 2) , D (4 , 5) , E (5 , 4) are rotated to 90° clockwise about the origin. Find the new co-ordinates.

A) C' (2 , − 3) , D' (5 , − 4) , E' (4 , − 5)

B) C' (− 2 , 3) , D' (− 5 , 4) , E' (− 4 , 5)

C) C' (4 , − 3) , E' (5 , − 2) , C' (3 , 0)

D) C' (− 3 , − 2) , D' (− 4 , − 5) , E' (− 5 , − 4)

5) Below points U (1 , − 5) , V (− 1 , − 2) , W (− 1 , − 1) , X (3 , − 3) are rotated to 180° about the origin. Find the new co-ordinates.

A) U' (5 , 1) , V' (2 , − 1) , W' (1 , − 1) , X' (3 , 3)

B) V' (− 1 , − 4) , W' (− 1 , − 5) , X' (3 , − 3) , U' (1 , − 1)

C) U' (− 1 , 5) , V' (1 , 2) , W' (1 , 1) , X' (− 3 , 3)

D) U' (− 5 , − 1) , V' (− 2 , 1) , W' (− 1 , 1) , X' (− 3 , − 3)

6) Below points J (− 5 , 0) , K (− 2 , 4) , L (3 , 4) , M (− 1 , 1) are rotated to 180° about the origin. Find the new co-ordinates.

A) J' (0 , 5) , K' (4 , 2) , L' (4 , − 3) , M' (1 , 1)

B) J' (0 , − 5) , K' (− 4 , − 2) , L' (− 4 , 3) , M' (− 1 , − 1)

C) K' (− 2 , 0) , L' (3 , 0) , M' (− 1 , 3) , J' (− 5 , 4)

D) J' (5 , 0) , K' (2 , − 4) , L' (− 3 , − 4) , M' (1 , − 1)

7) Below points A (– 1 , – 1) , B (0 , 3) , C (4 , – 1) , D (1 , – 5) are rotated to 180° about the origin.
Find the new co-ordinates.

A) A' (1 , – 1) , B' (– 3 , 0) , C' (1 , 4) , D' (5 , 1)

B) B' (0 , – 5) , C' (4 , – 1) , D' (1 , 3) , A' (– 1 , – 1)

C) A' (– 1 , 1) , B' (3 , 0) , C' (– 1 , – 4) , D' (– 5 , – 1)

D) A' (1 , 1) , B' (0 , – 3) , C' (– 4 , 1) , D' (– 1 , 5)

8) Below points E (– 3 , – 3) , F (0 , 1) , G (3 , 0) , H (1 , – 3) are rotated to 90° counterclockwise about the origin.
Find the new co-ordinates.

A) F' (2 , 1) , G' (– 1 , 0) , H' (1 , – 3) , E' (5 , – 3)

B) E' (3 , 3) , F' (0 , – 1) , G' (– 3 , 0) , H' (– 1 , 3)

C) E' (– 3 , 3) , F' (1 , 0) , G' (0 , – 3) , H' (– 3 , – 1)

D) E' (3 , – 3) , F' (– 1 , 0) , G' (0 , 3) , H' (3 , 1)

9) Below points H (– 2 , – 5) , I (– 3 , – 1) , J (– 2 , – 1) , K (– 1 , – 3) are rotated to 90° clockwise about the origin.
Find the new co-ordinates.

A) I' (– 3 , – 1) , J' (– 4 , – 1) , K' (– 5 , – 3) , H' (– 4 , – 5)

B) H' (5 , – 2) , I' (1 , – 3) , J' (1 , – 2) , K' (3 , – 1)

C) H' (– 5 , 2) , I' (– 1 , 3) , J' (– 1 , 2) , K' (– 3 , 1)

D) H' (2 , 5) , I' (3 , 1) , J' (2 , 1) , K' (1 , 3)

10) Below points B (– 5 , 2) , C (0 , 3) , D (– 1 , – 1) are rotated to 90° clockwise about the origin.
Find the new co-ordinates.

A) B' (0 , 3) , C' (5 , 4) , D' (4 , 0)

B) B' (5 , – 2) , C' (0 , – 3) , D' (1 , 1)

C) B' (2 , 5) , C' (3 , 0) , D' (– 1 , 1)

D) B' (– 2 , – 5) , C' (– 3 , 0) , D' (1 , – 1)

11) Below points B (– 2 , 1) , C (– 3 , 5) , D (0 , 4) , E (1 , 2) are rotated to 90° clockwise about the origin.
Find the new co-ordinates.

A) B' (1 , 2) , C' (5 , 3) , D' (4 , 0) , E' (2 , – 1)

B) B' (– 1 , – 2) , C' (– 5 , – 3) , D' (– 4 , 0) , E' (– 2 , 1)

C) B' (2 , – 1) , C' (3 , – 5) , D' (0 , – 4) , E' (– 1 , – 2)

D) B' (– 3 , 1) , C' (– 4 , 5) , D' (– 1 , 4) , E' (0 , 2)

12) Below points S (– 1 , – 2) , T (4 , – 1) , U (4 , – 3) are rotated to 90° clockwise about the origin.
Find the new co-ordinates.

A) S' (2 , – 1) , T' (1 , 4) , U' (3 , 4)

B) S' (– 4 , – 4) , T' (1 , – 3) , U' (1 , – 5)

C) S' (1 , 2) , T' (– 4 , 1) , U' (– 4 , 3)

D) S' (– 2 , 1) , T' (– 1 , – 4) , U' (– 3 , – 4)

13) Below points C (– 4 , – 4) , D (– 3 , – 2) ,
E (– 1 , – 1) , F (– 1 , – 5) are rotated
to 180° about the origin.
Find the new co-ordinates.

A) D' (– 3 , – 2) , E' (– 1 , – 3) ,
F' (– 1 , 1) , C' (– 4 , 0)

B) C' (4 , – 4) , D' (2 , – 3) ,
E' (1 , – 1) , F' (5 , – 1)

C) C' (4 , 4) , D' (3 , 2) ,
E' (1 , 1) , F' (1 , 5)

D) C' (– 4 , 4) , D' (– 2 , 3) ,
E' (– 1 , 1) , F' (– 5 , 1)

14) Below points K (– 3 , – 1) , L (– 3 , 4) ,
M (2 , 5) , N (1 , 1) are rotated to
90° counterclockwise about the origin.
Find the new co-ordinates.

A) K' (1 , – 3) , L' (– 4 , – 3) ,
M' (– 5 , 2) , N' (– 1 , 1)

B) K' (3 , 1) , L' (3 , – 4) ,
M' (– 2 , – 5) , N' (– 1 , – 1)

C) K' (– 1 , 3) , L' (4 , 3) ,
M' (5 , – 2) , N' (1 , – 1)

D) K' (– 5 , – 2) , L' (– 5 , 3) ,
M' (0 , 4) , N' (– 1 , 0)

15) Below points E (– 3 , – 2) , D (2 , – 1) ,
C (– 1 , – 5) are rotated to 90° clockwise
about the origin.
Find the new co-ordinates.

A) E' (2 , – 3) , D' (1 , 2) ,
C' (5 , – 1)

B) E' (0 , 2) , D' (5 , 3) ,
C' (2 , – 1)

C) E' (3 , 2) , D' (– 2 , 1) ,
C' (1 , 5)

D) E' (– 2 , 3) , D' (– 1 , – 2) ,
C' (– 5 , 1)

16) Below points Z (– 1 , 0) , Y (– 4 , 4) ,
X (– 2 , 5) , W (1 , 5) are rotated to
180° about the origin.
Find the new co-ordinates.

A) Z' (0 , – 1) , Y' (– 4 , – 4) ,
X' (– 5 , – 2) , W' (– 5 , 1)

B) Z' (1 , 0) , Y' (4 , – 4) ,
X' (2 , – 5) , W' (– 1 , – 5)

C) Z' (0 , 1) , Y' (4 , 4) ,
X' (5 , 2) , W' (5 , – 1)

D) Z' (– 2 , 0) , Y' (– 5 , 4) ,
X' (– 3 , 5) , W' (0 , 5)

17) Below points E (– 3 , – 4) , F (– 3 , 0) ,
G (2 , – 2) are rotated to 90°
clockwise about the origin.
Find the new co-ordinates.

A) E' (– 4 , 3) , F' (0 , 3) ,
G' (– 2 , – 2)

B) E' (3 , 4) , F' (3 , 0) ,
G' (– 2 , 2)

C) E' (– 3 , – 2) , F' (– 3 , 2) ,
G' (2 , 0)

D) E' (4 , – 3) , F' (0 , – 3) ,
G' (2 , 2)

18) Below points G (2 , – 4) , H (– 1 , 0) ,
I (0 , 1) , J (5 , 0) are rotated to 90°
counterclockwise about the origin.
Find the new co-ordinates.

A) G' (– 4 , – 2) , H' (0 , 1) ,
I' (1 , 0) , J' (0 , – 5)

B) G' (– 1 , – 1) , H' (– 4 , 3) ,
I' (– 3 , 4) , J' (2 , 3)

C) G' (– 2 , 4) , H' (1 , 0) ,
I' (0 , – 1) , J' (– 5 , 0)

D) G' (4 , 2) , H' (0 , – 1) ,
I' (– 1 , 0) , J' (0 , 5)

19) Below points D (− 4 , − 4) , E (− 5 , − 3) , F (− 1 , 0) , G (1 , − 4) are rotated to 180° about the origin.
Find the new co-ordinates.

A) D' (4 , − 4) , E' (3 , − 5) ,
F' (0 , − 1) , G' (4 , 1)

B) D' (4 , 4) , E' (5 , 3) ,
F' (1 , 0) , G' (− 1 , 4)

C) E' (1 , − 3) , F ' (− 3 , 0) ,
G' (− 5 , − 4) , D' (0 , − 4)

D) D' (− 4 , 4) , E' (− 3 , 5) ,
F' (0 , 1) , G' (− 4 , − 1)

20) Below points B (− 1 , − 3) , C (3 , − 2) , D (4 , − 5) are rotated to 180° about the origin.
Find the new co-ordinates.

A) B' (1 , 3) , C' (− 3 , 2) , D' (− 4 , 5)

B) B' (3 , − 1) , C' (2 , 3) , D' (5 , 4)

C) B' (0 , 3) , C' (4 , 4) , D' (5 , 1)

D) B' (− 3 , 1) , C' (− 2 , − 3) ,
D' (− 5 , − 4)

21) Below points W (1 , − 5) , V (0 , − 3) , U (4 , − 1) , T (5 , − 2) are rotated to 180° about the origin.
Find the new co-ordinates.

A) W' (− 1 , 5) , V' (0 , 3) ,
U' (− 4 , 1) , T' (− 5 , 2)

B) W' (− 5 , − 1) , V' (− 3 , 0) ,
U' (− 1 , − 4) , T' (− 2 , − 5)

C) V' (0 , 3) , U' (4 , 1) ,
T' (5 , 2) , W' (1 , 5)

D) W' (5 , 1) , V' (3 , 0) ,
U' (1 , 4) , T' (2 , 5)

22) Below points A (− 4 , 0) , B (− 4 , 1) , C (0 , 3) , D (1 , 1) are rotated to 90° counterclockwise about the origin.
Find the new co-ordinates.

A) A' (4 , 0) , B' (4 , − 1) ,
C' (0 , − 3) , D' (− 1 , − 1)

B) A' (0 , − 4) , B' (− 1 , − 4) ,
C' (− 3 , 0) , D' (− 1 , 1)

C) A' (− 1 , − 1) , B' (− 1 , 0) ,
C' (3 , 2) , D' (4 , 0)

D) A' (0 , 4) , B' (1 , 4) ,
C' (3 , 0) , D' (1 , − 1)

23) Below points F (− 3 , 1) , E (− 3 , 3) , D (0 , 4) , C (− 2 , − 1) are rotated to 180° about the origin.
Find the new co-ordinates.

A) F' (− 1 , − 3) , E' (− 3 , − 3) ,
D' (− 4 , 0) , C' (1 , − 2)

B) E' (1 , 3) , D' (− 2 , 4) ,
C' (0 , − 1) , F' (1 , 1)

C) F' (1 , 3) , E' (3 , 3) ,
D' (4 , 0) , C' (− 1 , 2)

D) F' (3 , − 1) , E' (3 , − 3) ,
D' (0 , − 4) , C' (2 , 1)

24) Below points S (2 , − 4) , R (0 , − 1) , Q (2 , 0) , P (4 , 0) are rotated to 180° about the origin.
Find the new co-ordinates.

A) S' (4 , 2) , R' (1 , 0) ,
Q' (0 , 2) , P' (0 , 4)

B) S' (− 2 , 4) , R' (0 , 1) ,
Q' (− 2 , 0) , P' (− 4 , 0)

C) S' (− 4 , − 2) , R' (− 1 , 0) ,
Q' (0 , − 2) , P' (0 , − 4)

D) S' (− 1 , − 3) , R' (− 3 , 0) ,
Q' (− 1 , 1) , P' (1 , 1)

Algebra 1

25) Below points C (2 , – 2) , D (1 , 1) , E (4 , 3) are rotated to 90° counterclockwise about the origin. Find the new co-ordinates.

A) C' (2 , 2) , D' (– 1 , 1) , E' (– 3 , 4)

B) D' (1 , – 3) , E' (4 , – 5) , C' (2 , 0)

C) C' (– 2 , 2) , D' (– 1 , – 1) , E' (– 4 , – 3)

D) C' (– 2 , – 2) , D' (1 , – 1) , E' (3 , – 4)

26) Below points Q (– 3 , 1) , R (2 , 3) , S (2 , 1) are rotated to 90° counterclockwise about the origin. Find the new co-ordinates.

A) Q' (1 , 3) , R' (3 , – 2) , S' (1 , – 2)

B) R' (2 , – 5) , S' (2 , – 3) , Q' (– 3 , – 3)

C) Q' (3 , – 1) , R' (– 2 , – 3) , S' (– 2 , – 1)

D) Q' (– 1 , – 3) , R' (– 3 , 2) , S' (– 1 , 2)

27) Below points H (– 4 , 1) , I (– 2 , 3) , J (– 1 , – 2) are rotated to 90° counterclockwise about the origin. Find the new co-ordinates.

A) H' (– 1 , – 4) , I' (– 3 , – 2) , J' (2 , – 1)

B) H' (0 , – 2) , I' (2 , 0) , J' (3 , – 5)

C) H' (1 , 4) , I' (3 , 2) , J' (– 2 , 1)

D) H' (4 , – 1) , I' (2 , – 3) , J' (1 , 2)

28) Find the rule of translation from E (– 3 , – 3) , F (– 1 , 2) , G (2 , – 2)
to
E' (3 , – 3) , F' (– 2 , – 1) , G' (2 , 2)

A) rotation 180° about the origin

B) rotation 90° clockwise about the origin

C) translation: 1 unit left and 1 unit down

D) rotation 90° counterclockwise about the origin

29) Find the rule of translation from A (– 2 , – 4) , B (– 3 , – 1) , C (– 2 , 0) , D (– 1 , – 2)
to
A' (4 , – 2) , B' (1 , – 3) , C' (0 , – 2) , D' (2 , – 1)

A) reflection across x = – 1

B) rotation 90° clockwise about the origin

C) rotation 90° counterclockwise about the origin

D) rotation 180° about the origin

30) Find the rule of translation from K (– 3 , 2) , L (– 3 , 3) , M (– 1 , 2)
to
K' (2 , 3) , L' (3 , 3) , M' (2 , 1)

A) rotation 90° clockwise about the origin

B) rotation 90° counterclockwise about the origin

C) rotation 180° about the origin

D) reflection across y = 3

Algebra 1

31) Find the rule of translation from
I (2 , − 5) , J (2 , 0) ,
K (4 , 0) , L (3 , − 4)
to
I' (− 2 , 5) , J' (− 2 , 0) ,
K' (− 4 , 0) , L' (− 3 , 4)

A) rotation 90° counterclockwise about the origin
B) rotation 90° clockwise about the origin
C) rotation 180° about the origin
D) translation: 4 units left and 5 units up

32) Find the rule of translation from
W (1 , − 5) , V (4 , − 3) ,
U (3 , − 5)
to
W' (− 1 , 5) , V' (− 4 , 3) ,
U' (− 3 , 5)

A) rotation 90° clockwise about the origin
B) rotation 90° counterclockwise about the origin
C) translation: 5 units left and 6 units up
D) rotation 180° about the origin

33) Find the rule of translation from
X (0 , 0) , W (0 , 2) , V (5 , 3) ,
U (3 , − 1)
to
X' (0 , 0) , W' (− 2 , 0) ,
V' (− 3 , 5) , U' (1 , 3)

A) rotation 90° counterclockwise about the origin
B) rotation 90° clockwise about the origin
C) translation: 5 units left and 2 units up
D) rotation 180° about the origin

34) Find the rule of translation from
G (4 , − 3) , F (4 , 2) , E (5 , 2) ,
D (5 , − 2)
to
G' (− 4 , 3) , F' (− 4 , − 2) ,
E' (− 5 , − 2) , D' (− 5 , 2)

A) rotation 90° clockwise about the origin
B) translation: 3 units left and 2 units down
C) rotation 90° counterclockwise about the origin
D) rotation 180° about the origin

35) Find the rule of translation from
M (1 , − 3) , L (3 , 2) ,
K (5 , 0) , J (2 , − 4)
to
M' (− 3 , − 1) , L' (2 , − 3) ,
K' (0 , − 5) , J' (− 4 , − 2)

A) rotation 90° counterclockwise about the origin
B) rotation 90° clockwise about the origin
C) translation: 5 units left and 1 unit down
D) rotation 180° about the origin

36) Find the rule of translation from
R (− 5 , − 2) , S (− 5 , 0) ,
T (− 1 , − 3)
to
R' (− 2 , 5) , S' (0 , 5) ,
T' (− 3 , 1)

A) rotation 90° clockwise about the origin
B) rotation 90° counterclockwise about the origin
C) reflection across y = 1
D) rotation 180° about the origin

www.math-knots.com | www.a4ace.com

37) Find the rule of translation from
K (− 4 , − 3) , L (− 2 , 0) ,
M (− 1 , 0) , N (− 1 , − 2)
to
K' (3 , − 4) , L' (0 , − 2) ,
M' (0 , − 1) , N' (2 , − 1)

A) rotation 180° about the origin

B) reflection across x = − 2

C) rotation 90° clockwise about
the origin

D) rotation 90° counterclockwise
about the origin

38) Find the rule of translation from
V (− 3 , 1) , W (− 1 , 4) ,
X (2 , 0) , Y (− 2 , − 3)
to
V' (3 , − 1) , W' (1 , − 4) ,
X' (− 2 , 0) , Y' (2 , 3)

A) reflection across the y-axis

B) rotation 180° about the origin

C) rotation 90° clockwise about
the origin

D) rotation 90° counterclockwise
about the origin

39) Find the rule of translation from
U (− 1 , − 4) , V (0 , − 2) ,
W (4 , − 3) , X (4 , − 5)
to
U' (− 4 , 1) , V' (− 2 , 0) ,
W' (− 3 , − 4) , X' (− 5 , − 4)

A) rotation 180° about the origin

B) reflection across y = − 2

C) rotation 90° counterclockwise
about the origin

D) rotation 90° clockwise about
the origin

40) Find the rule of translation from
S (− 3 , 1) , T (− 3 , 5) ,
U (− 2 , 3)
to
S' (− 1 , − 3) , T' (− 5 , − 3) ,
U' (− 3 , − 2)

A) rotation 180° about the origin

B) reflection across x = − 3

C) rotation 90° clockwise about
the origin

D) rotation 90° counterclockwise
about the origin

41) Find the rule of translation from
J (− 5 , 0) , I (− 4 , 4) ,
H (0 , 1) , G (− 4 , − 3)
to
J' (0 , − 5) , I' (− 4 , − 4) ,
H' (− 1 , 0) , G' (3 , − 4)

A) rotation 90° clockwise about
the origin

B) rotation 90° counterclockwise
about the origin

C) rotation 180° about the origin

D) reflection across the x-axis

42) Find the rule of translation from
P (3 , 0) , Q (2 , 3) , R (4 , 4) ,
S (4 , 2)
to
P' (− 3 , 0) , Q' (− 2 , − 3) ,
R' (− 4 , − 4) , S' (− 4 , − 2)

A) rotation 90° counterclockwise
about the origin

B) rotation 180° about the origin

C) translation: 5 units left and
1 unit up

D) rotation 90° clockwise about
the origin

114

43) Find the rule of translation from
F (– 4 , – 1) , G (– 4 , 3) ,
H (0 , 2) , I (– 2 , – 1)
to
F' (4 , 1) , G' (4 , – 3) ,
H' (0 , – 2) , I' (2 , 1)

A) rotation 180° about the origin

B) **reflection across** the y-axis

C) rotation 90° counterclockwise
about the origin

D) rotation 90° clockwise about
the origin

44) Find the rule of translation from
J (– 5 , – 1) , K (– 1 , 0) ,
L (– 1 , – 4)
to
J' (5 , 1) , K' (1 , 0) , L' (1 , 4)

A) translation: 1 unit right and
3 units up

B) rotation 90° counterclockwise
about the origin

C) rotation 180° about the origin

D) rotation 90° clockwise about
the origin

45) Find the rule of translation from
T (3 , – 5) , U (4 , – 3) ,
V (5 , – 5)
to
T' (– 3 , 5) , U' (– 4 , 3) ,
V' (– 5 , 5)

A) translation: 6 units left and
4 units up

B) rotation 90° counterclockwise
about the origin

C) rotation 180° about the origin

D) rotation 90° clockwise about
the origin

46) Find the rule of translation from
A (– 5 , – 1) , B (– 5 , 3) ,
C (– 3 , 0)
to
A' (1 , – 5) , B' (– 3 , – 5) ,
C' (0 , – 3)

A) rotation 180° about the origin

B) rotation 90° counterclockwise
about the origin

C) translation: 6 units right and
1 unit up

D) rotation 90° clockwise about
the origin

47) Find the rule of translation from
Y (2 , 3) , X (2 , 4) ,
W (4 , 4) , V (4 , 3)
to
Y' (– 2 , – 3) , X' (– 2 , – 4) ,
W' (– 4 , – 4) , V' (– 4 , – 3)

A) translation: 7 units left and 6
units down

B) rotation 90° counterclockwise
about the origin

C) rotation 90° clockwise about
the origin

D) rotation 180° about the origin

48) Find the rule of translation from
V (1 , – 4) , U (1 , – 2) ,
T (3 , – 4)
to
V' (– 1 , 4) , U' (– 1 , 2) ,
T' (– 3 , 4)

A) rotation 180° about the origin

B) rotation 90° clockwise about
the origin

C) rotation 90° counterclockwise
about the origin

D) translation: 1 unit right and
5 units up

49) Find the rule of translation from
C (− 1 , 0) , D (− 3 , 3) ,
E (1 , 4) , F (3 , 1)
to
C' (0 , 1) , D' (3 , 3) ,
E' (4 , − 1) , F' (1 , − 3)

A) rotation 90° clockwise about the origin
B) rotation 180° about the origin
C) reflection across x = − 1
D) rotation 90° counterclockwise about the origin

50) Find the rule of translation from
P (− 3 , 4) , Q (2 , 5) , R (− 1 , 2)
to
P' (4 , 3) , Q' (5 , − 2) , R' (2 , 1)

A) rotation 180° about the origin
B) rotation 90° counterclockwise about the origin
C) translation: 2 units right
D) rotation 90° clockwise about the origin

51) Find the rule of translation from
D (0 , − 5) , C (− 1 , − 2) ,
B (2 , − 1) , A (3 , − 5)
to
D' (− 5 , 0) , C' (− 2 , 1) ,
B' (− 1 , − 2) , A' (− 5 , − 3)

A) rotation 180° about the origin
B) rotation 90° clockwise about the origin
C) rotation 90° counterclockwise about the origin
D) reflection across the y-axis

52) Find the rule of translation from
B (0 , − 5) , C (3 , − 1) ,
D (3 , − 3)
to
B' (− 5 , 0) , C' (− 1 , − 3) ,
D' (− 3 , − 3)

A) rotation 180° about the origin
B) rotation 90° counterclockwise about the origin
C) translation: 4 units left and 3 units up
D) rotation 90° clockwise about the origin

53) Find the rule of translation from
X (− 3 , − 2) , W (0 , 2) ,
V (4 , 0) , U (0 , − 3)
to
X' (3 , 2) , W' (0 , − 2) ,
V' (− 4 , 0) , U' (0 , 3)

A) rotation 90° counterclockwise about the origin
B) rotation 90° clockwise about the origin
C) rotation 180° about the origin
D) translation: 1 unit left and 1 unit down

54) Find the rule of translation from
S (− 5 , − 5) , R (− 4 , − 2) ,
Q (− 3 , − 3)
to
S' (− 5 , 5) , R' (− 2 , 4) ,
Q' (− 3 , 3)

A) rotation 90° counterclockwise about the origin
B) rotation 90° clockwise about the origin
C) rotation 180° about the origin
D) reflection across the x-axis

Algebra 1

55) Find the rule of translation from
J (– 1 , – 3) , K (– 1 , 2) ,
L (2 , 3) , M (4 , – 2)
to
J' (1 , 3) , K' (1 , – 2) ,
L' (– 2 , – 3) , M' (– 4 , 2)

A) rotation 180° about the origin

B) rotation 90° counterclockwise
about the origin

C) rotation 90° clockwise about
the origin

D) reflection across x = 1

56) Graph the transformation of the figure
rotation 180° about the origin.

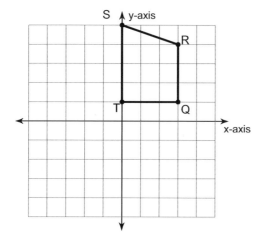

57) Graph the transformation of the figure
rotation 90° about the origin.

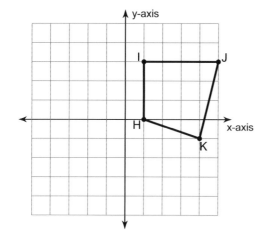

58) Graph the transformation of the figure
rotation 90° counterclockwise about the origin.

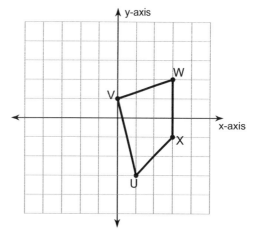

59) Graph the transformation of the figure
rotation 90° clockwise about the origin.

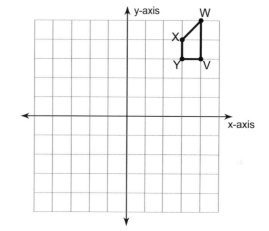

60) Graph the transformation of the figure
rotation 180° about the origin.

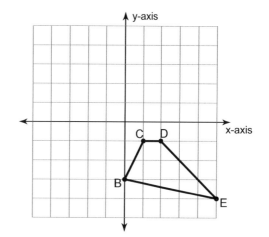

www.math-knots.com | www.a4ace.com

61) Graph the transformation of the figure rotation 90° clockwise about the origin.

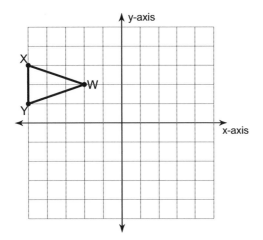

62) Graph the transformation of the figure rotation 90° clockwise about the origin.

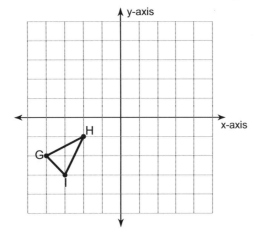

63) Graph the transformation of the figure rotation 90° clockwise about the origin.

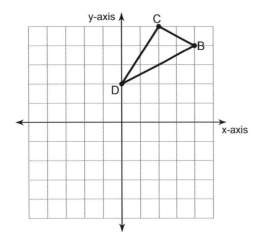

64) Graph the transformation of the figure rotation 180° about the origin.

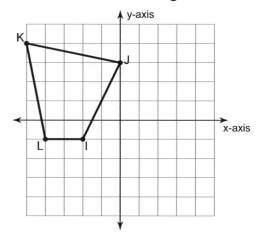

65) Graph the transformation of the figure rotation 180° about the origin.

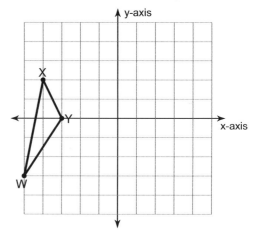

66) Graph the transformation of the figure rotation 180° about the origin.

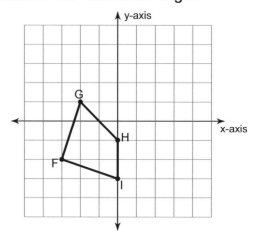

Algebra 1

67) Graph the transformation of the figure rotation 90° clockwise about the origin.

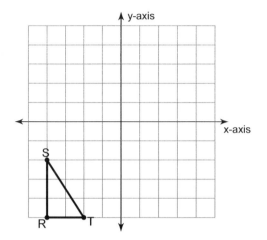

68) Graph the transformation of the figure rotation 180° about the origin.

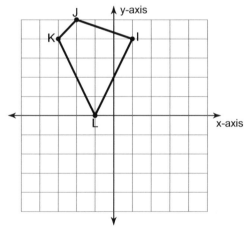

69) Graph the transformation of the figure rotation 90° clockwise about the origin.

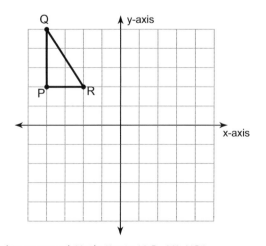

70) Graph the transformation of the figure rotation 180° about the origin.

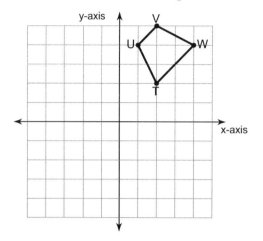

71) Graph the transformation of the figure rotation 90° clockwise about the origin.

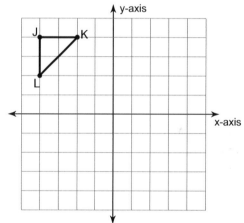

72) Graph the transformation of the figure rotation 180° about the origin.

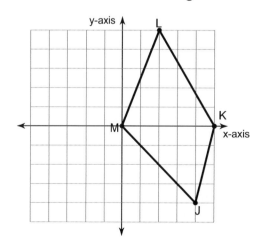

www.math-knots.com | www.a4ace.com

Algebra 1

73) Graph the transformation of the figure rotation 90° counterclockwise about the origin.

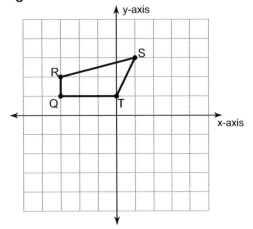

74) Graph the transformation of the figure rotation 180° about the origin.

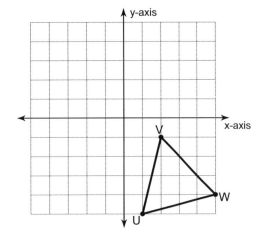

75) Graph the transformation of the figure rotation 90° counterclockwise about the origin.

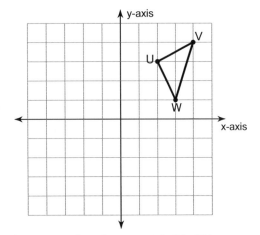

76) Graph the transformation of the figure rotation 180° about the origin.

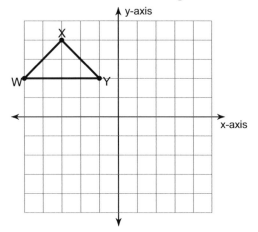

77) Graph the transformation of the figure rotation 180° about the origin.

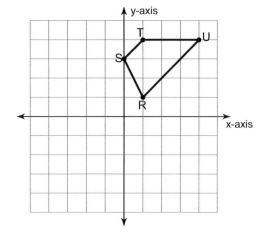

78) Graph the transformation of the figure rotation 180° about the origin.

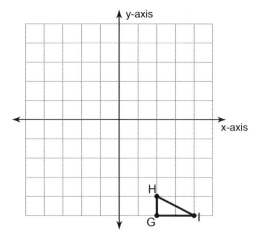

Algebra 1

79) Graph the transformation of the figure rotation 180° about the origin.

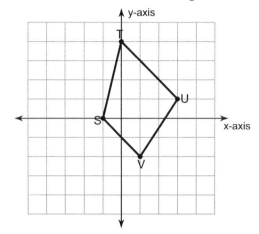

80) Graph the transformation of the figure rotation 90° counterclockwise about the origin.

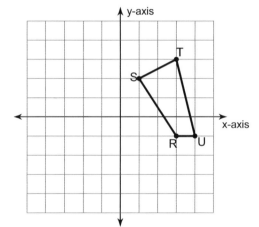

81) Graph the transformation of the figure rotation 180° about the origin.

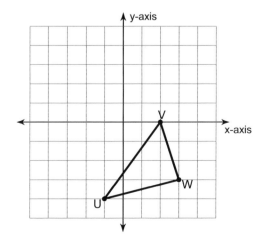

82) Graph the transformation of the figure rotation 180° about the origin.

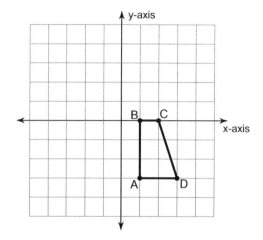

83) Graph the transformation of the figure rotation 90° counterclockwise about the origin.

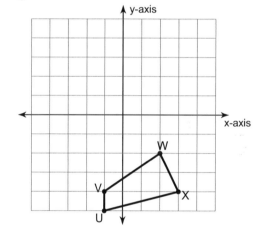

www.math-knots.com | www.a4ace.com

The height and weight of few players from Supersonic basketball club are given below

Height (ft)	Weight (lbs)
4.7	97.6
4.85	106
5.7	145
5.95	164
6.1	163
6.4	187

The data is modeled by the equation $y = 50.7 x - 141$ where x is height in feet and y is weight in pounds.

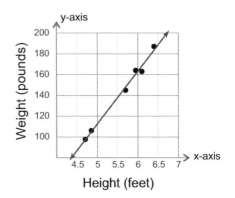

84)   What does the slope of the line represent?

85)   What does the y-intercept of this function represent?

86)   Using this model, what would be the weight of a player with a height of 5.45 ft ? Round your answer to the nearest tenth.

87)   According to the model, what would be the weight of a player with a height of 6 ft ? Round your answer to the nearest tenth.

88)   Based on the model, what height corresponds to a weight of 120 pounds ? Round your answer to the nearest hundredth.

www.math-knots.com | www.a4ace.com

The weather department took the following sample data from a bulbul cyclone in 2019.

Air Pressure (kPa)	Wind Speed (knots)
930	132
936	
945	88.3
984	54.6
993	46.1
1,000	34.8

The weather department found a relation between the air pressure and wind speed which is modeled as below
$y = -1.21 x + 1250$ where x is the air pressure in millibars (kPa) and y is the maximum sustained wind speed in knots (nautical miles per hour).

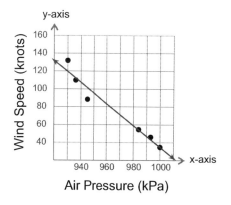

89) What does the slope of the line represent?

90) What does the y-intercept of this function represent?

91) Using this model, find the wind speed of a cyclone with an air pressure of 962 kPa? Round your answer to the nearest knot.

92) Based on the model, find the wind speed of a cyclone with an air pressure of 895 kPa ? Round your answer to the nearest knot.

93) Based on the model , find the air pressure when the wind speed is 74 knots. Round your answer to the nearest millibar.

The weather department took the following sample data from the Asani cyclone in 2022.

Air Pressure (kPa)	Wind Speed (knots)
927	122
941	
956	79.2
973	65
990	40.7
1,002	33

The weather found the relation between the air pressure and wind speed. They modeled the below equation.
$y = -1.24x + 1270$ where x is the air pressure in millibars (kPa) and y is the maximum sustained wind speed in knots (nautical miles per hour).

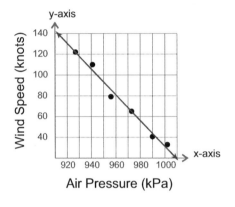

94) What does the slope of the line represent?

95) What does the y-intercept of this function represent?

96) Based on the model, find the wind speed of a cyclone when the air pressure is 944 kPa? Round your answer to the nearest knot.

97) Bases on the model, find the speed of a cyclone when the air pressure is 889 kPa. Round your answer to the nearest knot.

98) Based on the model, find the air pressure of a cyclone when the wind speed is 56 knots. Round your answer to the nearest millibar.

Residents of Arizona use more electricity during summer times. A sample of electricity usage data is taken as below.

Temperature	Electricity (kWh)
58	32.1
60	33.9
69	37.7
71	38.7
80	42
83	45

The relation between the temperature and the electricity usage of the households is modeled as below.

$y = 0.473 x + 5.04$ where x is the average daily temperature in °F and y is the average amount of electricity consumed in kilowatt-hours (kWh).

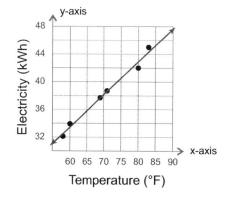

99)   What does the slope of the line represent?

100)   What does the y-intercept of this function represent?

101)   Based on the model, how much electricity would be consumed if the average daily temperature was 65 °F ? kPa? Round your answer to the nearest knot.

102)   Based on the model, how much electricity would be consumed if the average daily temperature was 104 °F ? Round your answer to the nearest kilowatt-hour.

103)   Based on the model, At what temperature will a house use 41 kWh elceticity? Round your answer to the nearest degree.

www.math-knots.com | www.a4ace.com

Residents of Florida use more electricity during summer times. A sample of electricity usage data is taken as below.

Temperature	Electricity (kWh)
58	32.6
64	35.4
73	38.2
74	39.3
78	42.7
82	44

The relation between the temperatures and the electricity usage is modeled by the below equation
$y = 0.474 x + 4.78$ where x is the average daily temperature in °F and y is the average amount of electricity consumed in kilowatt-hours (kWh).

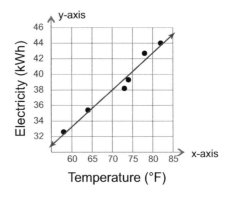

104) What does the slope of the line represent?

105) What does the y-intercept of this function represent?

106) Based on the model, how much electricity will be consumed if the average daily temperature is 69 °F ? Round your answer to the nearest kilowatt-hour.

107) Based on the model, find the electricity usage if the average daily temperature is 94 °F ? Round your answer to the nearest kilowatt-hour.

108) Based on the model, at what temperature will a household use 33 kWh. Round your answer to the nearest degree.

Algebra 1

Vol 2
Week 21
Rotations

The weather department took the following sample data from the Nisarga cyclone in 2020.

Air Pressure (kPa)	Wind Speed (knots)
925	108
933	109
973	81.9
979	56.5
991	45.1
1,006	29.5

Based on the relation between the air pressure and wind speed the following equation is modeled.
$y = -x + 1040$ where x is the air pressure in millibars (kPa) and y is the maximum sustained wind speed in knots (nautical miles per hour).

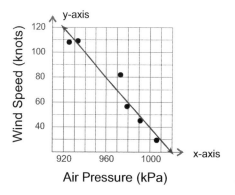

109) What does the slope of the line represent?

110) What does the y-intercept of this function represent?

111) Based on the model, find the wind speed of a cyclone with an air pressure of 949 kPa? Round your answer to the nearest knot.

112) Based the model, find the wind speed of a cyclone with an air pressure of 899 kPa. Round your answer to the nearest knot.

113) Based on the model, find the air pressure of a cyclone when the wind speed is 80 knots. Round your answer to the nearest millibar.

Algebra 1

The cost of a flight is related to the distance it travels

Miles	Cost ($)
450	85
550	76
1,125	
1,850	156
2,150	247
2,300	257

The relation between the cost and the miles a flight travels is modeled by the equation $y = 0.0931 x + 25.7$ where x is distance in miles and y is cost in dollars.

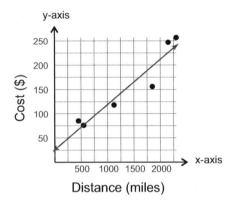

114)   What does the slope of the line represent?

115)   What does the y-intercept of this function represent?

116)   Based on the model, find the cost of a flight that travels 1550 miles.
        Round your answer to the nearest dollar.

117)   Based on the model, find the cost of a flight travelling 3725 miles.
        Round your answer to the nearest dollar.

118)   Based on the model, how many miles will a flight travel with a cost of $120?
        Round your answer to the nearest mile.

www.math-knots.com | www.a4ace.com

Algebra 1

The height and weight of flying club members are collected and a sample data is taken as below

Height (ft)	Weight (lbs)
4.7	98.1
4.95	
5.45	135
5.65	144
6.05	167
6.4	184

The relation between the height and weight of flying club members is modeled as below.
y = 50.5 x − 140 where x is height in feet and y is weight in pounds.

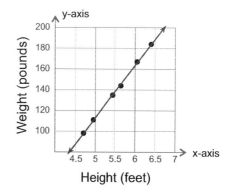

119)   What does the slope of the line represent?

120)   What does the y-intercept of this function represent?

121)   Based on the model, find the weight of a member with a height of 5.95 ft. Round your answer to the nearest tenth.

122)   Based on the model, find the weight of a member with a height of 7 ft. Round your answer to the nearest tenth.

123)   Based on the model, find the height of a club member with a weight of 120 pounds. Round your answer to the nearest hundredth.

 **Algebra 1**

124) Find the GCF for the below.

$48xy, 24$

A) $48xy$ B) $24y$

C) 24 D) $24x$

125) Find the GCF for the below.

$52uv, 76uv$

A) $4uv$ B) $988uv$

C) $2uv$ D) $4u$

126) Find the GCF for the below.

$80uv, 100u^2v$

A) $2uv$ B) $20uv$

C) $100uv$ D) $400u^2v$

127) Find the GCF for the below.

$68y, 64x^2$

A) 3 B) 4

C) 2 D) $1088yx^2$

128) Find the GCF for the below.

$100, 40uv$

A) 20 B) 10

C) 4 D) $200uv$

129) Find the GCF for the below.

$48y^3x, 96xy$

A) $48x$ B) $48yx$

C) $96y^3x$ D) $21yx$

130) Find the GCF for the below.

$57x^2, 57$

A) $57x^2$ B) 57

C) $57x$ D) 3

131) Find the GCF for the below.

$38y^2x, 57y^2x$

A) $19yx$ B) $19y^2x$

C) $114y^2x$ D) $5y^2x$

132) Find the GCF for the below.

$18, 15x$

A) $90x$ B) 3

C) 6 D) $3x$

133) Find the GCF for the below.

$45a^2, 99a^3$

A) $495a^3$ B) $9a$

C) $9a^2$ D) $45a^2$

134) Find the GCF for the below.

$50, 75, 100b^2$

A) $300b^2$ B) 5

C) 25   D) $25a$

135) Find the GCF for the below.

$84x^2, 15y, 93x$

A) $3y$ B) $13020x^2y$

C) 15   D) 3

136) Find the GCF for the below.

$72x^4, 60x^3y, 48x^3$

A) $720x^4y$ B) $3x^3$

C) $12x^3$ D) $6x^3$

137) Find the GCF for the below.

$26x^2, 26x, 78yx$

A) $78yx^2$ B) $13x$

C) $2x$ D) $26x$

138) Find the GCF for the below.

$61v, 51u, 26u^2$

A) $80886u^2v$ B) 1

C) 5   D) v

139) Find the GCF for the below.

$64, 80, 48y^2$

A) $16y$ B) $960y^2$

C) 16   D) 11

140) Find the GCF for the below.

$51u, 85u^2, 51u^2$

A) $4u$ B) $14u$

C) $17u$ D) $255u^2$

141) Find the GCF for the below.

$80n^3, 90n^2, 30n^3$

A) $2n^2$ B) $10n^2$

C) $10n^2m$ D) $720n^3$

142) Find the GCF for the below.

$32ab, 92b^2, 48$

A) 16   B) $2208ab^2$

C) 2   D) 4

143) Find the GCF for the below.

$93x^2, 76x^2, 63y^2$

A) 1   B) y

C) $148428y^2x^2$ D) 5

144) Find the LCM for the below.

$24x^2, 18$

A) $432x^2$ B) 6

C) $72x^2$ D) $72x^3$

145) Find the LCM for the below.

$54m^2, 39m^2$

A) $702m$ B) $2106m^4$

C) $3m^2$ D) $702m^2$

146) Find the LCM for the below.

$90y, 54x^2$

A) $4860yx^2$ B) $810yx^2$

C) 18   D) $270yx^2$

147) Find the LCM for the below.

$27x^2y, 24y^3x$

A) $3xy$ B) $9x^2y^3$

C) $648x^3y^4$ D) $216x^2y^3$

148) Find the LCM for the below.

$90y^2, 99$

A) 9   B) $8910y^2$

C) $990y^2$ D) $6y^2$

149) Find the LCM for the below.

$80, 60xy$

A) 20   B) $42xy$

C) $240xy$ D) $4800xy$

150) Find the LCM for the below.

$91x^2, 78y$

A) $7098x^2y$ B) $1092x^2y$

C) 13   D) $546x^2y$

151) Find the LCM for the below.

$98y^2, 42y^2$

A) $49y^2$ B) $14y^2$

C) $294y^2$ D) $4116y^4$

152) Find the LCM for the below.

$45, 54x^2$

A) $270x$ B) 9

C) $270x^2$ D) $2430x^2$

153) Find the LCM for the below.

$70b^3, 80b$

A) $10b$ B) $2800b^3$

C) $560b^3$ D) $5600b^4$

154) Find the LCM for the below.

$42y^2, 98y, 56xy^2$

A) $2352xy^2$ B) $14y$

C) $230496y^5x$ D) $1176xy^2$

155) Find the LCM for the below.

$19n, 24mn, 5m$

A) $2280m$ B) $2280n^2m^2$

C) 1 D) $2280mn$

156) Find the LCM for the below.

$45xy, 90y, 75y^2$

A) $303750xy^4$ B) $225y^2x$

C) $450y^2x$ D) $15y$

157) Find the LCM for the below.

$49, 83xy, 79$

A) $321293xy$ B) 1

C) $314117xy$ D) $4067xy$

158) Find the LCM for the below.

$39yx^2, 52x^2, 78y^2x^2$

A) $158184y^3x^6$ B) $156y^2x^2$

C) $780y^2x^2$ D) $13x^2$

159) Find the LCM for the below.

$96x^2, 72x^4, 96x^3y$

A) $663552x^9y$ B) $288yx^4$

C) $92yx^4$ D) $24x^2$

160) Find the LCM for the below.

$90, 74x^2, 36$

A) $239760x^2$ B) $18x^2$

C) $6660x^2$ D) 2

161) Find the LCM for the below.

$16n^2, 24, 16m^2$

A) $96m^2n^2$ B) $48m^2n^2$

C) $6144n^2m^2$ D) 8

162) Find the LCM for the below.

$96m^2, 84m^4, 78m^2$

A) $6m^2$ B) $628992m^8$

C) $8736m^4$ D) $7287m^4$

163) Find the LCM for the below.

$45a, 36a^2, 63a$

A) $1260a^2$ B) $843a^2$

C) $102060a^4$ D) $9a$

1) Find the slope of the straight line passing through the points

$$A (-3 , -18) , B (19 , -10)$$

A) $\dfrac{11}{4}$

B) $\dfrac{4}{11}$

C) $-\dfrac{4}{11}$

D) $-\dfrac{11}{4}$

2) Write the standard form of the equation of the straight line passing through the point $A (-4 , 5)$ and parallel to

$$y = -\dfrac{6}{7} x - 5.$$

A) $6x + 7y = 11$

B) $11x - 7y = -6$

C) $6x + 3y = -7$

D) $6x + 7y = -3$

3) Find the slope of the straight line from the below graph.

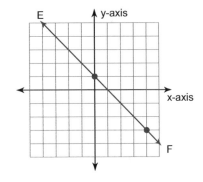

A) $\dfrac{5}{2}$

B) -1

C) $-\dfrac{5}{2}$

D) 1

4) Plot the straight line in the graph below with x-intercept as 3 and y-intercept as 3.

A)

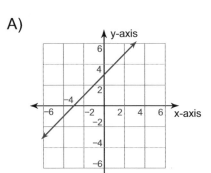

B)

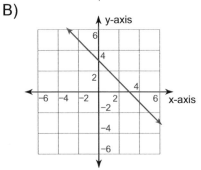

C)

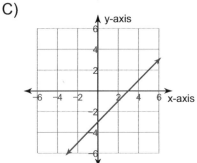

D)

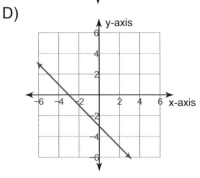

The weather department collected the following sample data during the Asani cyclone in 2022.

Air Pressure (kPa)	Wind Speed (knots)
922	117
949	86.3
961	96.8
979	52.1
996	41.9
1,008	26.3

The weather department related the air pressure and wind speed. The information is modeled as below.
$y = -1.07x + 1110$ where x is the air pressure in millibars(kPa) and y is the maximum sustained wind speed in knots (nautical miles per hour).

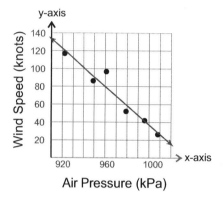

5) What does the slope of the line represent?

6) What does the y-intercept of this function represent?

7) Based on the model, find the wind speed of a cyclone with an air pressure of 968 kPa? Round your answer to the nearest knot.

8) Based on the model, find the wind speed of a cyclone with an air pressure of 899 kPa. Round your answer to the nearest knot.

9) Based on the model, find the air pressure when wind speed is 110 knots Round your answer to the nearest millibar.

www.math-knots.com | www.a4ace.com

Algebra 1

10) Plot the straight-line $y = -\dfrac{1}{3}x - 4$ in the graph below.

A)

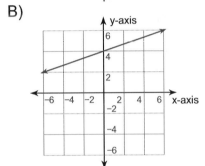

B)

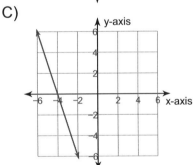

C)

D)

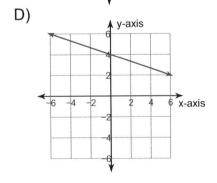

11) Find the slope of the straight line from the below graph.

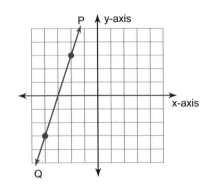

A) −3

B) $-\dfrac{1}{3}$

C) $\dfrac{1}{3}$

D) 3

12) Find the slope of the straight line passing through the points

P (− 1 , 7) , Q (15 , 4)

A) $-\dfrac{16}{3}$

B) $\dfrac{16}{3}$

C) $-\dfrac{3}{16}$

D) $\dfrac{3}{16}$

13) Write the standard form of the equation of the straight line passing through the point A (− 5 , 3) and perpendicular to $y = \dfrac{5}{6}x - 4$.

A) 6 x + 5 y = 15

B) 6 x + 15 y = 5

C) 6 x − 5 y = − 15

D) 6 x + 5 y = − 15

www.math-knots.com | www.a4ace.com

14) Write the equation of the straight line passing through the points A (0 , − 5) and B (− 1 , 4) in slope intercept form.

A) $y = -9x - 5$ B) $y = 9x - 5$

C) $y = -5x - 9$ D) $y = 3x - 9$

15) Find the slope of the straight line from the below graph.

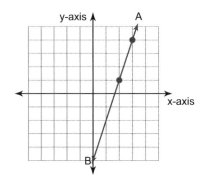

A) $-\dfrac{1}{3}$ B) 3

C) -3 D) $\dfrac{1}{3}$

16) Find the slope of the straight line passing through the points

M (19 , 5) , N (− 4 , − 9)

A) $\dfrac{14}{23}$ B) $\dfrac{23}{14}$

C) $-\dfrac{14}{23}$ D) $-\dfrac{23}{14}$

17) Find the slope of the straight line passing through the points

A (1 , 35) and B (x , 41) ; slope: $-\dfrac{1}{4}$

A) 30   B) − 39

C) − 30   D) − 23

18) Write the standard form of the equation of the straight line passing through the point A (− 5 , 5) and perpendicular to $y = \dfrac{1}{2}x - 1.$

A) $5x + y = 1$

B) $5x - y = 1$

C) $5x + y = -1$

D) $2x + y = -5$

19) Find the slope intercept form of the straight line from the below graph.

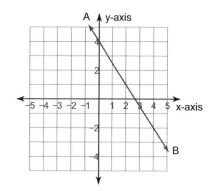

A) $y = \dfrac{1}{2}x + 4$ B) $y = \dfrac{3}{2}x + 4$

C) $y = -\dfrac{5}{2}x + 4$ D) $y = -\dfrac{3}{2}x + 4$

20) What is the slope of the straight line given below?

$$y = -3x + 4$$

A) $\dfrac{1}{3}$ B) $-\dfrac{1}{3}$

C) 3   D) -3

21) Find the slope of the straight line from the below graph.

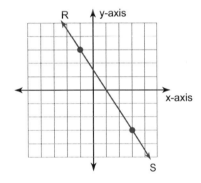

A) $-\dfrac{3}{2}$ B) $\dfrac{3}{2}$

C) $\dfrac{2}{3}$ D) $-\dfrac{2}{3}$

22) Write the equation of the straight line passing through the points A (-1 , 1) and B (1 , -2) in slope intercept form.

A) $y = -\dfrac{1}{2}x - \dfrac{3}{2}$ B) $y = -x - \dfrac{1}{2}$

C) $y = -\dfrac{3}{2}x - \dfrac{1}{2}$ D) $y = \dfrac{3}{2}x - \dfrac{1}{2}$

23) Plot the straight-line $y = -\dfrac{8}{5}x + 5$ in the graph below.

A)

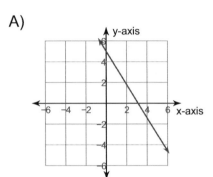

B)

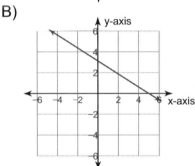

C)

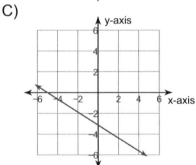

D)

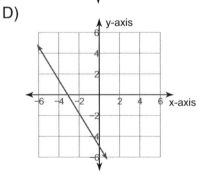

www.math-knots.com | www.a4ace.com

 **Algebra 1**

24) Find the slope of the straight line passing through the points

R (12 , 4) , S (14 , 12)

A) $\dfrac{1}{4}$ B) $-\dfrac{1}{4}$

C) 4   D) −4

25) Write the equation of the straight line passing through the points A (2 , 1) and B (2 , − 4) in standard form.

A) − y = 0   B) x = 0

C) x = 2   D) y = 0

26) Find the slope of the straight line passing through the points

R (x , 19) and S (− 26 , − 7); slope: $-\dfrac{13}{10}$

A) 12   B) −14

C) −31   D) −46

27) What is the slope of the straight line given below?

4 x − y = 1

A) 4   B) − 4

C) $\dfrac{1}{4}$ D) $-\dfrac{1}{4}$

28) What is the slope of the straight line given below?

$$y = -\dfrac{3}{5}x + 2$$

A) $\dfrac{5}{3}$ B) $-\dfrac{5}{3}$

C) $-\dfrac{3}{5}$ D) $\dfrac{3}{5}$

29) Rewrite the below equation of straight line into slope intercept form of the straight line.

$$y - 2 = 4 (x - 1)$$

A) y = − 4 x − 2   B) y = 4 x − 2

C) y = − x − 4   D) y = − 2 x − 4

30) Find the slope of the straight line passing through the points

M (20 , y) and N (11 , 10); slope: $-\dfrac{5}{9}$

A) 5   B) 32

C) 16   D) 8

31) What is the slope of the straight line given below?

3 x + 2 y = 4

A) $\dfrac{2}{3}$ B) $-\dfrac{3}{2}$

C) $-\dfrac{2}{3}$ D) $\dfrac{3}{2}$

Algebra 1

32) Find the slope intercept form of the straight line from the below graph.

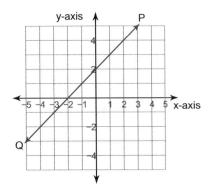

A) $y = 2x + 1$ B) $y = x + 2$

C) $y = -x + 2$ D) $y = -5x + 1$

33) What is the slope of the straight line given below?

$$y = -\frac{1}{4}x + 4$$

A) $\frac{1}{4}$ B) 4

C) $-\frac{1}{4}$ D) -4

34) Which of the below straight lines has the highest slope ?

A) $y = x + 2$ B) $y = 3x + 2$

C) $y = 2x - 3$ D) $y = -3x + 2$

35) Find the slope of the straight line passing through the points

P (-6 , y) and Q (36 , -45); slope: $-\frac{3}{7}$

A) -27 B) 0

C) 46   D) 16

36) What is the slope of the straight line given below?

$$4x - 5y = -25$$

A) $-\frac{4}{5}$ B) $\frac{5}{4}$

C) $-\frac{5}{4}$ D) $\frac{4}{5}$

37) Write the equation of the straight line passing through the points A (4 , -4) and B (3 , -1) in standard form.

A) $8x - y = 3$ B) $3x + y = 8$

C) $3x + y = 2$ D) $3x + 5y = 1$

38) What is the slope of the straight line given below?

$$x - 5y = 0$$

A) $\frac{1}{5}$ B) -5

C) 5   D) $-\frac{1}{5}$

39) Plot the straight-line $-1 = -y + 2x$ in the graph below.

A)
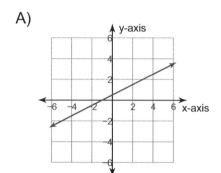

B)
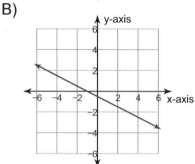

C)
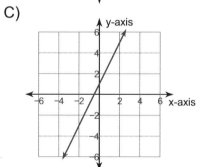

D)
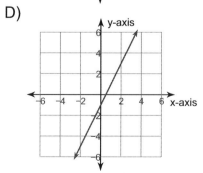

40) Find the rule of translation from
$H(1,0), I(3,4), J(4,3), K(1,-1)$
to
$H'(-3,-2), I'(-1,2), J'(0,1), K'(-3,-3)$

A) translation: 4 units left and 2 units down

B) translation: 4 units left and 4 units down

C) translation: 1 unit right and 1 unit down

D) translation: 5 units left

41) Find the slope intercept form of the straight line from the below point slope form of the straight line.

through: $(2,1)$, slope $= -1$

A) $y = 4x + 3$ B) $y = -2x + 3$

C) $y = 2x + 3$ D) $y = -x + 3$

42) Find the slope intercept form of the straight line from the below point slope form of the straight line.

through: $(-4,4)$, slope $= -\dfrac{1}{2}$

A) $y = -\dfrac{1}{2}x + 2$ B) $y = \dfrac{1}{2}x + 2$

C) $y = \dfrac{5}{2}x + 2$ D) $y = -\dfrac{5}{2}x + 2$

www.math-knots.com | www.a4ace.com

43) Find the slope intercept form of the straight line from the below point standard form of the straight line .

$$x - 2y = 4$$

A) $y = -2x + \dfrac{1}{2}$

B) $y = \dfrac{1}{2}x - 2$

C) $y = \dfrac{1}{2}x + 2$

D) $y = 2x + \dfrac{1}{2}$

44) What is the slope of the straight line given below?

$$x + y = -5$$

A) -1

B) -2

C) 2

D) 1

45) Write the equation of the straight line passing through the points A (4 , 4) and B (− 1 , − 2) in standard form.

A) $5x - 6y = -4$

B) $6x - 5y = 4$

C) $5x + 6y = -24$

D) $6x + 5y = 24$

46) Which of the below straight lines has the least slope ?

$$y + 1 = 0$$

A) $x = -1$

B) $x = 1$

C) $y = -\dfrac{1}{4}x$

D) $y = -1$

47) Find the slope intercept form of the straight line from the below point standard form of the straight line .

$$y - 1 = -(x + 1)$$

A) $y = -x$

B) $y = 2$

C) $y = 2x$

D) $y = -2x$

48) Find the slope intercept form of the straight line from the below point standard form of the straight line.

$$y + 3 = -\dfrac{7}{4}(x - 3)$$

A) $y = -\dfrac{1}{2}x + \dfrac{9}{4}$

B) $y = \dfrac{5}{4}x + \dfrac{9}{4}$

C) $y = \dfrac{3}{4}x + \dfrac{9}{4}$

D) $y = -\dfrac{7}{4}x + \dfrac{9}{4}$

49) Find the rule of translation from
A (4 , − 1) , B (3 , 2) , C (4 , 3) , D (5 , − 1)
to
A' (− 2 , − 4) , B' (− 3 , − 1) , C' (− 2 , 0) , D' (− 1 , − 4)

A) translation: 2 units left and 4 units down

B) translation: 7 units left and 4 units down

C) translation: 3 units left and 1 unit down

D) translation: 6 units left and 3 units down

50) Write the equation of the straight line passing through the points A (− 5 , 0) and B (0 , − 2) in point slope form.

A) $y - 5 = 5x$

B) $y = -\dfrac{2}{5}(x + 5)$

C) $y = -\dfrac{1}{5}(x + 5)$

D) $y = 5(x + 5)$

51) Write the equation of the straight line passing through the point A (2 , 2) and parallel to $y = \dfrac{1}{2}x + 5$.

A) $y + 2 = \dfrac{1}{2}(x + 2)$

B) $y - 2 = -\dfrac{1}{2}(x - 2)$

C) $y - 2 = \dfrac{1}{2}(x - 2)$

D) $y - 2 = \dfrac{1}{2}(x + 2)$

52) Find the slope intercept form of the straight line from the below point slope form of the straight line.

through: (− 4 , 1) , slope = $-\dfrac{3}{2}$

A) $y = x - 5$ B) $y = 2x + 1$

C) $y = -\dfrac{3}{2}x - 5$ D) $y = -5x + 1$

53) Plot the straight-line $-4y + 4 + x = 0$ in the graph below.

A)
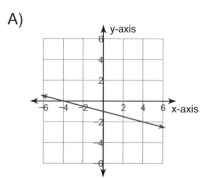

B)
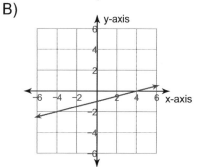

C)
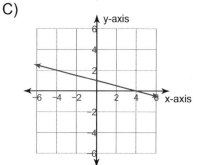

D)
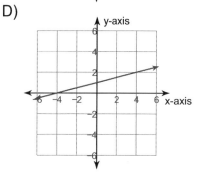

54) Rewrite the below equation of straight line into slope intercept form of the straight line.

$$y - 4 = \frac{1}{3}(x + 3)$$

A) $y = \frac{1}{3}x + 5$

B) $y = 5x - 1$

C) $y = x + 5$

D) $y = -x + 5$

55) Below points Q (− 3 , − 1) , R (− 1 , 4) , S (2 , 0), T (2 , − 1) are translated to 1 unit right and 1 unit down. Find the new coordinates.

A) Q' (− 2 , − 2) , R' (0 , 3) , S' (3 , − 1) , T' (3 , − 2)

B) Q' (− 5 , − 3) , R' (− 3 , 2) , S' (0 , − 2) , T' (0 , − 3)

C) Q' (− 5 , 0) , R' (− 3 , 5) , S' (0 , 1) , T' (0 , 0)

D) Q' (− 1 , 0) , R' (1 , 5) , S' (4 , 1) , T' (4 , 0)

56) Write the equation of the straight line passing through the points A (0 , − 5) and B (− 5 , − 2) in slope intercept form.

A) $y = -\frac{1}{5}x - 5$

B) $y = \frac{1}{5}x - 5$

C) $y = \frac{3}{5}x - 5$

D) $y = -\frac{3}{5}x - 5$

57) Write the equation of the straight line passing through the point A (− 3 , 1) and parallel to $y = \frac{2}{3}x - 4$.

A) $y - 1 = \frac{2}{3}(x + 3)$

B) $y + 3 = \frac{3}{2}(x - 1)$

C) $y - 1 = \frac{3}{2}(x + 3)$

D) $y + 3 = -\frac{3}{2}(x - 1)$

58) Write the standard form of the equation of the straight line passing through the point A (− 2 , 2) and perpendicular to $y = -x - 1$.

A) $2x - y = 2$

B) $2x + y = 2$

C) $x - y = -2$

D) $x - y = -4$

59) Write the equation of the straight line passing through the points A (3 , − 2) and B (0 , 4) in slope intercept form.

A) $y = -5x - 2$

B) $y = -4x - 2$

C) $y = 4x - 2$

D) $y = -2x + 4$

60) Write the equation of the straight line passing through the points A (5 , − 4) and B (− 2 , − 2) in point slope form.

A) $y - 5 = -\dfrac{4}{7}(x + 4)$

B) $y - 5 = \dfrac{4}{7}(x - 4)$

C) $y - 4 = \dfrac{2}{7}(x + 5)$

D) $y + 4 = -\dfrac{2}{7}(x - 5)$

61) Below points J (1 , − 3) , K (− 1 , 0) , L (3 , 1) , M (2 , − 3) are translated to 2 units left and 1 unit up. Find the new coordinates.

A) J' (− 3 , − 5) , K' (− 5 , − 2) , L' (− 1 , − 1) , M' (− 2 , − 5)

B) J' (− 3 , − 4) , K' (− 5 , − 1) , L' (− 1 , 0) , M' (− 2 , − 4)

C) J' (− 1 , − 2) , K' (− 3 , 1) , L' (1 , 2) , M' (0 , − 2)

D) J' (− 2 , − 3) , K' (− 4 , 0) , L' (0 , 1) , M' (− 1 , − 3)

62) Write the equation of the straight line passing through the point A (− 2 , 4) and perpendicular to $y = -\dfrac{3}{8}x - 5$.

A) $y = \dfrac{8}{3}x + \dfrac{1}{3}$

B) $y = \dfrac{8}{3}x + \dfrac{28}{3}$

C) $y = \dfrac{28}{3}x + \dfrac{8}{3}$

D) $y = \dfrac{1}{3}x + \dfrac{8}{3}$

63) Write the standard form of the equation of the straight line passing through the point A (1 , − 3) and parallel to y = 4.

A) y = 1

B) y = − 3

C) −y = − 4

D) − y = 4

64) Rewrite the below equation of straight line into slope intercept form of the straight line.

$$y - 3 = \dfrac{1}{7}(x - 3)$$

A) $y = \dfrac{1}{7}x + \dfrac{18}{7}$

B) $y = -\dfrac{5}{7}x + \dfrac{18}{7}$

C) $y = -\dfrac{2}{7}x + \dfrac{18}{7}$

D) $y = -\dfrac{3}{7}x + \dfrac{18}{7}$

65) Graph the transformation of the figure plotted below to 1 unit down.

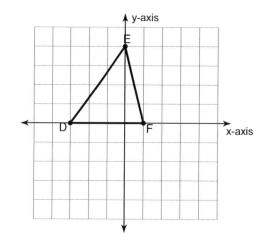

66) Graph the transformation of the figure reflected across y-axis.

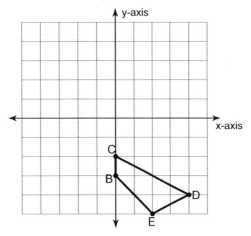

67) Below points N (– 2 , – 4) , M (– 2 , 0) , L (3 , – 1) , K (2 , – 5) are reflected across x = 1. Find the new coordinates.

A) M' (4 , 0) , L' (– 1 , – 1) ,
   K' (0 , – 5) , N' (4 , – 4)

B) M' (– 2 , – 4) , L' (3 , – 3) ,
   K' (2 , 1) , N' (– 2 , 0)

C) M' (– 2 , 0) , L' (3 , 1) ,
   K' (2 , 5) , N' (– 2 , 4)

D) M' (2 , 0) , L' (– 3 , – 1) ,
   K' (– 2 , – 5) , N' (2 , – 4)

68) Write the equation of the straight line passing through the points A (0 , 1) and B (5 , – 1) in point slope form.

A) $y = -\dfrac{3}{5}(x-1)$

B) $y - 1 = -\dfrac{2}{5}x$

C) $y = 0$

D) $y = -5(x-1)$

69) Find the rule of translation from

D (4 , – 3) , E (1 , 1) , F (3 , 4) , G (5 , 1)
                    to
E' (1 , – 1) , F' (3 , – 4) ,
G' (5 , – 1) , D' (4 , 3)

A) reflection across the x-axis

B) reflection across x = 2

C) reflection across x = 1

D) reflection across y = 1

70) Graph the transformation of the figure plotted below to 5 units right and 2 units up.

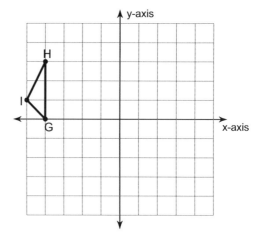

71) Find the x-intercept for the below straight line

$8x - 7y = -35$

Algebra 1

72) Find the y-intercept and
    x-intercept of the below

$$y = x^2 + 4x - 1$$

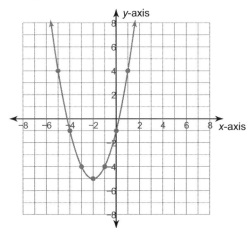

73) Write the standard form of the equation
    of the straight line passing through the
    point A (2 , – 3) and perpendicular to

$$y = -\frac{1}{2}x + 2.$$

   A) $x + 4y = -2$

   B) $2x + y = -7$

   C) $2x - y = 7$

   D) $2x + y = 2$

74) Find the y-intercept for the
    below straight line

$$y = \frac{1}{4}x - 4$$

75) Write the point slope of the equation of
     the straight line passing through the
     point A (2 , 4) and perpendicular to

$$y = -\frac{2}{7}x - 5.$$

   A) $y - 4 = -\frac{2}{5}(x - 2)$

   B) $y - 2 = \frac{5}{2}(x + 4)$

   C) $y - 4 = \frac{7}{2}(x - 2)$

   D) $y + 4 = -\frac{5}{2}(x - 2)$

76) Graph the transformation of the figure
    rotation 180° about the origin.

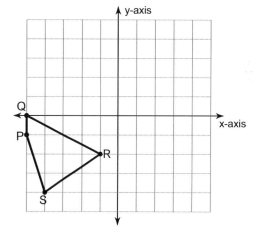

77) Find the y-intercept for the
    below straight line

$$y = -4x - 5$$

78) Below points K (– 5 , – 1) , J (– 5 , 0) , I (– 1 , – 1) , H (– 4 , – 3) are rotated to 90° counterclockwise about the origin. Find the new co-ordinates.

A) K' (– 3 , 0) , J' (– 3 , 1) , I' (1 , 0) , H' (– 2 , – 2)

B) K' (1 , – 5) , J' (0 , – 5) , I' (1 , – 1) , H' (3 , – 4)

C) K' (– 1 , 5) , J' (0 , 5) , I' (– 1 , 1) , H' (– 3 , 4)

D) K' (5 , 1) , J' (5 , 0) , I' (1 , 1) , H' (4 , 3)

79) Graph the transformation of the figure rotation 180° about the origin.

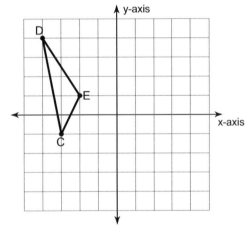

80) Find the rule of translation from

Q (– 3 , – 2) , R (– 4 , 3) , S (– 1 , 0)
to
R' (2 , 3) , S' (– 1 , 0) , Q' (1 , – 2)

A) reflection across the y-axis

B) reflection across x = – 1

C) reflection across x = – 2

D) reflection across y = – 1

81) Plot the straight line in the graph below with x-intercept as – 5 and y-intercept as – 3.

A)
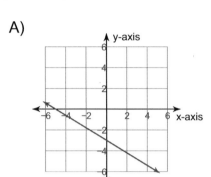

B)
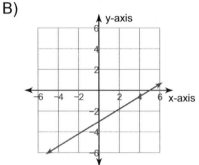

C)
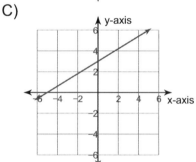

D)
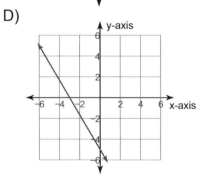

82) Find the y-intercept and x-intercept of the below

$$-3 = -x + y$$

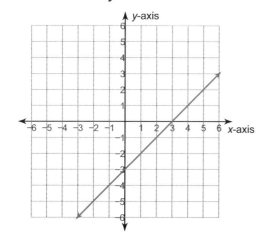

83) Graph the transformation of the figure reflected across x = 1.

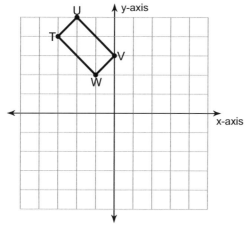

84) Below points C (− 3 , 1) , D (− 3 , 4) , E (− 1 , 3) , F (0 , − 2) are reflected across x-axis. Find the new coordinates.

A) D' (5 , 4) , E' (3 , 3) , F' (2 , − 2) , C' (5 , 1)

B) D' (− 3 , − 2) , E' (− 1 , − 1) , F' (0 , 4) , C' (− 3 , 1)

C) D' (− 3 , − 4) , E' (− 1 , − 3) , F' (0 , 2) , C' (− 3 , − 1)

D) D' (1 , 4) , E' (− 1 , 3) , F' (− 2 , − 2) , C' (1 , 1)

85) Find the rule of translation from

D (− 5 , − 3) , E (− 3 , 2) , F (− 2 , 1)
to
E' (− 3 , − 4) , F' (− 2 , − 3) , D' (− 5 , 1)

A) reflection across the y-axis

B) reflection across x = − 3

C) reflection across y = − 1

D) reflection across y = 1

86) Find the rule of translation from I (0 , 2) , H (3 , 5) , G (2 , 0) to I' (0 , − 2) , H' (− 3 , − 5) , G' (− 2 , 0)

A) reflection across y = 2

B) rotation 90° clockwise about the origin

C) rotation 90° counterclockwise about the origin

D) rotation 180° about the origin

87) Find the x-intercept for the below straight line

$$6x + 5y = -5$$

88) Below points H (− 3 , − 4) , I (− 2 , 1) , J (1 , − 2) , K (0 , − 4) are rotated to 180° about the origin.
Find the new co-ordinates.

A) I' (− 2 , − 3) , J' (1 , 0) , K' (0 , 2) , H' (− 3 , 2)

B) H' (4 , − 3) , I' (− 1 , − 2) , J' (2 , 1) , K' (4 , 0)

C) H' (3 , 4) , I' (2 , − 1) , J' (− 1 , 2) , K' (0 , 4)

D) H' (− 4 , 3) , I' (1 , 2) , J' (− 2 , − 1) , K' (− 4 , 0)

89) Below points M (− 4 , 2) , L (− 3 , 4) , K (0 , 3) , J (2 , − 2) are rotated to 90° counterclockwise about the origin.
Find the new co-ordinates.

A) M' (− 2 , − 4) , L' (− 4 , − 3) , K' (− 3 , 0) , J' (2 , − 2)

B) M' (4 , − 2) , L' (3 , − 4) , K' (0 , − 3) , J' (2 , 2)

C) M' (2 , 4) , L' (4 , 3) , K' (3 , 0) , J' (− 2 , 2)

D) L' (− 3 , − 4) , K' (0 , − 3) , J' (− 2 , 2) , M' (− 4 , − 2)

90) Find the rule of translation from

E (− 5 , − 2) , F (− 3 , 3) , G (− 1 , 3) , H (0 , − 2)
to
F' (3 , 3) , G' (1 , 3) , H' (0 , − 2) , E' (5 , − 2)

A) reflection across x = −2

B) reflection across the x-axis

C) reflection across y = −1

D) reflection across the y-axis

91) Find the y-intercept and x-intercept of the below

$$y = x^2 + 8x + 14$$

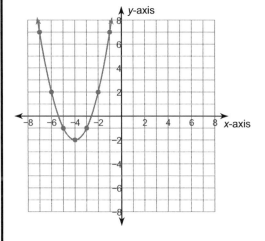

92) Below points G (1 , − 2) , F (1 , − 1) , E (4 , − 2) , D (0 , − 4) are reflected across x = 2. Find the new coordinates.

A) F' (1 , − 1) , E' (− 2 , − 2) , D' (2 , − 4) , G' (1 , − 2)

B) F' (1 , − 5) , E' (4 , − 4) , D' (0 , − 2) , G' (1 , − 4)

C) F' (1 , − 3) , E' (4 , − 2) , D' (0 , 0) , G' (1 , − 2)

D) F' (3 , − 1) , E' (0 , − 2) , D' (4 , − 4) , G' (3 , − 2)

93) Write the equation of the straight line passing through the points A(1 , -2) with a slope of undefined in point slope form.

A) 0 = x − 1

B) y − 2 = − 3 (x − 1)

C) y − 2 = 3 (x − 1)

D) y + 2 = 0

www.math-knots.com | www.a4ace.com

94) Below points R (− 5 , − 4) , S (− 4 , − 1) , T (− 4 , − 4) are reflected across the x = − 1. Find the new coordinates.

A) S' (− 4 , − 1) , T' (− 4 , 2) , R' (− 5 , 2)

B) S' (− 4 , − 1) , T' (− 4 , − 4) , R' (− 3 , − 4)

C) S' (2 , − 1) , T' (2 , − 4) , R' (3 , − 4)

D) S' (− 4 , − 3) , T' (− 4 , 0) , R' (− 5 , 0)

95) Write the equation of the straight line passing through the points A (− 5 , 5) with a slope of slope $= -\dfrac{4}{5}$ in point slope form.

A) $y - 5 = - 5 (x - 5)$

B) $y - 5 = - (x - 5)$

C) $y - 5 = -\dfrac{4}{5} (x + 5)$

D) $y + 5 = 5 (x + 5)$

96) Find the rule of translation from
D (− 4 , 2) , C (− 2 , 3) , B (− 2 , 1)
to
D' (4 , − 2) , C' (2 , − 3) , B' (2 , − 1)

A) rotation 90° counterclockwise about the origin

B) translation: 5 units right and 6 units down

C) rotation 180° about the origin

D) rotation 90° clockwise about the origin

97) Find the y-intercept and x-intercept of the below

$$5x = 6 + 3y$$

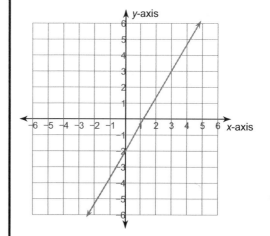

www.math-knots.com | www.a4ace.com

1) Find the solution for the below system of equations using method of substitution
$$x - y = -5$$
$$-2x - 7y = 19$$

A) $(-6, -1)$

B) Infinite number of solutions

C) $(-6, -10)$

D) $(-6, 10)$

2) Find the solution for the below system of equations using method of substitution
$$-2x - 6y = -10$$
$$x + 11y = -3$$

A) $(-3, 6)$ B) $(8, -1)$

C) $(-3, -1)$ D) No solution

3) Find the solution for the below system of equations using method of substitution
$$x + 6y = -3$$
$$12x - 2y = -36$$

A) $(-3, 0)$ B) $(0, 4)$

C) $(0, -3)$ D) $(3, 0)$

4) Find the solution for the below system of equations using method of substitution
$$-11x + y = -1$$
$$5x - 4y = 4$$

A) $(1, 2)$ B) $(1, -10)$

C) $(0, 1)$ D) $(0, -1)$

5) Find the solution for the below system of equations using method of substitution
$$-12x - 3y = -6$$
$$-7x + y = 24$$

A) $(-2, 10)$ B) $(11, -10)$

C) $(7, -10)$ D) $(-2, -10)$

6) Find the solution for the below system of equations using method of substitution
$$-2x - 5y = 27$$
$$x - 5y = 9$$

A) No solution   B) $(-6, -3)$

C) $(10, -3)$ D) $(4, 6)$

7) Find the solution for the below system of equations using method of substitution
$$9x + 10y = 12$$
$$-x + y = -14$$

A) $(-8, -2)$ B) $(8, -6)$

C) $(-8, 2)$ D) $(8, -2)$

8) Find the solution for the below system of equations using method of substitution
$$-3x + 2y = 28$$
$$-3x + y = 32$$

A) $(9, -4)$ B) $(-9, -4)$

C) $(-12, -4)$ D) $(-4, -9)$

9) Find the solution for the below system of equations using method of substitution

$9x + 3y = 1$
$3x + y = 7$

A) No solution

B) (1 , 10)

C) (– 1 , – 10)

D) (– 1 , 10)

10) Find the solution for the below system of equations using method of substitution

$y = -11$
$4x - y = -29$

A) (– 10 , 5)

B) (– 10 , – 3)

C) (– 10 , 11)

D) (– 10 , – 11)

11) Find the solution for the below system of equations using method of substitution

$3x - 5y = 0$
$x - 2y = 0$

A) (0 , 10)

B) (– 10 , 0)

C) (0 , 0)

D) (0 , – 10)

12) Find the solution for the below system of equations using method of substitution

$-6x - 2y = -24$
$y = 0$

A) (4 , – 5)

B) (– 4 , – 5)

C) (4 , 0)

D) (4 , 5)

13) Find the solution for the below system of equations using method of substitution

$x - 3y = 4$
$-2x + 8y = -10$

A) (– 1 , 5)

B) (1 , 5)

C) (1 , – 1)

D) (– 1 , – 5)

14) Find the solution for the below system of equations using method of substitution

$-6x - 3y = 3$
$2x + y = -1$

A) (– 12 , – 4)

B) (– 12 , 4)

C) (4 , – 12)

D) Infinite number of solutions

15) Find the solution for the below system of equations using method of substitution

$9x + y = -14$
$11x + 5y = -36$

A) (– 9 , – 1)

B) (– 9 , 3)

C) (– 1 , – 5)

D) No solution

16) Find the solution for the below system of equations using method of substitution

$4x - 11y = 20$
$x + 9y = 5$

A) (– 5 , 0)

B) Infinite number of solutions

C) No solution

D) (5 , 0)

17) Find the solution for the below system of equations using method of substitution
$$-9x + 7y = -11$$
$$-11x + y = -21$$

A) (1 , 2)

B) Infinite number of solutions

C) (1 , - 2)

D) (2 , 1)

18) Find the solution for the below system of equations using method of substitution
$$-x + y = 0$$
$$5x - 2y = -27$$

A) (9 , - 9)   B) (- 9 , - 9)

C) (- 9 , 1)   D) (- 9 , - 8)

19) Find the solution for the below system of equations using method of substitution
$$y = 0$$
$$3x + 2y = -12$$

A) (- 10 , 4)   B)   4 , - 10)

C) (4 , - 10)   D)   4 , 0)

20) Find the solution for the below system of equations using method of substitution
$$9x + 11y = 14$$
$$x - y = -14$$

A) No solution

B) (- 7 , 7)

C) Infinite number of solutions

D) (- 7 , - 7)

21) Find the solution for the below system of equations using method of substitution
$$x + 4y = -5$$
$$8x + 7y = 35$$

A) (- 2 , 7)   B) (8 , 7)

C) (- 3 , 7)   D) (7 , - 3)

22) Find the solution for the below system of equations using method of substitution
$$-3x + 21y = -99$$
$$x - 7y = 33$$

A) (- 12 , 2)

B) (- 12 , - 2)

C) (12 , - 2)

D) Infinite number of solutions

23) Find the solution for the below system of equations using method of substitution
$$y = -1$$
$$6x - 11y = -7$$

A) (3 , 1)   B) (- 1 , 3)

C) (- 3 , - 1)   D) (1 , 3)

24) Find the solution for the below system of equations using method of substitution
$$x - 9y = -13$$
$$6x + 10y = -14$$

A) (1 , 4)   B) No solution

C) (4 , 1)   D) (- 4 , 1)

Algebra 1

25) Find the solution for the below system of equations using method of substitution
$$x + 4y = -19$$
$$3x + 12y = -57$$

A) $(11, -4)$

B) $(-12, 4)$

C) $(-11, -4)$

D) Infinite number of solutions

26) Find the solution for the below system of equations using method of substitution
$$-12x - 12y = -36$$
$$-12x - 5y = 34$$

A) $(-8, 7)$ B) $(10, -7)$

C) $(-5, 7)$ D) $(-7, 10)$

27) Find the solution for the below system of equations using method of substitution
$$7x + 11y = 11$$
$$-3x + 2y = 2$$

A) $(6, 8)$ B) $(0, 1)$

C) $(1, 0)$ D) $(-1, 0)$

28) Find the solution for the below system of equations using method of substitution
$$10x - 3y = 2$$
$$-11x + 6y = 14$$

A) $(2, -6)$ B) $(-7, -2)$

C) $(-6, -2)$ D) $(2, 6)$

29) Find the solution for the below system of equations using method of substitution
$$-4x - 8y = 36$$
$$y = -7$$

A) $(5, -7)$

B) $(-7, -3)$

C) $(-7, 5)$

D) Infinite number of solutions

30) Find the solution for the below system of equations using method of substitution
$$-2x + 10y = 8$$
$$4x - 20y = 4$$

A) No solution

B) $(-3, -12)$

C) $(-3, 12)$

D) $(3, 12)$

31) Find the solution for the below system of equations using method of substitution
$$4x + 5y = 24$$
$$8x + 4y = 0$$

A) No solution   B) $(-8, 4)$

C) $(-4, -8)$ D) $(-4, 8)$

32) Find the solution for the below system of equations using method of substitution
$$-10x - 5y = -10$$
$$-2x + 2y = 28$$

A) $(-10, -4)$ B) $(-4, 10)$

C) $(4, 10)$ D) $(10, 4)$

 **Algebra 1**

33) Find the solution for the below system of equations using method of substitution

$$-5x + 3y = 0$$
$$12x - y = -31$$

A) $(-7, 3)$ B) $(-5, -3)$

C) $(-7, -3)$ D) $(-3, -5)$

34) Find the solution for the below system of equations using method of substitution

$$-9x + 7y = 24$$
$$5x - y = -22$$

A) $(-5, -3)$ B) $(3, -5)$

C) $(-5, 3)$ D) $(5, -3)$

35) Find the solution for the below system of equations using method of substitution

$$-6x - 5y = -32$$
$$12x + 3y = 36$$

A) $(2, 4)$ B) $(-12, 1)$

C) $(-12, -8)$ D) $(-4, 2)$

36) Find the solution for the below system of equations using method of substitution

$$-12x - 12y = 12$$
$$-5x - 9y = -3$$

A) $(-3, 2)$ B) $(-3, -12)$

C) $(-3, 12)$ D) $(3, -12)$

37) Find the solution for the below system of equations using method of substitution

$$4x + 11y = 29$$
$$-8x - 22y = -58$$

A) No solution

B) $(-12, -5)$

C) Infinite number of solutions

D) $(-12, -10)$

38) Find the solution for the below system of equations using method of substitution

$$9x + 5y = 0$$
$$-2x - 2y = -8$$

A) $(-8, 9)$

B) Infinite number of solutions

C) $(8, 5)$

D) $(-5, 9)$

39) Find the solution for the below system of equations using method of substitution

$$-9x + 4y = 23$$
$$y = -1$$

A) $(-3, -2)$ B) $(-2, -6)$

C) $(-3, -1)$ D) $(-6, -2)$

40) Find the solution for the below system of equations using method of substitution

$$3x + 12y = 96$$
$$-x - 4y = -32$$

A) $(-5, 10)$

B) Infinite number of solutions

C) No solution

D) $(-5, -9)$

41) Find the solution for the below system
of equations using method of substitution
$$-4x - 9y = -33$$
$$2x + 4y = 14$$

A) (3 , 12)   B) (3 , 5)

C) (−3 , 12)   D) (−3 , 5)

42) Find the solution for the below system
of equations using method of substitution
$$y = -2$$
$$5x + 9y = 7$$

A) Infinite number of solutions

B) (7 , 9)

C) (−2 , 7)

D) (7 , −2)

43) Find the solution for the below system
of equations using method of substitution
$$-9x + 4y = 20$$
$$5x - 3y = -15$$

A) (0 , 5)   B) (−9 , −5)

C) (5 , 5)   D) (0 , −5)

44) Find the solution for the below system
of equations using method of substitution
$$-8x - 4y = -20$$
$$y = -3$$

A) (4 , −3)   B) (3 , 8)

C) (−4 , −3)   D) (−4 , 3)

45) Find the solution for the below system
of equations using method of substitution
$$-2x + 4y = 16$$
$$6x - y = -4$$

A) (−4 , 0)

B) (0 , 4)

C) (4 , 0)

D) Infinite number of solutions

46) The sum of two numbers is 22. Their
difference is 6. What are the numbers?

A) 4 and 12

B) 8 and 14

C) 13 and 21

D) 5 and 13

47) Find the value of two numbers if their
sum is 18 and their difference is 2.

A) 4 and 14

B) 6 and 14

C) 7 and 14

D) 8 and 10

48) Find the value of two numbers if their sum is 15 and their difference is 1.

A) 3 and 13

B) 10 and 7

C) 8 and 6

D) 7 and 8

49) Find the value of two numbers if their sum is 24 and their difference is 4.

A) 8 and 12

B) 10 and 14

C) 4 and 17

D) 7 and 6

50) Find the value of two numbers if their sum is 20 and their difference is 2.

A) 11 and 7

B) 9 and 11

C) 11 and 4

D) 5 and 14

51) Find the value of two numbers if their sum is 23 and their difference is 3.

A) 4 and 6

B) 7 and 20

C) 13 and 21

D) 10 and 13

52) The difference of two numbers is 5. Their sum is 23. What are the numbers?

A) 14 and 7

B) 12 and 21

C) 9 and 14

D) 6 and 17

53) Find the value of two numbers if their sum is 23 and their difference is 1.

A) 12 and 10

B) 18 and 11

C) 8 and 10

D) 11 and 12

54) The difference of two numbers is 1. Their sum is 19. Find the numbers.

A) 8 and 14

B) 8 and 15

C) 9 and 10

D) 7 and 5

55) Find the value of two numbers if their sum is 27 and their difference is 1.

A) 15 and 22

B) 10 and 15

C) 8 and 19

D) 13 and 14

56) The difference of two numbers is 2. Their sum is 22. What are the numbers?

A) 15 and 6

B) 12 and 13

C) 14 and 15

D) 10 and 12

57) The difference of two numbers is 5. Their sum is 19. Find the numbers.

A) 9 and 16

B) 9 and 10

C) 7 and 12

D) 3 and 8

58) The sum of two numbers is 16. Their difference is 2. Find the numbers.

A) 7 and 9

B) 8 and 13

C) 8 and 14

D) 3 and 11

59) The sum of two numbers is 21. Their difference is 3. What are the numbers?

A) 14 and 8

B) 11 and 17

C) 9 and 12

D) 5 and 8

60) Find the value of two numbers if their sum is 19 and their difference is 3.

A) 3 and 18

B) 12 and 16

C) 3 and 8

D) 8 and 11

61) The sum of two numbers is 25. Their difference is 1. Find the numbers.

A) 12 and 13

B) 17 and 5

C) 18 and 21

D) 5 and 19

62) Find the value of two numbers if their sum is 21 and their difference is 5.

A) 8 and 13

B) 7 and 14

C) 3 and 7

D) 13 and 20

63) The difference of two numbers is 1. Their sum is 21. What are the numbers?

A) 10 and 11

B) 14 and 16

C) 11 and 13

D) 15 and 15

 **Algebra 1**

64) The sum of two numbers is 22. Their difference is 4. Find the numbers.

A) 12 and 14

B) 8 and 20

C) 9 and 13

D) 7 and 10

65) The difference of two numbers is 4. Their sum is 18. Find the numbers.

A) 9 and 7

B) 5 and 14

C) 8 and 9

D) 7 and 11

66) The difference of two numbers is 1. Their sum is 17. Find the numbers.

A) 7 and 4

B) 3 and 8

C) 7 and 12

D) 8 and 9

67) Grade 9 and Grade 10 students of MK Academy are going to STEM fair in Washington DC. Grade 9 students traveled in 9 vans and 4 buses with 171 students in total. Grade 10 students traveled in 11 vans and 4 buses with 185 students in total. If each van and bus have the same number of students seating, then how many students are seated in a van? How many students are seated in a bus?

A) Van : 27 , Bus : 7

B) Van : 7 , Bus : 27

C) Van : 3 , Bus : 35

D) Van : 4 , Bus : 19

68) Grade 10 and Grade 11 students of MK Academy are going to field trip to Bush gardens. Grade 10 students traveled in 8 vans and 8 buses with 320 students in total. Grade 11 students traveled in 9 vans and 8 buses with 330 students in total. If each van and bus have the same number of students seating, then how many students are seated in a van? How many students are seated in a bus?

A) Van : 15 , Bus : 23

B) Van : 16 , Bus : 25

C) Van : 10 , Bus : 30

D) Van : 8 , Bus : 26

69) Grade 7 and Grade 8 students of MK Academy are going to STEM fair in Washington DC. Grade 7 students traveled in 2 vans and 7 buses with 237 students in total. Grade 8 students traveled in 13 vans and 7 buses with 347 students in total. If each van and bus have the same number of students seating, then how many students are seated in a van? How many students are seated in a bus?

A) Van : 31 , Bus : 10

B) Van : 10 , Bus : 31

C) Van : 6 , Bus : 35

D) Van : 16 , Bus : 21

70) Grade 7 and Grade 8 students of MK Academy are going to Hershey's park for musical performance. Grade 7 students traveled in 13 vans and 13 buses with 923 students in total. Grade 8 students traveled in 2 vans and 13 buses with 725 students in total. If each van and bus have the same number of students seating, then how many students are seated in a van? How many students are seated in a bus?

A) Van : 23 , Bus : 25

B) Van : 11 , Bus : 22

C) Van : 53 , Bus : 18

D) Van : 18 , Bus : 53

There are 18 animals in the barn. There are chickens and goats, with 56 legs in all.

71) How many chickens are there?

72) How many goats are there?

A farmhouse shelters 20 animals. There are oxen and chickens. Altogether there are 62 legs.

73) How many chickens are there?

74) How many oxen are there?

A farmhouse shelters 18 animals. There are horses and geese. Altogether there are 54 legs.

75) How many horses are there?

76) How many geese are there?

A farmhouse shelters 12 animals.
There are sheep and chickens.
Altogether there are 30 legs.

77) How many chickens are there?

78) How many sheeps are there?

There are 25 animals in the barn.
There are ducks and sheep.
There are 86 legs in all.

79) How many ducks are there?

80) How many sheeps are there?

There are 8 animals in the barn.
There are ducks and buffalo.
There are 26 legs in all.

81) How many ducks are there?

82) How many buffalo are there?

There are 5 animals in the field.
There are buffalos and chickens.
There are 16 legs in all.

83) How many chickens are in the field?

84) How many buffalo are in the field?

There are 27 animals in the field.
There are goats and geese.
There are 90 legs in all.

85) How many geese are in the field?

86) How many goats are in the field?

There are 21 animals in the barn.
There are geese and sheep.
There are 68 legs in all.

87) How many geese are there?

88) How many sheep are there?

A farmhouse shelters 28 animals.
There are horses and chickens.
Altogether there are 94 legs.

89) How many chickens are there?

90) How many horses are there?

There are 14 animals in the barn.
There are geese and goats.
There are 50 legs in all.

91) How many geese are there?

92) How many goats are there?

There are 14 animals in the field.
There are sheep and ducks.
There are 46 legs in all.

93) How many ducks are in the field?

94) How many sheeps are in the field?

A farmhouse shelters 25 animals.
There are buffalos and chickens.
Altogether there are 80 legs.

95) How many buffalos are there?

96) How many chickens are there?

There are 20 animals in the field.
There are cows and ducks.
There are 60 legs in all.

97) How many ducks are in the field?

98) How many cows are in the field?

There are 17 animals in the field.
There are goats and chickens.
There are 54 legs in all.

99) How many chickens are in the field?

100) How many goats are in the field?

A farmhouse shelters 14 animals.
There are buffalos and chickens.
Altogether there are 44 legs.

101) How many chickens are there?

102) How many buffalos are there?

There are 16 animals in the field.
There are sheeps and ducks.
There are 50 legs in all.

103) How many ducks are in the field?

104) How many sheeps are in the field?

A farmhouse shelters 10 animals.
There are horses and ducks.
Altogether there are 28 legs.

105) How many ducks are there?

106) How many horses are there?

There are 23 animals in the barn.
There are ducks and goats.
There are 76 legs in all.

107) How many ducks are there?

108) How many goatss are there?

There are 15 animals in the barn.
There are ducks and horses.
There are 54 legs in all.

109) How many ducks are there?

110) How many horses are there?

The school cafeteria sells oranges for
$0.35 and tangerines for $0.95. On friday
they sold 8 pieces of fruit earning a total
of $4.

111) How many oranges did they sell?

112) How many tangerines did they sell?

Bella spent $322 on a pair of pants. Dress pants cost $88 and jeans cost $29. If she bought a total of 5.

113) How many pants did she buy?

114) How many jeans did she buy?

Rik spent $798 on pairs of pants.
Dress pants cost $79 and jeans cost $81. If he bought a total of 10 pair of pants.

115) How many pants did he buy?

116) How many jeans did he buy?

The book store sells graph paper and lined paper. Graph paper cost $0.70 and lined paper cost $1.25. On Friday they sold 8 pieces of paper earning a total of $7.80.

117) How many graph paper did they sell?

118) How many lined paper did they sell?

Mary spent $123.40 on shirts. Fancy shirts cost $20.10 and plain shirts cost $8.60. If she bought a total of 9.

119) How many fancy shirts did she buy?

120) How many plain shirts did she buy?

The cafeteria sells corn chips for $1.25 and potato chips for $1.15. On Thursday they sold 10 bags of chips earning a total of $12.10.

121) How many corn chips did they sell?

122) How many potato chips did they sell?

Tom bought 7 eating utensils for a total of $26.60. Spoons cost $5.30 and forks cost $3.20.

123) How many spoons did he buy?

124) How many forks did he buy?

The grocery store sells apples for $0.25 and pears for $0.15. On Tuesday they sold 11 pieces of fruit earning a total of $2.05.

125) How many apples did they sell?

126) How many pears did they sell?

Lola bought 6 eating utensils for a total of $33. Spoons cost $5.90 and forks cost $4.70.

127) How many spoons did she buy?

128) How many forks did she buy?

Helen spent $29.70 on eating utensils. Spoons cost $2.80 and forks cost $3.70. If she bought a total of 9.

129) How many spoons did she buy?

130) How many forks did she buy?

Fiona bought 10 items for a total of $26.40. Pens cost $3.20 and pencils cost $1.80.

131) How many pens did she buy?

132) How many pencils did she buy?

Ian spent $31 on schools supplies. Pens cost $4 and pencils cost $2.20. If he bought a total of 10.

133) How many pens did he buy?

134) How many pencils did he buy?

The fruit vendor sells apples for $0.90 and pears for $1.15. On Thursday they sold 9 pieces of fruit earning a total of $9.85.

135) How many apples did he sell?

136) How many pears did he sell?

 **Algebra 1**

The book store sells correction ink for $0.95 and correction tape for $0.80. On Thursday they sold 10 of these items earning a total of $9.20.

137) How many correction ink did they sell?

138) How many correction tape did they sell?

The book store sells graph paper and lined paper. Graph paper cost $0.25 and lined paper cost $0.50. On Monday they sold 8 pieces of paper earning a total of $3.

139) How many graph paper did they sell?

140) How many lined paper did they sell?

Matt bought 4 writing tools for a total of $7.60. Pens cost $2.30 and pencils cost $1.50.

141) How many pens did he buy?

142) How many pencils did he buy?

Zara spent $141 on shirts. Fancy shirts cost $20.10 and plain shirts cost $10.20. If she bought a total of 8.

143) How many fancy shirts did she buy?

144) How many plain shirts did she buy?

The store sells corn chips for $0.95 and potato chips for $0.75. On Monday they sold 16 bags of chips earning a total of $13.20.

145) How many corn chips did they sell?

146) How many potato chips did they sell?

Elsa spent $67.60 on shirts. Tee shirts cost $2.40 and long sleeve shirts cost $11.60. If she bought a total of 9.

147) How many tee shirts did she buy?

148) How many sleeve shirts did she buy?

The cafeteria sells corn chips for $0.90 and potato chips for $0.35. On Wednesday they sold 16 bags of chips earning a total of $10.55.

149) How many corn chips did they sell?

150) How many potato chips did they sell?

Lola's Custom Kitchen Supplies sells handmade forks and spoons. It costs the store $2 to buy the supplies to make a fork and $1 to buy the supplies to make a spoon. The store sells forks for $4 and spoons for $6. This month they spent $41 on materials for forks and spoons. They sold the finished products for a total of $118.

151) How many forks did they make this month?

152) How many spoons did they make this month?

Sam's Custom Kitchen Supplies sells edible forks and spoons. It costs the store $2 to buy the supplies to make a fork and $1 to buy the supplies to make a spoon. The store sells forks for $4 and spoons for $4. This month they spent $41 on materials for forks and spoons. They sold the finished products for a total of $108.

153) How many forks did they sell this month?

154) How many spoons did they sell this month?

Rik's Custom Kitchen Supplies sells edible forks and spoons. It costs the store $2 to buy the supplies to make a fork and $1 to buy the supplies to make a spoon. The store sells forks for $6 and spoons for $5. This month Rik's Custom Kitchen Supplies spent $24 on materials for forks and spoons. They sold the finished products for a total of $88.

155) How many forks did they make this month?

156) How many spoons did they make this month?

Rak's Custom Kitchen Supplies sells edible plates and cups. It costs the store $2 to buy the supplies to make a plate and $1 to buy the supplies to make a cup. The store sells plates for $6 and cups for $4. This month Maria's Custom Kitchen Supplies spent $33 on materials for plates and cups. They sold the finished products for a total of $112.

157) How many plates did they make this month?

158) How many cups did they make this month?

Sara's Custom Kitchen Supplies sells handmade forks and spoons. It costs the store $1 to buy the supplies to make a fork and $2 to buy the supplies to make a spoon. The store sells forks for $6 and spoons for $6. This month Sara's Custom Kitchen Supplies spent $42 on materials for forks and spoons. They sold the finished products for a total of $156.

159) How many forks did they make this month?

160) How many spoons did they make this month?

Ishan's Custom Kitchen Supplies sells edible cups and spoons. It costs the store $2 to buy the supplies to make a cup and $1 to buy the supplies to make a spoon. The store sells cups for $5 and spoons for $5. This month Ishan's Custom Kitchen Supplies spent $25 on materials for cups and spoons. They sold the finished products for a total of $90.

161) How many cups did they make this month?

162) How many spoons did they make this month?

MK's Custom Kitchen Supplies sells edible plates and spoons. It costs the store $2 to buy the supplies to make a plate and $2 to buy the supplies to make a spoon. The store sells plates for $6 and spoons for $4. This month MK's Custom Kitchen Supplies spent $48 on materials for plates and spoons. They sold the finished products for a total of $128.

163) How many plates did they make this month?

164) How many spoons did they make this month?

Amy's Custom Kitchen Supplies sells edible forks and spoons. It costs the store $2 to buy the supplies to make a fork and $1 to buy the supplies to make a spoon. The store sells forks for $6 and spoons for $5. This month Amy's Custom Kitchen Supplies spent $28 on materials for forks and spoons. They sold the finished products for a total of $96.

165) How many forks did they
      make this month?

166) How many spoons did they
      make this month?

Violet's Custom Kitchen Supplies sells edible cups and bowls. It costs the store $2 to buy the supplies to make a cup and $1 to buy the supplies to make a bowl. The store sells cups for $4 and bowls for $4. This month violet's Custom Kitchen Supplies spent $48 on materials for cups and bowls. They sold the finished products for a total of $120.

167) How many cups did they make this?

168) How many bowls did they make this?

Mk's Custom Kitchen Supplies sells edible bowls and spoons. It costs the store $1 to buy the supplies to make a bowl and $2 to buy the supplies to make a spoon. The store sells bowls for $5 and spoons for $4. This month Julio's Custom Kitchen Supplies spent $17 on materials for bowls and spoons. They sold the finished products for a total of $49.

169) How many bowls did they
      make this month?

170) How many spoons did they
      make this month?

At MK's Printing Company there are two kinds of printers: Model A can print 80 books per day and Model B can print 40 books per day. The company owns 13 total printers to print 640 books per day.

171) How many model A printers
      do they have?

172) How many model B printers
      do they have?

MK's Printing Inc. has two type of
printers: Model A and Model B.
Model A can print 60 books per day and
Model B can print 70 books per day.
Altogether they have 11 printers.
If they can print 700 books in a day, then
answer the below

173) How many Model A printers
does they have?

174) How many Model B printers
does they have?

At KK's Printing Company there are
two kinds of printers: Model A
can print 70 books per day and
Model B can print 35 books per
day. The company owns 9 total printers
and this allows them to print 560
books per day.

175) How many Model A printers
do they have?

176) How many Model B printers
do they have?

At JJ's Printing Company there are
two kinds of printers Model A
can print 70 books per day and
Model B can print 30 books per
day. The company owns 13 total
printers and this allows them to print
550 books per day.

177) How many Model A printers
do they have?

178) How many Model B printers
do they have?

Mk's Printing Inc. has two types of
printers Model A and Model B.
Model A can print 50 books per day
and Model B can print 55 books per
day. Altogether they have 17 printers.
If he can print 885 books in a day,
then answer the below questions

179) How many Model A printer
does they have?

180) How many Model B printer
does they have?

At RP's Printing Company LLC there are two kinds of printers: Model A can print 50 books per day and Model B can print 45 books per day. The company owns 14 total printers and this allows them to print 660 books per day.

181) How many Model A printers do they have?

182) How many Model B printers do they have?

At JJ's Printing Company there are two kinds of printers: Model A can print 50 books per day and Model B can print 35 books per day. The company owns 8 total printing presses and this allows them to print 370 books per day.

183) How many Model A printers do they have?

184) How many Model B printers do they have?

GV's Printing Inc. has two type of printers: Model A and Model B. Model A can print 50 books per day and Model B can print 55 books per day. Altogether they have 11 printers. If he can print 580 books in a day, then answer the below questions.

185) How many Model A printers do they have?

186) How many Model B printers do they have?

At MK's Printing Company there are two kinds of printers: Model A can print 60 books per day and Model B can print 55 books per day. The company owns 16 total printers and this allows them to print 930 books per day.

187) How many Model A printers do they have?

188) How many Model B printers do they have?

MK' Printing Inc. has two type of printers: Model A and Model B.
Model A can print 50 books per day and Model B can print 75 books per day.
Altogether they have 7 printers.
If they can print 475 books in a day, then answer the below questions.

189) How many Model A printers do they have?

190) How many Model B printers do they have?

Grade 9 class of MK academy used vans and buses to go on a field trip. They used 12 vehicles to go on the trip. Each van holds 11 students and each bus holds 55 students. If 484 students went on the trip, then

191) How many vans did the class use?

192) How many buses did the class use?

The grade 8 class of MK Academy trip because all of the buses were already in use. They used 14 vehicles to go on the trip. Each car holds 5 students and each van holds 9 students. If 102 students went on the trip, then

193) How many cars did the class use?

194) How many vans did the class use?

The grade 7 class of MK Academy used cars and vans to go on a field trip because all of the buses were already in use. They used 6 vehicles to go on the trip. Each car holds 5 students and each van holds 9 students. If 46 students went on the trip, then

195) How many cars did the class use?

196) How many vans did the class use?

The grade 9 class of MK Academy used cars and buses to go on afield trip. They used 16 vehicles to go on the trip. Each car holds 5 students and each bus holds 35 students. If 320 students went on the trip, then

197) How many cars did the class use?

198) How many buses did the class use?

The grade 10 class of MK Academy used cars and vans to go on a field
trip because all of the buses were already in use. They used 10 vehicles to go on the trip. Each car holds 4 students and each van holds 10 students. If 70 students went on the trip, then

199) How many cars did the class use?

200) How many vans did the class use?

The grade 6 class of MK Academy used cars and vans to go on a field trip because all of the buses were already in use. They used 10 vehicles to go on the trip. Each car holds 5 students and each van holds 6 students. If 55 students went on the trip, then

201) How many cars did the class use?

202) How many vans did the class use?

The grade 7 class of MK Academy used vans and buses to go on a field trip. They used 10 vehicles to go on the trip. Each van holds 6 students and each bus holds 45 students. If 255 students went on the trip, then how many of each type of vehicle did the class use?

203) How many vans did the class use?

204) How many buses did the class use?

The grade 9 class of MK Academy 157 students went on a field
trip. They took 13 vehicles, some cars and some buses. Find the number of cars and the number of buses they took if each car holds 4 students and each bus hold 25 students.

205) How many cars did the class use?

206) How many buses did the class use?

The grade 7 class of MK Academy used cars and vans to go on a field
trip because all of the buses were already in use. They used 11 vehicles to go on the trip. Each car holds 5 students and each van holds 7 students. If 61 students went on the trip, then

207) How many cars did the class use?

208) How many vans did the class use?

The grade 8 class of MK Academy used cars and buses to go on a field trip. They used 9 vehicles to go on the trip. Each car holds 3 students and each bus holds 30 students.
If 108 students went on the trip, then

209) How many cars did the class use?

210) How many buses did the class use?

The grade 10 class of 225 students went on a field trip. They took 12 vehicles, some cars and some buses. Find the number of cars and the number of buses they took if each car holds 3 students and each bus hold 30 students.

211) How many cars did the class use?

212) How many buses did the class use?

All 421 students in the Math Club went on a field trip. Some students rode in vans which hold 9 students each and some students rode in buses which hold 55 students each.
If there are 11 vehicles in total answer the below?

213) How many cars did the class use?

214) How many buses did the class use?

1) Find the solution for the below system of equations using method of elimination.

$$5x - 4y = 12$$
$$9x + 4y = -12$$

A) $(6, 5)$ B) $(-3, 0)$

C) $(-3, 5)$ D) $(0, -3)$

2) Find the solution for the below system of equations using method of elimination.

$$x + 2y = 12$$
$$-x - 4y = -28$$

A) $(-4, 8)$ B) $(4, 8)$

C) $(4, 6)$ D) $(4, -8)$

3) Find the solution for the below system of equations using method of elimination.

$$x + 4y = -8$$
$$-x + 10y = -6$$

A) $(-4, 1)$ B) $(-4, -1)$

C) $(1, -4)$ D) $(-1, 4)$

4) Find the solution for the below system of equations using method of elimination.

$$-8x + 5y = 20$$
$$8x - 3y = -28$$

A) $(-9, -7)$ B) $(-7, -9)$

C) $(-5, -4)$ D) $(-9, -4)$

5) Find the solution for the below system of equations using method of elimination.

$$2x + 5y = -28$$
$$-2x - 2y = 4$$

A) Infinite number of solutions

B) $(6, -8)$

C) $(-6, -2)$

D) $(-6, 2)$

6) Find the solution for the below system of equations using method of elimination.

$$10x + 6y = 0$$
$$-10x + 6y = 0$$

A) $(0, 6)$ B) $(0, -6)$

C) $(5, -7)$ D) $(0, 0)$

7) Find the solution for the below system of equations using method of elimination.

$$-7x + 5y = 28$$
$$7x + 10y = -28$$

A) $(-4, 0)$ B) $(0, 7)$

C) $(-7, 0)$ D) $(7, 0)$

8) Find the solution for the below system of equations using method of elimination.

$$2x + 10y = -6$$
$$2x - 10y = -26$$

A) $(-8, 1)$ B) No solution

C) $(-9, 1)$ D) $(9, 1)$

Algebra 1

9) Find the solution for the below system of equations using method of elimination.

$$-x - 10y = 24$$
$$x + 4y = -12$$

A) (4 , 6)   B) (-4 , -6)

C) (4 , -6)   D) (-4 , -2)

10) Find the solution for the below system of equations using method of elimination.

$$7x + 10y = -9$$
$$-7x - 7y = 21$$

A) (-7 , -1)   B) (-7 , 4)

C) (7 , 4)   D) (-5 , -1)

11) Find the solution for the below system of equations using method of elimination.

$$6x - 7y = 18$$
$$-6x + 7y = -18$$

A) (4 , -3)

B) (-4 , -3)

C) (4 , 3)

D) Infinite number of solutions

12) Find the solution for the below system of equations using method of elimination.

$$-9x - 6y = -12$$
$$-5x + 6y = -16$$

A) (-2 , -1)

B) (6 , 2)

C) Infinite number of solutions

D) (2 , -1)

13) Find the solution for the below system of equations using method of elimination.

$$-10x - y = 13$$
$$10x - 4y = 2$$

A) (-1 , -3)   B) (1 , -4)

C) (-1 , -4)   D) (1 , -3)

14) Find the solution for the below system of equations using method of elimination.

$$9x - 6y = 21$$
$$-8x + 6y = -16$$

A) No solution   B) (5 , 4)

C) (5 , -4)   D) (-5 , 4)

15) Find the solution for the below system of equations using method of elimination.

$$-x - 2y = -20$$
$$x - y = 5$$

A) (10 , -5)   B) (5 , 10)

C) (-10 , 5)   D) (10 , 5)

16) Find the solution for the below system of equations using method of elimination.

$$4x + y = 18$$
$$-x - y = 0$$

A) (6 , -6)   B) (-4 , -6)

C) (9 , -2)   D) (9 , -6)

17) Find the solution for the below system of equations using method of elimination.

$$-3x + y = 7$$
$$-10x - y = -7$$

A) (9 , 0)   B) (9 , 7)

C) (0 , 7)   D) (7 , 0)

18) Find the solution for the below system of equations using method of elimination.

$$10x + 2y = -6$$
$$-10x - 7y = -4$$

A) (-4 , -1)   B) (4 , -1)

C) (4 , 4)   D) (-1 , 2)

19) Find the solution for the below system of equations using method of elimination.

$$6x - 2y = -24$$
$$-6x + 5y = 15$$

A) (5 , -3)

B) Infinite number of solutions

C) (-5 , -3)

D) (5 , 3)

20) Find the solution for the below system of equations using method of elimination.

$$3x + 3y = 24$$
$$-3x - 4y = -22$$

A) (2 , 1)

B) (1 , 2)

C) (-2 , -2)

D) (10 , -2)

21) Find the solution for the below system of equations using method of elimination.

$$-6x - 5y = 15$$
$$x + 5y = 10$$

A) (7 , 5)

B) (-5 , -7)

C) Infinite number of solutions

D) (-5 , 3)

22) Find the solution for the below system of equations using method of elimination.

$$10x + 5y = 5$$
$$-10x - 5y = -5$$

A) (-6 , -10)

B) Infinite number of solutions

C) (-10 , -6)

D) (-10 , 6)

23) Find the solution for the below system of equations using method of elimination.

$$3x - 4y = 11$$
$$-3x + 6y = -15$$

A) (-2 , -7)
B) (-2 , -6)
C) (1 , -2)
D) Infinite number of solutions

24) Find the solution for the below system of equations using method of elimination.

$$8x + 6y = 24$$
$$4x - 6y = 12$$

A) No solution   B) (-3 , 9)

C) (3 , 0)   D) (-3 , -9)

25) Find the solution for the below system of equations using method of elimination.

$$6x - y = -5$$
$$-6x + y = 5$$

A) $(9, -6)$

B) No solution

C) Infinite number of solutions

D) $(-6, -9)$

26) Find the solution for the below system of equations using method of elimination.

$$9x + 8y = -14$$
$$9x + 6y = -6$$

A) $(-2, -4)$ B) $(6, -4)$

C) $(2, -4)$ D) $(4, 6)$

27) Find the solution for the below system of equations using method of elimination.

$$-2x + 2y = -6$$
$$6x + 2y = -6$$

A) $(8, 2)$

B) $(0, -3)$

C) $(8, 0)$

D) No solution

28) Find the solution for the below system of equations using method of elimination.

$$9x - 6y = 6$$
$$9x - 5y = 11$$

A) $(8, 4)$ B) $(4, -5)$

C) $(4, 5)$ D) $(4, -8)$

29) Find the solution for the below system of equations using method of elimination.

$$-x + 2y = 15$$
$$2x + 2y = 12$$

A) $(-1, 7)$ B) $(7, 1)$

C) $(2, 1)$ D) $(7, -1)$

30) Find the solution for the below system of equations using method of elimination.

$$-10x - 10y = 20$$
$$-9x - 10y = 18$$

A) $(7, 10)$

B) $(-7, 10)$

C) Infinite number of solutions

D) $(-2, 0)$

31) Find the solution for the below system of equations using method of elimination.

$$2x + 10y = -28$$
$$2x - 8y = -10$$

A) $(-9, -1)$ B) $(-1, -4)$

C) $(-4, -1)$ D) $(4, -1)$

32) Find the solution for the below system of equations using method of elimination.

$$3x - 7y = -8$$
$$3x - 9y = -18$$

A) $(9, 5)$ B) $(9, -5)$

C) $(9, -4)$ D) $(9, 4)$

33) Find the solution for the below system of equations using method of elimination.

$$3x - 4y = -30$$
$$8x - 4y = -20$$

A) (9 , 2) B) (– 1 , 5)

C) (9 , 9) D) (2 , 9)

34) Find the solution for the below system of equations using method of elimination.

$$x - y = -17$$
$$x + 3y = 23$$

A) (– 3 , 9) B) (– 9 , – 3)

C) (– 7 , 10) D) (9 , – 3)

35) Find the solution for the below system of equations using method of elimination.

$$8x - 4y = 16$$
$$8x + 3y = 30$$

A) (8 , 4) B) (– 10 , – 4)

C) (3 , 2) D) (10 , – 4)

36) Find the solution for the below system of equations using method of elimination.

$$-x - 8y = -25$$
$$3x - 8y = 11$$

A) (– 9 , – 1) B) (9 , – 2)

C) (9 , 2) D) (– 9 , 2)

37) Find the solution for the below system of equations using method of elimination.

$$5x - 7y = -23$$
$$5x + 5y = 25$$

A) (– 4 , 1) B) (– 1 , 4)

C) (1 , 4) D) (– 4 , – 1)

38) Find the solution for the below system of equations using method of elimination.

$$-4x + 10y = -12$$
$$x + 10y = 28$$

A) (8 , – 1) B) (2 , – 8)

C) (2 , 8) D) (8 , 2)

39) Find the solution for the below system of equations using method of elimination.

$$7x - 2y = 22$$
$$7x - 4y = 16$$

A) (– 4 , – 3) B) (4 , – 3)

C) (4 , 3) D) (– 4 , 3)

40) Find the solution for the below system of equations using method of elimination.

$$-7x + 10y = 17$$
$$-7x + 10y = 26$$

A) (– 1 , – 9) B) (– 9 , – 1)

C) No solution D) (6 , 1)

41) Find the solution for the below system of equations using method of elimination.

$$10 x - 5 y = 5$$
$$10 x + 10 y = 20$$

A) $(-1, 1)$ B) $(1, 1)$

C) $(1, -5)$ D) $(1, -1)$

42) Find the solution for the below system of equations using method of elimination.

$$-8 x - y = -25$$
$$5 x - y = 27$$

A) $(4, -7)$ B) $(7, 3)$

C) $(-7, 4)$ D) $(-4, -7)$

43) Find the solution for the below system of equations using method of elimination.

$$4 x - 8 y = -16$$
$$10 x - 8 y = 8$$

A) $(4, 4)$

B) $(4, -4)$

C) Infinite number of solutions

D) $(-4, 4)$

44) Find the solution for the below system of equations using method of elimination.

$$-9 x + 10 y = 2$$
$$x + 10 y = 22$$

A) $(-2, -2)$ B) $(-2, 2)$

C) $(2, -2)$ D) $(2, 2)$

45) Find the solution for the below system of equations using method of elimination.

$$-2 x - 7 y = -28$$
$$-9 x - 7 y = -28$$

A) $(0, 4)$ B) $(-4, 2)$

C) $(3, 4)$ D) $(3, -2)$

46) Find the solution for the below system of equations using method of elimination.

$$2 x - 2 y = -8$$
$$5 x - 2 y = -14$$

A) $(5, 2)$ B) $(-5, 2)$

C) $(-5, -2)$ D) $(-2, 2)$

47) Find the solution for the below system of equations using method of elimination.

$$4 x - 4 y = 16$$
$$4 x + 7 y = -17$$

A) $(1, 6)$ B) $(1, -3)$

C) $(6, 1)$ D) $(1, 3)$

48) Find the solution for the below system of equations using method of elimination.

$$-6 x - 2 y = -2$$
$$-7 x - 2 y = -4$$

A) $(5, -2)$ B) $(5, 2)$

C) $(-2, -5)$ D) $(2, -5)$

49) Find the solution for the below system of equations using method of elimination.

$$-3x + 9y = 6$$
$$4x + 9y = -29$$

A) $(-1, -4)$ B) $(-5, -1)$

C) $(4, -1)$ D) $(-1, 4)$

50) Find the solution for the below system of equations using method of elimination.

$$8x - 3y = 14$$
$$8x - 6y = 20$$

A) $(1, -2)$ B) $(-2, -1)$

C) $(-2, 1)$ D) $(2, -1)$

51) Find the solution for the below system of equations using method of elimination.

$$-9x + 9y = 36$$
$$-18x - 2y = 32$$

A) $(-2, -2)$ B) $(-2, 2)$

C) $(12, 2)$ D) $(2, 12)$

52) Find the solution for the below system of equations using method of elimination.

$$4x + 7y = -8$$
$$-2x - 2y = -8$$

A) Infinite number of solutions

B) $(12, -8)$

C) $(-12, 8)$

D) $(8, -12)$

53) Find the solution for the below system of equations using method of elimination.

$$10x + 3y = -3$$
$$2x + 12y = -12$$

A) $(-9, -8)$ B) $(0, -9)$

C) $(0, -1)$ D) $(9, 0)$

54) Find the solution for the below system of equations using method of elimination.

$$x - 7y = -30$$
$$-9x + 14y = -24$$

A) $(-6, -12)$ B) $(6, 12)$

C) $(-12, -6)$ D) $(12, 6)$

55) Find the solution for the below system of equations using method of elimination.

$$-x + 12y = -7$$
$$-5x - 6y = -35$$

A) $(-7, -1)$ B) $(7, 0)$

C) $(7, 1)$ D) $(-3, -7)$

56) Find the solution for the below system of equations using method of elimination.

$$2x - 8y = 4$$
$$-x - 10y = -2$$

A) $(0, 2)$ B) $(-12, 2)$

C) $(2, 0)$ D) $(-5, 12)$

57) Find the solution for the below system of equations using method of elimination.

$$-4x + 8y = 8$$
$$-8x + 16y = 16$$

A) $(2, 11)$

B) Infinite number of solutions

C) $(-2, 11)$

D) $(-2, -11)$

58) Find the solution for the below system of equations using method of elimination.

$$6x + 2y = -28$$
$$7x + 6y = -29$$

A) $(1, 3)$ B) $(-5, 1)$

C) $(3, 1)$ D) $(-3, 1)$

59) Find the solution for the below system of equations using method of elimination.

$$7x - 16y = 18$$
$$-9x + 8y = 2$$

A) $(-5, 12)$ B) $(-2, -5)$

C) $(-5, 7)$ D) $(-2, -2)$

60) Find the solution for the below system of equations using method of elimination.

$$-6x - 6y = 6$$
$$12x + 4y = 4$$

A) $(-1, -8)$ B) $(1, -2)$

C) $(-1, -2)$ D) $(1, -8)$

61) Find the solution for the below system of equations using method of elimination.

$$-9x + 14y = 8$$
$$-8x + 7y = 18$$

A) $(-4, -11)$ B) $(4, 11)$

C) $(-4, -2)$ D) $(4, 5)$

62) Find the solution for the below system of equations using method of elimination.

$$-12x + 8y = 36$$
$$7x - 2y = -29$$

A) $(5, -9)$ B) $(-5, -3)$

C) $(-5, 3)$ D) $(-5, -9)$

63) Find the solution for the below system of equations using method of elimination.

$$-10x + 6y = 34$$
$$-7x + 18y = 10$$

A) No solution   B) $(1, 4)$

C) $(4, -1)$ D) $(-4, -1)$

64) Find the solution for the below system of equations using method of elimination.

$$3x + 5y = 4$$
$$12x + 10y = -4$$

A) $(2, 5)$ B) $(-2, 5)$

C) No solution   D) $(-2, 2)$

65) Find the solution for the below system of equations using method of elimination.

$$3x - 11y = 3$$
$$x - y = 1$$

A) No solution

B) $(0, -1)$

C) $(1, 0)$

D) $(0, 1)$

66) Find the solution for the below system of equations using method of elimination.

$$16x + 9y = -21$$
$$-8x - 2y = 18$$

A) $(3, -3)$ B) No solution

C) $(-3, -3)$ D) $(-3, 3)$

67) Find the solution for the below system of equations using method of elimination.

$$5x + 8y = -36$$
$$-15x - 10y = 10$$

A) $(4, -7)$

B) $(-8, -7)$

C) Infinite number of solutions

D) $(-8, 7)$

68) Find the solution for the below system of equations using method of elimination.

$$-10x + 11y = 31$$
$$5x - y = 34$$

A) $(-9, -11)$ B) $(11, -9)$

C) $(-9, 11)$ D) $(9, 11)$

69) Find the solution for the below system of equations using method of elimination.

$$10x + 6y = 10$$
$$20x + y = -35$$

A) $(2, -1)$ B) $(2, 5)$

C) $(2, -5)$ D) $(-2, 5)$

70) Find the solution for the below system of equations using method of elimination.

$$-5x + 9y = 11$$
$$-10x + 18y = 22$$

A) $(-12, -2)$

B) No solution

C) Infinite number of solutions

D) $(-12, 2)$

71) Find the solution for the below system of equations using method of elimination.

$$-x - 2y = 22$$
$$3x - 3y = -12$$

A) $(6, -10)$ B) $(-10, 6)$

C) $(10, -6)$ D) $(-10, -6)$

72) Find the solution for the below system of equations using method of elimination.

$$x + 15y = 22$$
$$-4x - 5y = 22$$

A) $(-8, 2)$ B) $(12, 8)$

C) $(12, 5)$ D) $(-12, 8)$

 **Algebra 1**

73) Find the solution for the below system of equations using method of elimination.

$$- 10 x - y = 0$$
$$- 20 x - 5 y = - 30$$

A) $(- 1 , 10)$

B) Infinite number of solutions

C) $(10 , - 1)$

D) $(1 , 10)$

74) Find the solution for the below system of equations using method of elimination.

$$- 8 x - 10 y = 0$$
$$- 9 x - 5 y = 0$$

A) $(- 6 , 8)$ B) $(- 7 , 8)$

C) $(0 , 0)$ D) $(- 8 , - 7)$

75) Find the solution for the below system of equations using method of elimination.

$$9 x + 2 y = 9$$
$$- 3 x + 3 y = - 3$$

A) $(1 , - 6)$

B) No solution

C) $(- 1 , 0)$

D) $(1 , 0)$

76) Find the solution for the below system of equations using method of elimination.

$$11 x + 5 y = 28$$
$$- 2 x - 13 y = 7$$

A) $(3 , - 1)$ B) $(1 , 3)$

C) $(1 , - 13)$ D) $(- 1 , 3)$

77) Find the solution for the below system of equations using method of elimination.

$$8 x - 3 y = - 30$$
$$- 13 x + 14 y = - 6$$

A) $(- 6 , - 2)$ B) $(- 6 , - 6)$

C) $(- 6 , 2)$ D) $(6 , 2)$

78) Find the solution for the below system of equations using method of elimination.

$$- 5 x - 6 y = 27$$
$$2 x + 11 y = - 28$$

A) $(- 3 , 2)$ B) $(- 3 , - 2)$

C) $(2 , - 3)$ D) $(3 , 2)$

79) Find the solution for the below system of equations using method of elimination.

$$108 x + 9 y = - 22$$
$$60 x + 5 y = - 20$$

A) No solution   B) $(3 , 5)$

C) $(3 , 12)$ D) $(- 5 , 12)$

80) Find the solution for the below system of equations using method of elimination.

$$4 x - 3 y = - 4$$
$$6 x + 7 y = 40$$

A) $(2 , 13)$ B) $(2 , - 13)$

C) $(- 2 , - 13)$ D) $(2 , 4)$

81) Find the solution for the below system of equations using method of elimination.

$$-6x + 16y = -6$$
$$-21x + 56y = -21$$

A) $(8, -7)$

B) $(8, 7)$

C) $(8, 14)$

D) Infinite number of solutions

82) Find the solution for the below system of equations using method of elimination.

$$12x + 13y = -17$$
$$10x + 3y = 25$$

A) $(-5, 4)$ B) $(-5, -4)$

C) $(4, -5)$ D) $(-5, -12)$

83) Find the solution for the below system of equations using method of elimination.

$$-5x - 12y = 6$$
$$-9x - 11y = 32$$

A) $(4, 2)$

B) $(-6, 2)$

C) $(-4, 2)$

D) Infinite number of solutions

84) Find the solution for the below system of equations using method of elimination.

$$-3x + 3y = 3$$
$$-11x - 10y = -31$$

A) $(-1, 2)$

B) $(1, 2)$

C) $(2, -1)$

D) Infinite number of solutions

85) Find the solution for the below system of equations using method of elimination.

$$7x + 5y = -15$$
$$5x - 3y = -37$$

A) $(-5, 4)$ B) $(5, 8)$

C) $(8, 5)$ D) $(8, -5)$

86) Find the solution for the below system of equations using method of elimination.

$$-5x + 3y = -28$$
$$-7x - 13y = -22$$

A) $(-12, -5)$

B) $(5, -1)$

C) Infinite number of solutions

D) $(5, -12)$

87) Find the solution for the below system of equations using method of elimination.

$$-10x + 13y = -3$$
$$8x + 5y = -13$$

A) $(1, -1)$ B) $(-1, -1)$

C) No solution   D) $(1, 1)$

88) Find the solution for the below system of equations using method of elimination.

$$11x - 5y = 30$$
$$-6x - 2y = -40$$

A) $(-5, -12)$ B) $(12, -5)$

C) $(-12, -5)$ D) $(5, 5)$

89) Find the solution for the below system of equations using method of elimination.

$$7x - 12y = -28$$
$$8x - 10y = -32$$

A) $(-11, 0)$

B) $(4, 0)$

C) $(0, -4)$

D) $(-4, 0)$

90) Find the solution for the below system of equations using method of elimination.

$$-12x - 9y = 6$$
$$11x + 11y = 33$$

A) $(-6, -11)$ B) $(6, -11)$

C) $(-11, 6)$ D) $(-11, 14)$

91) Find the solution for the below system of equations using method of elimination.

$$12x - 11y = 29$$
$$10x - 9y = 25$$

A) $(7, 5)$ B) $(-7, 4)$

C) $(4, 7)$ D) $(4, -7)$

92) Find the solution for the below system of equations using method of elimination.

$$7x - 7y = 21$$
$$8x - 4y = -24$$

A) $(-9, -12)$

B) $(-12, 10)$

C) $(11, 10)$

D) Infinite number of solutions

93) Find the solution for the below system of equations using method of elimination.

$$12x + 4y = 36$$
$$9x + 7y = -21$$

A) $(12, -7)$ B) $(12, 7)$

C) $(-7, -12)$ D) $(7, -12)$

94) Find the solution for the below system of equations using method of elimination.

$$-11x + 8y = 10$$
$$14x - 14y = 14$$

A) $(6, 5)$ B) $(-5, 6)$

C) $(6, -5)$ D) $(-6, -7)$

MK Academy is selling tickets for a musical concert. On Monday they sold 6 senior citizen ticket and 1 child ticket for a total of $96. On Tuesday they sold 3 senior citizen tickets and 2 child tickets for a total of $57.

95) Find the price of senior citizen ticket

96) Find the price of child ticket

www.math-knots.com | www.a4ace.com

MK Academy is selling tickets for a
STEM fair. On Monday they sold
14 senior citizen ticket and 3 student tickets
for a total of $126. On Tuesday they sold
7 senior citizen tickets and 11 student tickets
for a total of $196.

97) Find the price of senior citizen ticket

98) Find the price of student ticket

Lola is selling tickets to a local fair.
On Friday she sold 3 senior citizen ticket
and 6 child tickets for a total of $60.
On Saturday she sold 12 senior citizen
tickets and 3 child tickets for a total of $135.

99) Find the price of senior citizen ticket

100) Find the price of child ticket

Bella is selling tickets to a local fair.
On Friday she sold 4 adult tickets
and 12 student tickets for a total of $152.
On Saturday she sold 3 adult
tickets and 6 student tickets for a total of $81.

101) Find the price of adult ticket

102) Find the price of student ticket

Nancy is selling tickets to a local
spring festival. On Friday she sold
12 senior citizen tickets and 10 student
tickets for a total of $246. On Saturday
she sold 7 senior citizen tickets and 5
student tickets for a total of $136.

103) Find the price of adult ticket

104) Find the price of student ticket

105) The sum of the digits of a certain
two digit number is 4. Reversing its
digits decreases the number by 36.
Find the number.

106) Lola reversed the digits of a two
digit number. The value of the
two digit number is decreased
by 27. What is the number if the
sum of its digits is 7 ?

107) The sum of the digits of a certain
two digit number is 11. Reversing its
digits decreases the number by 63.
Find the number.

108) Kelly reversed the digits of a two digit number. The value of the two digit number is increased by 36. What is the number if the sum of its digits is 6 ?

109) The sum of the digits of a certain two-digit number is 11. Reversing its digits decreases the number by 45. What is the number?

110) Ron reversed the digits of a two digit number. The value of the two digit number is increased by 9. What is the number if the sum of its digits is 15 ?

111) Tom reversed the digits of a two digit number. The value of the two digit number is decreased by 72. What is the number if the sum of its digits is 8 ?

112) The sum of the digits of a certain two-digit number is 13. Reversing its digits increases the number by 9. Find the number.

113) The sum of the digits of a certain two digit number is 4. Reversing its digits decreases the number by 18. Find the number.

114) Beth reversed the digits of a two digit number. The value of the two digit number is decreased by 18. What is the number if the sum of its digits is 16 ?

115) Stella reversed the digits of a two digit number. The value of the two digit number is increased by 9. What is the number if the sum of its digits is 3 ?

Ron and Shawn are selling flower bulbs at a local farm. Ron sold 6 packs of Tulip bulbs and 5 packs of Daffodil bulbs for a total of $153. Shawn sold 7 packs of Tulip bulbs and 3 packs of Daffodil bulbs for a total of $136.

116) Find the cost of 1 pack of Tulip bulb

117) Find the cost of 1 pack of Daffodil bulb

Sara and Zara are selling gift wrapping paper for a school fund riser. Sara sold 2 rolls of plain gift paper and 12 rolls of shiny gift paper for a total of $212. Zara sold 11 rolls of plain gift paper and 13 rolls of shiny gift paper for a total of $318.

120) Find the cost of 1 roll of plain gift paper

121) Find the cost of 1 roll of shiny gift paper

MK Academy is selling tickets for a STEM fair. On Monday they sold 9 senior citizen ticket and 6 student tickets for a total of $102. On Tuesday they sold 6 senior citizen tickets and 5 child tickets for a total of $79.

118) Find the price of senior citizen ticket

119) Find the price of student ticket

Bella and Stella are selling fruit boxes at a local fair. Bella sold 3 small boxes of oranges and 7 large boxes of mangoes for a total of $99. Stella sold 13 small boxes of oranges and 6 large boxes of mangoes for a total of $137.

122) Find the cost of 1 small box of Oranges

123) Find the cost of 1 large box of mangoes

124) A train left Washington DC and traveled towards New York at an average speed of 80 mph. Sometime later a diesel train left traveling in the same direction but at an average speed of 90 mph. After traveling for 16 hours the diesel train caught up with the train. How long did the train travel before the diesel train caught up?

A) 18 hours   B) 30 hours

C) 20 hours   D) 29 hours

125) Train X left Delaware and traveled towards Boston. Train Y left three hours later traveling at 51 mph in an effort to catch up to train X. After traveling for 14 hours the train Y finally caught up. Find the speed of train X.

A) 45 mph   B) 44 mph

C) 42 mph   D) 35 mph

126) An aircraft left Baltimore and traveled towards Dallas at an average speed of 25 mph due to heavy winds. Sometime later a boat left traveling in the opposite direction with an average speed of 25 mph. After the aircraft had traveled for 13 hours the ships were 650 mi. apart. Find the number of hours the boat traveled.

A) 3 hours   B) 8 hours

C) 5 hours   D) 13 hours

127) A jet flight left New York and flew towards London at an average speed of 450 mph. Sometime later a passenger plane left flying in the same direction but at an average speed of 495 mph. After flying for ten hours the passenger plane caught up with the jet. How long did the jet fly before the passenger plane caught up?

A) 17 hours   B) 14 hours

C) 11 hours   D) 19 hours

128) A jet flight X left London and flew towards Delhi at an average speed of 225 km/h. Sometime later a passenger plane Y left flying in the same direction but at an average speed of 270 km/h. After flying for ten hours the passenger plane caught up with the jet. Find the number of hours the jet flew before the passenger plane caught up.

A) 15 hours   B) 13 hours

C) 20 hours   D) 12 hours

129) A cruise ship left Baltimore and traveled north. Two hours later a plane left traveling at 7 mph in an effort to catch up to the cruise ship. After traveling for 12 hours the plane finally caught up. What was the cruise ship's average speed?

A) 4 mph   B) 2 mph

C) 5 mph   D) 6 mph

130) Dora left her office and traveled towards her home. One hour later Lola left traveling at 25 km/h in an effort to catch up to Dora. After traveling for four hours Lola finally caught up. What was Dora's average speed?

A) 5 km/h   B) 20 km/h

C) 12 km/h   D) 15 km/h

131) Zara traveled to Virginia beach and back. The trip there took five hours and the trip back took three hours. She averaged 40 mi/h on the return trip. Find the average speed of the trip there.

A) 24 mi/h   B) 10 mi/h

C) 20 mi/h   D) 30 mi/h

132) A cargo train made a trip to Florida and back. The trip there took 15 hours and the trip back took 12 hours. What was the cargo train's average speed on the trip there if it averaged 20 mi/h on the return trip?

A) 4 mi/h   B) 18 mi/h

C) 16 mi/h   D) 11 mi/h

133) Tom made a trip to his beach cabin in Florida and back. On the trip there he drove 24 mph and on the return trip he went 30 mph. How long did the trip there take if the return trip took four hours?

A) 5 hours   B) 8 hours

C) 6 hours   D) 9 hours

134) Dan left his home and drove towards his friend Ron's house. One hour later Maria left driving at 60 mi/h in an effort to catch up to Dan. After driving for two hours Maria finally caught up. Find Dan's average speed.

A) 50 mi/h   B) 40 mi/h

C) 51 mi/h   D) 31 mi/h

135) Train A left Chicago and traveled east at an average speed of 35 mph. After sometime train B left traveling in the opposite direction with an average speed of 75 mph. After the train A had traveled for 15 hours the trains were 1650 mi. apart. How long did the train B travel?

A) 9 hours   B) 6 hours

C) 13 hours   D) 15 hours

136) A train X left New Jersey and traveled south at an average speed of 24 mi/h. A train Y left sometime later traveling in the same direction at an average speed of 40 mi/h. After traveling for nine hours the train Y caught up with the train X. How long did the train X travel before the train Y caught up?

A) 12 hours   B) 16 hours

C) 15 hours   D) 21 hours

137) A plane X left London and flew towards Italy. One hour later plane Y left flying at 260 mph in an effort to catch up to plane X. After flying for three hours the plane Y finally caught up. Find the plane X's average speed.

A) 195 mph   B) 210 mph

C) 110 mph   D) 190 mph

138) A plane X left Torrento, Canada and flew towards New York. Four hours later a plane Y left flying at 240 mph in an effort to catch up to the plane X. After flying for eight hours the plane Y finally caught up. Find the average speed of plane X.

A) 160 mph   B) 98 mph

C) 105 mph   D) 165 mph

139) Plane X left Dubai and flew north. Two hours later plane Y left flying at 470 mph in an effort to catch up to plane X. After flying for three hours the plane Y plane finally caught up. Find the average speed of plane X.

A) 165 mph   B) 385 mph

C) 290 mph   D) 282 mph

140) Fiona traveled to the science fair and back. On the trip there she traveled 36 km/h and on the return trip she went 60 km/h. How long did the trip there take if the return trip took three hours?

A) 5 hours   B) 9 hours

C) 4 hours   D) 7 hours

141) Jet X left Maryland and flew south at an average speed of 210 km/h. Jet Y left sometime later flying in the same direction at an average speed of 280 km/h. After flying for six hours Jet Y caught up with the jet X. Find the number of hours the jet X flew before the jet Y caught up.

A) 13 hours   B) 9 hours

C) 14 hours   D) 8 hours

142) Ivanka left Fairfax, Virginia and traveled towards Richmond, Virginia at an average speed of 44 mph. Mary left sometime later traveling in the same direction at an average speed of 55 mph. After traveling for four hours Mary caught up with Ivanka. How long did Ivanka traveled before Mary caught up?

A) 5 hours   B) 6 hours

C) 9 hours   D) 8 hours

143) A plane flew to Detroit and back from Virginia. On the trip there it flew 225 mph and on the return trip it went 315 mph. How long did the trip there take if the return trip took five hours?

A) 10 hours   B) 7 hours

C) 11 hours   D) 8 hours

144) Train X left Delaware and traveled east. Train Y left two hours later traveling at 55 mph in an effort to catch up to the train X. After traveling for nine hours the train Y finally caught up. Find the average speed of train X.

A) 40 mph   B) 53 mph

C) 50 mph   D) 45 mph

145) Samantha left school and drove north. One hour later Pamela left driving at 80 mph in an effort to catch up to Samantha. After driving for three hours Pamela finally caught up. Find the average speed of Samantha.

A) 30 mph   B) 75 mph

C) 25 mph   D) 60 mph

146) Working together, Sara and Sam can weed a garden in 21.26 minutes. Had he done it alone it would have taken Sam 37 minutes. Find how long it would take Sara to do it alone.

A) 55.78 minutes   B) 50.36 minutes

C) 49.98 minutes   D) 42.78 minutes

147) Edward can blow fifty balloons in 40 minutes. Dora can blow the same fifty balloons in 30 minutes. If they worked together how long would it take them?

A) 18.09 minutes   B) 19.71 minutes

C) 17.14 minutes   D) 13.14 minutes

148) Arya can wash a car in 38 minutes. Joe can wash the same car in 35 minutes. Find how long it would take them if they worked together.

A) 18.22 minutes

B) 15.92 minutes

C) 17.26 minutes

D) 14.07 minutes

149) John can blow fifty balloons in 47 minutes. Julia can blow the same fifty balloons in 27 minutes. If they worked together how long would it take them?

A) 20.26 minutes   B) 22.03 minutes

C) 13.74 minutes   D) 17.15 minutes

150) Working together, Tom and Beth can weed a garden in 18.75 minutes. Had she done it alone it would have taken Beth 30 minutes. Find how long it would take Tom to do it alone.

A) 53.71 minutes   B) 50 minutes

C) 54.44 minutes   D) 64.89 minutes

151) Jade can weed a garden in 45 minutes. One day her friend Helen helped her and it only took 18 minutes. Find how long it would take Helen to do it alone.

A) 37.54 minutes   B) 32.48 minutes

C) 21.65 minutes   D) 30 minutes

152) It takes Andy 26 minutes to mow a lawn. Kelly can mow the same lawn in 50 hours. How long would it take them if they worked together?

A) 17.11 hours   B) 17.4 hours

C) 19.45 hours   D) 16.34 hours

153) Working together, Dora and Bella can wax a floor in 19.69 hours. Had she done it alone it would have taken Bella 45 hours. Find how long it would take Dora to do it alone.

A) 40.94 hours   B) 35.01 hours

C) 32.56 hours   D) 44.11 hours

154) Wilson can polish a set of silverware in 45 minutes. One day his friend John helped him and it only took 21.18 minutes. How long would it take John to do it alone?

A) 40.01 minutes   B) 46.64 minutes

C) 51.63 minutes   D) 51.37 minutes

155) Working together, Peter and Oliver can install a new deck in 15.94 hours. Had he done it alone it would have taken Oliver 37 hours. Find how long it would take Peter to do it alone.

A) 21.97 hours   B) 28.52 hours

C) 28 hours   D) 31.69 hours

156) It takes Rik 30 minutes to wash a car. Rak can wash the same car in 38 minutes. How long would it take them if they worked together?

A) 19.35 minutes   B) 15.27 minutes

C) 18.75 minutes   D) 16.76 minutes

157) Zara can inflate twenty balloons in 30 minutes. Sara can inflate the same twenty balloons in 33 minutes. How long would it take them if they worked together?

A) 20.16 minutes   B) 11.36 minutes

C) 11.19 minutes   D) 15.71 minutes

158) For his birthday party Beth mixed together 2 L of Brand A fruit punch and 8 L of Brand B. Brand A contains 10% fruit juice and Brand B contains 20% fruit juice. What percent of the mixture is fruit juice?

A) 15%   B) 18%

C) 22%   D) 26%

159) For his birthday party Ron mixed together 5 gal. of Brand A fruit punch and 4 gal. of Brand B. Brand A contains 50% fruit juice and Brand B contains 5% fruit juice. What percent of the mixture is fruit juice?

A) 40%   B) 36%

C) 30%   D) 55%

www.math-knots.com | www.a4ace.com

160) For her birthday party Fiona mixed together 7 gal. of Brand A fruit punch and 9 gal. of grape juice. If Brand A contains 52% fruit juice, then what percent of the final mixture is fruit juice?

    A) 79%   B) 30%

    C) 87%   D) 95%

161) For his birthday party Lola mixed together 8 gal. of Brand A fruit punch and 2 gal. of grape juice. If Brand A contains 40% fruit juice, then what percent of the final mixture is fruit juice?

    A) 66%   B) 52%

    C) 79%   D) 80%

162) A metal alloy weighing 7 oz. and containing 78% copper is melted and mixed with 1 oz. of a different alloy which contains 14% copper. What percent of the resulting alloy is copper?

    A) 70%   B) 30%

    C) 15%   D) 95%

163) A metal alloy weighing 10 lb. and containing 60% platinum is melted and mixed with 6 lb. of a different alloy which contains 20% platinum. What percent of the resulting alloy is platinum?

    A) 33%   B) 50%

    C) 45%   D) 14%

164) A metal alloy weighing 3 kg and containing 55% gold is melted and mixed with 5 kg of a different alloy which contains 15% gold. What percent of the resulting alloy is gold?

    A) 30%   B) 16%

    C) 25%   D) 33%

165) A metal alloy weighing 9 lb. and containing 80% iron is melted and mixed with 1 lb. of pure iron. What percent of the resulting alloy is iron?

    A) 60%   B) 48%

    C) 16%   D) 82%

166) A metal alloy weighing 9 lb. and containing 35% gold is melted and mixed with 1 lb. of a different alloy which contains 85% gold. What percent of the resulting alloy is gold?

A) 30%   B) 10%

C) 60%   D) 40%

167) A metal alloy weighing 6 oz. and containing 20% platinum is melted and mixed with 5 oz. of a different alloy which contains 75% platinum. What percent of the resulting alloy is platinum?

A) 70%   B) 10%

C) 40%   D) 45%

 **Algebra 1**

1) Find the solution for the below system of equations by graphing them below

$$y = 5x - 3$$
$$y = x + 1$$

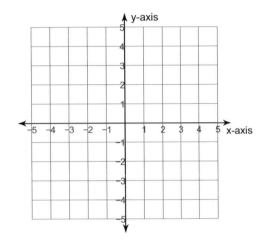

A) $(1, -2)$ B) $(-2, 1)$

C) $(1, 2)$ D) No solution

2) Find the solution for the below system of equations by graphing them below

$$y = -\frac{1}{2}x + 3$$
$$x = -2$$

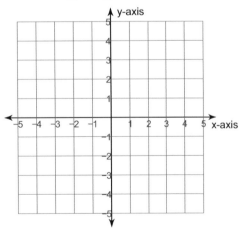

A) $(1, 4)$ B) $(-4, 4)$

C) $(-2, 4)$ D) $(4, 4)$

3) Find the solution for the below system of equations by graphing them below

$$y = \frac{1}{3}x + 2$$
$$y = 2x - 3$$

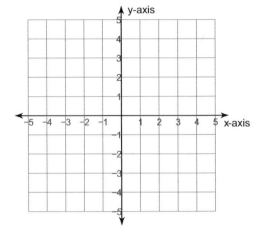

A) $(-3, -1)$ B) $(1, 3)$

C) $(3, 3)$ D) $(-1, 3)$

4) Find the solution for the below system of equations by graphing them below

$$y = \frac{3}{4}x - 2$$
$$y = -\frac{1}{2}x + 3$$

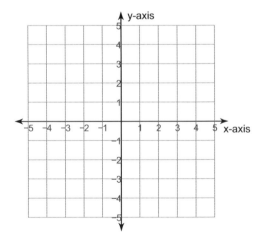

A) $(-4, 1)$ B) $(4, 1)$

C) $(4, -2)$ D) $(2, -2)$

www.math-knots.com | www.a4ace.com

Algebra 1

**Vol 2
Week 25**
Graphing
Simultaneous
equations

5) Find the solution for the below system of equations by graphing them below

$$y = -\frac{3}{4}x + 2$$

$$y = \frac{1}{2}x - 3$$

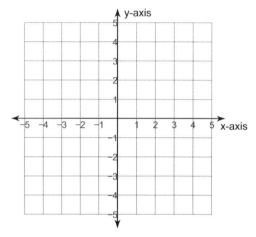

A) $(4, -1)$ B) $(-2, 4)$

C) $(4, 1)$ D) $(4, 2)$

6) Find the solution for the below system of equations by graphing them below

$$y = 3x + 2$$
$$y = -x - 2$$

A) $(1, -1)$

B) $(1, 1)$

C) $(-1, -1)$

D) Infinite number of solutions

7) Find the solution for the below system of equations by graphing them below

$$y = -\frac{3}{2}x - 1$$

$$y = -4$$

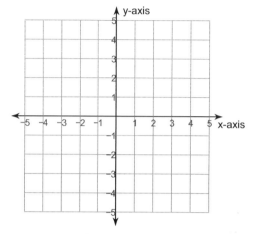

A) $(2, 4)$ B) $(-2, 4)$

C) No solution D) $(2, -4)$

8) Find the solution for the below system of equations by graphing them below

$$y = -\frac{5}{2}x - 2$$

$$y = -\frac{1}{2}x + 2$$

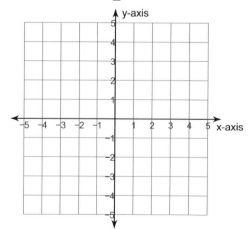

A) No solution

B) $(-2, 3)$

C) Infinite number of solutions

D) $(-2, -3)$

Algebra 1

Vol 2
Week 25
Graphing
Simultaneous
equations

9) Find the solution for the below system of equations by graphing them below

$$y = \frac{1}{2}x - 4$$

$$y = -\frac{5}{2}x + 2$$

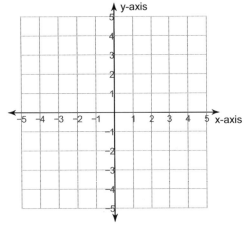

A) Infinite number of solutions

B) (2 , 4)

C) (4 , 2)

D) (2 , − 3)

10) Find the solution for the below system of equations by graphing them below

$$y = \frac{3}{2}x + 2$$

$$y = \frac{1}{2}x - 2$$

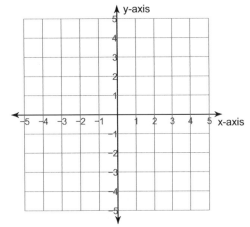

A) (3 , 4)   B) (− 4 , − 4)

C) (− 3 , 4)   D) (− 3 , − 4)

11) Find the solution for the below system of equations by graphing them below

$$y = -\frac{7}{3}x + 3$$

$$y = -\frac{7}{3}x + 4$$

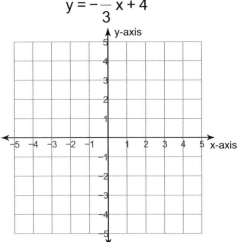

A) (4 , 4)   B) (− 4 , 4)

C) No solution   D) (4 , − 4)

12) Find the solution for the below system of equations by graphing them below

$$y = -\frac{1}{2}x + 2$$

$$y = -2x - 1$$

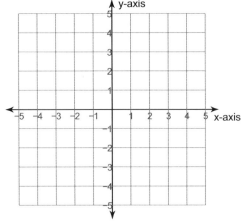

A) (3 , − 2)

B) (3 , − 4)

C) (− 2 , 3)

D) Infinite number of solutions

www.math-knots.com | www.a4ace.com

Algebra 1

Vol 2
Week 25
Graphing
Simultaneous
equations

13) Find the solution for the below system of equations by graphing them below

$$y = 2x - 1$$
$$y = -x - 4$$

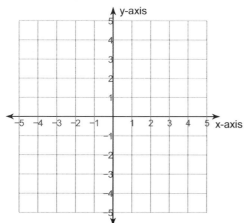

A) $(3, 1)$ B) $(-1, -3)$

C) $(3, -1)$ D) $(4, 1)$

14) Find the solution for the below system of equations by graphing them below

$$y = -\frac{8}{3}x - 4$$
$$y = -\frac{8}{3}x - 2$$

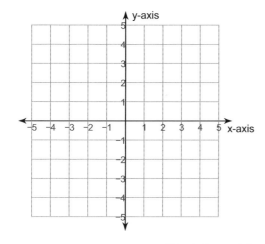

A) $(-4, 1)$ B) $(4, -1)$

C) No solution   D) $(4, 1)$

15) Find the solution for the below system of equations by graphing them below

$$y = x + 4$$
$$y = -\frac{3}{2}x - 1$$

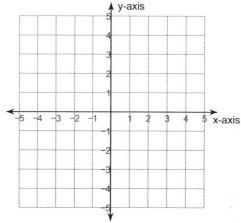

A) $(2, 2)$ B) $(-2, 2)$

C) $(-2, -2)$ D) $(2, -2)$

16) Find the solution for the below system of equations by graphing them below

$$-4 = -x + 2y$$
$$8 = 3x - 2y$$

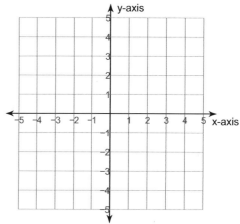

A) Infinite number of solutions

B) $(-3, -2)$

C) $(2, -1)$

D) $(-3, 2)$

www.math-knots.com | www.a4ace.com

Algebra 1

Vol 2
Week 25
Graphing
Simultaneous
equations

17) Find the solution for the below system of equations by graphing them below

$$0 = x - 2 - y$$
$$0 = y + 2 - x$$

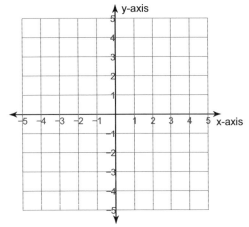

A) Infinite number of solutions

B) (5 , 3)

C) (3 , 5)

D) (3 , − 5)

18) Find the solution for the below system of equations by graphing them below

$$x + \frac{1}{2} = \frac{1}{8} y$$
$$9 = -3y + 3x$$

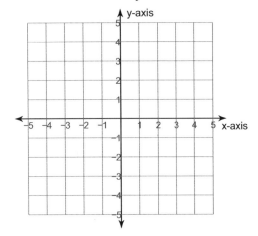

A) (− 1 , − 4)   B) (− 1 , 4)

C) (1 , − 4)   D) (1 , 4)

19) Find the solution for the below system of equations by graphing them below

$$0 = y + 4 + \frac{1}{2}x$$
$$12 = -5x + 4y$$

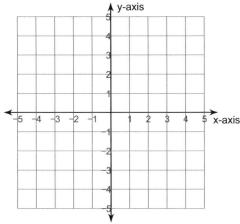

A) (− 4 , − 2)

B) (− 2 , 4)

C) (− 2 , − 4)

D) Infinite number of solutions

20) Find the solution for the below system of equations by graphing them below

$$-y - 6x = -2$$
$$-3 - y = x$$

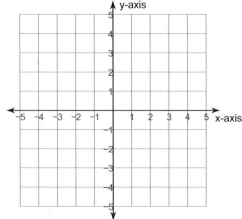

A) (1 , − 4)   B) (− 1 , − 5)

C) (− 1 , − 4)   D) (− 1 , 5)

Algebra 1

Vol 2
Week 25
Graphing
Simultaneous
equations

21) Find the solution for the below system
of equations by graphing them below

$$0 = y + x - 2$$

$$-y = x + 3$$

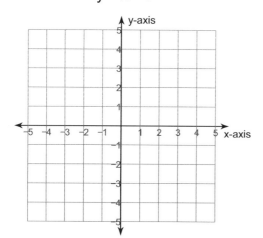

A) No solution   B) (3 , - 4)

C) (3 , 4)   D) (- 3 , 4)

22) Find the solution for the below system
of equations by graphing them below

$$-4y + 16 = 2x$$

$$-x - 1 = -y$$

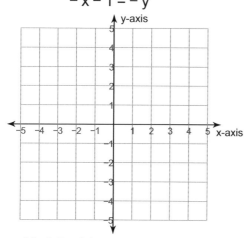

A) (5 , 1)

B) (2 , 3)

C) (5 , 2)

D) (2 , - 3)

23) Find the solution for the below system
of equations by graphing them below

$$0 = 7x + 4y - 16$$

$$x = -8 - 4y$$

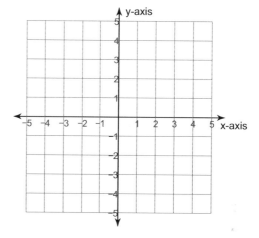

A) (- 4 , 3)   B) (3 , - 4)

C) (4 , - 3)   D) No solution

24) Find the solution for the below system
of equations by graphing them below

$$-4 + 3x - 2y = 0$$

$$-8 = -2y - 3x$$

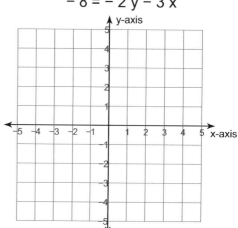

A) (- 2 , - 1)   B) (2 , - 1)

C) (2 , 1)   D) (- 4 , - 1)

www.math-knots.com | www.a4ace.com

Algebra 1

Vol 2
Week 25
Graphing
Simultaneous
equations

25) Find the solution for the below system of equations by graphing them below

$$-6 = -4x - 3y$$

$$-x = -9 - 3y$$

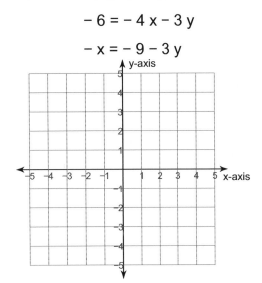

A) No solution

B) (3 , 3)

C) (3 , -2)

D) (-3 , -2)

26) Find the solution for the below system of equations by graphing them below

$$-y = -1 - x$$

$$y = -x + 3$$

A) (-2 , -2)

B) (1 , 2)

C) (1 , -2)

D) (2 , 2)

27) Find the solution for the below system of inequalities by graphing them below

$$y > -2x + 2$$

$$y \le \frac{1}{2}x - 3$$

A)

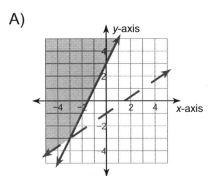

B)

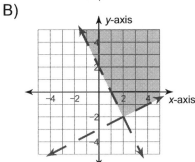

C)

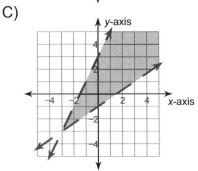

D)

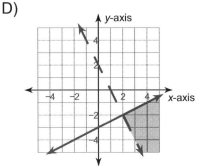

www.math-knots.com | www.a4ace.com

Algebra 1

Vol 2
Week 25
Graphing
Simultaneous
equations

28) Find the solution for the below system of inequalities by graphing them below

$$y > -\frac{2}{3}x - 1$$
$$x > -3$$

A)
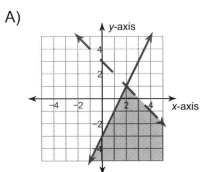

B)
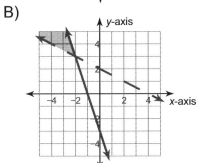

C)
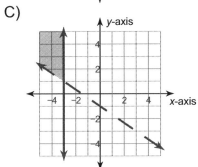

D)
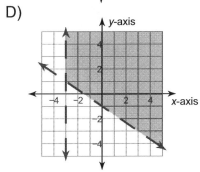

29) Find the solution for the below system of inequalities by graphing them below

$$y \leq 1$$
$$y < \frac{3}{2}x - 2$$

A)
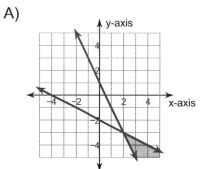

B)
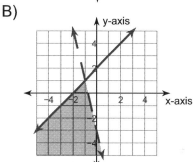

C)
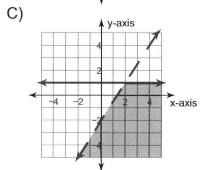

D)
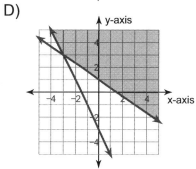

www.math-knots.com | www.a4ace.com

Algebra 1

Vol 2
Week 25
Graphing
Simultaneous
equations

30) Find the solution for the below system of inequalities by graphing them below

$$y < 5x + 2$$
$$y < x - 2$$

A)

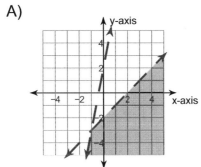

B)

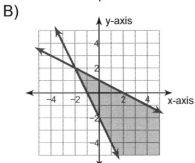

C)

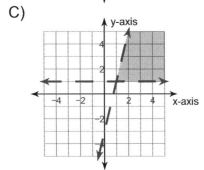

D)

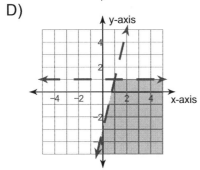

31) Find the solution for the below system of inequalities by graphing them below

$$y < -x + 3$$
$$y \geq -x - 3$$

A)

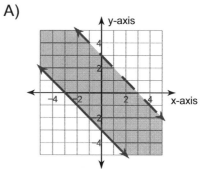

B)

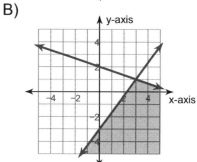

C)

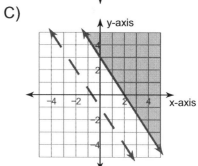

D)

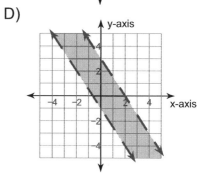

www.math-knots.com | www.a4ace.com

Algebra 1

Vol 2
Week 25
Graphing
Simultaneous
equations

32) Find the solution for the below system of inequalities by graphing them below

$$y > -\frac{1}{3}x + 3$$

$$y \geq \frac{1}{3}x + 1$$

A)

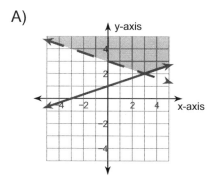

B)

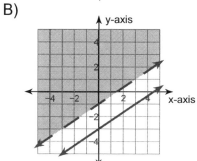

C)

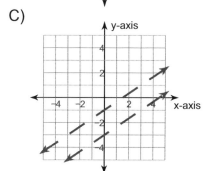

D)

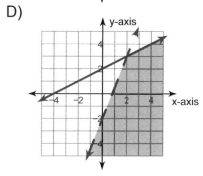

33) Find the solution for the below system of inequalities by graphing them below

$$x \geq 2$$

$$y \geq -\frac{3}{2}x + 2$$

A)

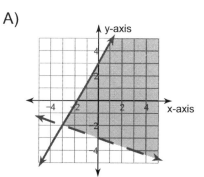

B)

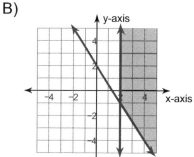

C)

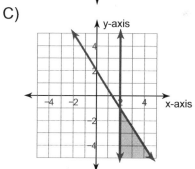

D)

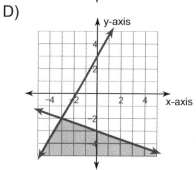

www.math-knots.com | www.a4ace.com

Algebra 1

34) Find the solution for the below system of inequalities by graphing them below

$$y < \frac{2}{3}x + 1$$

$$y \geq -\frac{2}{3}x - 3$$

A)
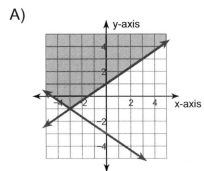

B)
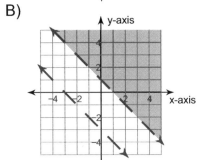

C)
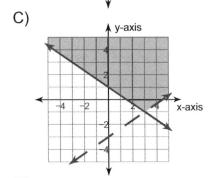

D)
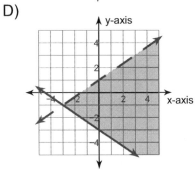

35) Find the solution for the below system of inequalities by graphing them below

$$y \geq x - 3$$

$$y \geq -4x + 2$$

A)
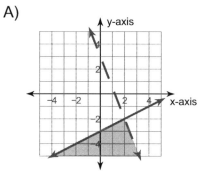

B)
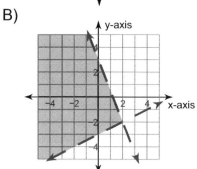

C)
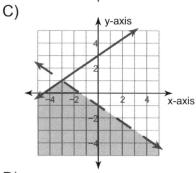

D)
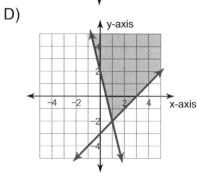

www.math-knots.com | www.a4ace.com

Algebra 1

Vol 2
Week 25
Graphing
Simultaneous
equations

36) Find the solution for the below system of inequalities by graphing them below

$$y \geq 4x - 1$$
$$y < 4x - 3$$

A)
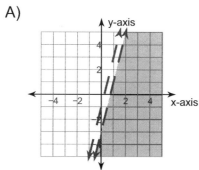

B)
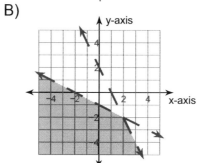

C)
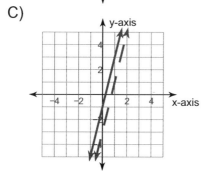

D)
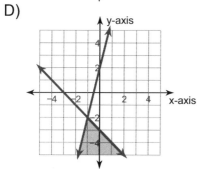

37) Find the solution for the below system of inequalities by graphing them below

$$y \geq -\frac{1}{3}x + 2$$
$$y \leq -\frac{5}{3}x - 2$$

A)
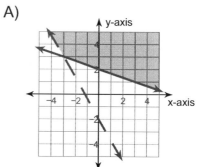

B)
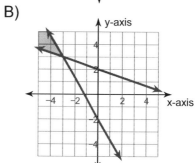

C)
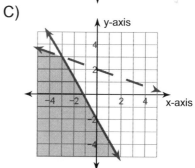

D)
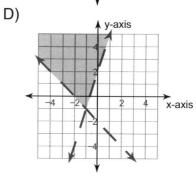

Algebra 1

Vol 2
Week 25
Graphing
Simultaneous
equations

38) Find the solution for the below system of inequalities by graphing them below

$$y \leq -2x + 3$$

$$y \geq \frac{1}{2}x - 2$$

A)

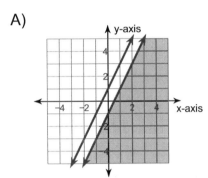

B)

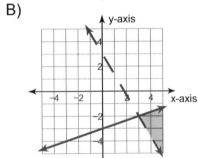

C)

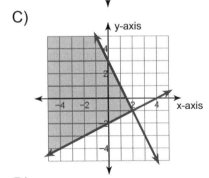

D)

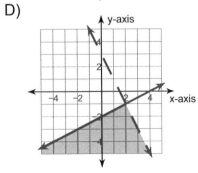

39) Find the solution for the below system of inequalities by graphing them below

$$y > -3x + 3$$

$$y \leq -3$$

A)

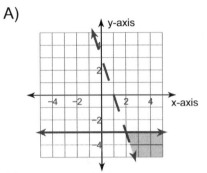

B)

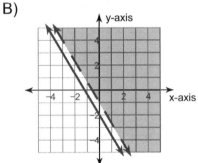

C)

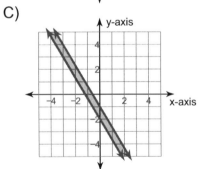

D)

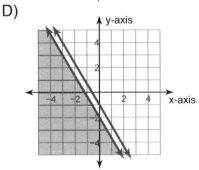

Algebra 1

Vol 2
Week 25
Graphing
Simultaneous
equations

40) Find the solution for the below system of inequalities by graphing them below

$$2x + y \geq -3$$
$$2x + 3y < 3$$

A)

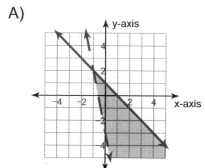

B)

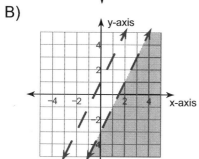

C)

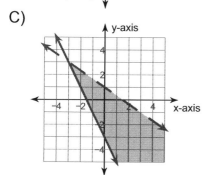

D)

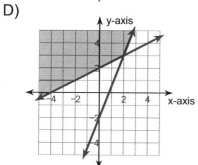

41) Find the solution for the below system of inequalities by graphing them below

$$2x - y \leq -1$$
$$2x + y \leq -3$$

A)

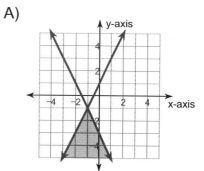

B)

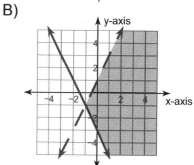

C)

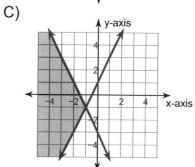

D)

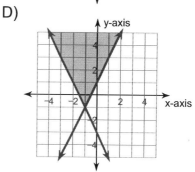

Algebra 1

Vol 2
Week 25
Graphing
Simultaneous
equations

42) Find the solution for the below system of inequalities by graphing them below

$$3x + 2y \le 4$$
$$y \ge -1$$

A)
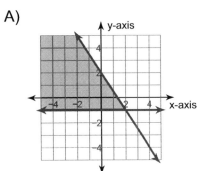

B)
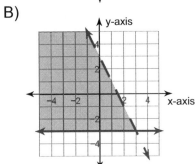

C)
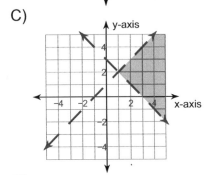

D)
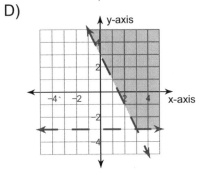

43) Find the solution for the below system of inequalities by graphing them below

$$5x + 3y < 6$$
$$x + 3y < -6$$

A)
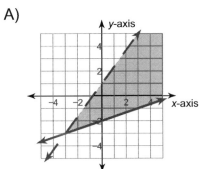

B)
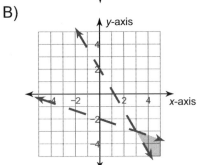

C)
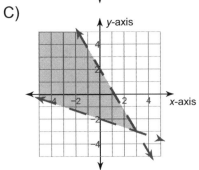

D)
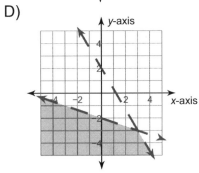

www.math-knots.com | www.a4ace.com

Algebra 1

44) Find the solution for the below system of inequalities by graphing them below

$$x - 3y > 6$$
$$x - 3y \leq -6$$

A)

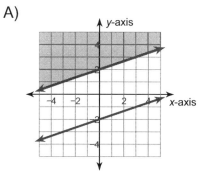

B)

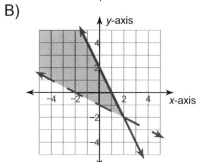

C)

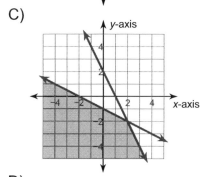

D)

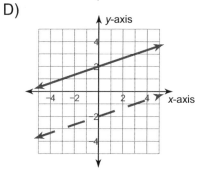

45) Find the solution for the below system of inequalities by graphing them below

$$5x - y < -2$$
$$x - y \geq 2$$

A)

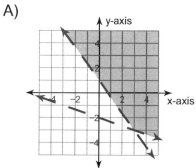

B)

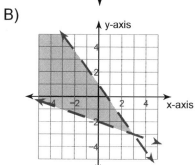

C)

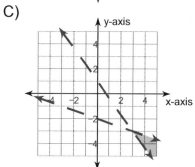

D)

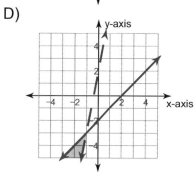

www.math-knots.com | www.a4ace.com

 Algebra 1

Vol 2
Week 25
Graphing
Simultaneous
equations

46) Find the solution for the below system of inequalities by graphing them below

$$2x + 3y < -3$$
$$x - 3y > -6$$

A)
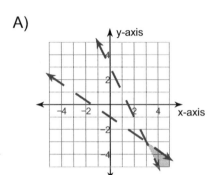

B)
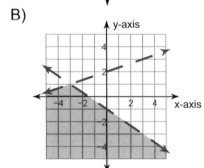

C)
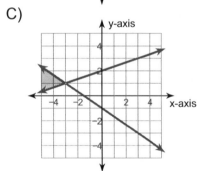

D)
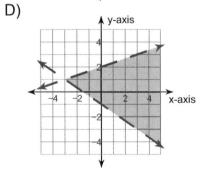

47) Find the solution for the below system of inequalities by graphing them below

$$5x - y < 3$$
$$x + y > 3$$

A)
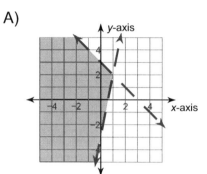

B)
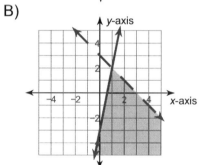

C)
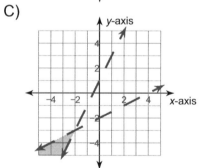

D)
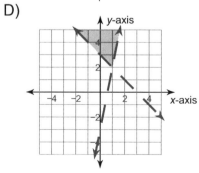

www.math-knots.com | www.a4ace.com

Algebra 1

Vol 2
Week 25
Graphing
Simultaneous
equations

48) Find the solution for the below system of inequalities by graphing them below

$$x - 3y > -9$$
$$x + y \le -1$$

A)

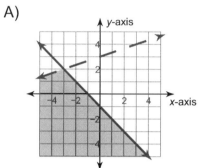

B)

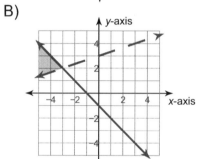

C)

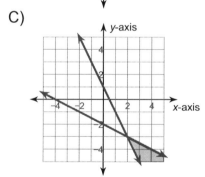

D)

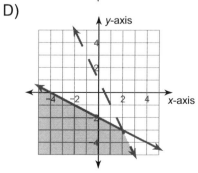

49) Find the solution for the below system of inequalities by graphing them below

$$x + 2y < -4$$
$$2x + y \ge 1$$

A)

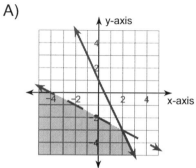

B)

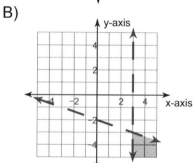

C)

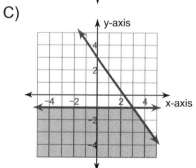

D)

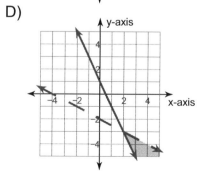

www.math-knots.com | www.a4ace.com

Algebra 1

Vol 2
Week 25
Graphing
Simultaneous
equations

50) Find the solution for the below system of inequalities by graphing them below

$$5x - y \geq -2$$
$$5x - y > 2$$

A)
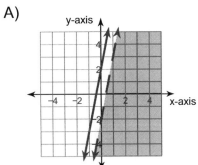

B)
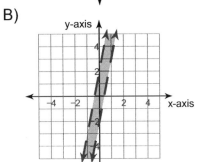

C)
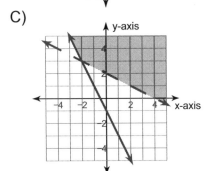

D)
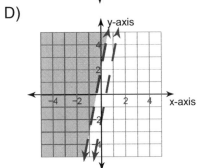

51) Find the solution for the below system of inequalities by graphing them below

$$x < 1$$
$$x + y \geq -2$$

A)
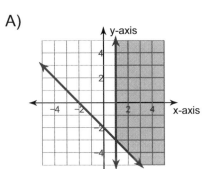

B)
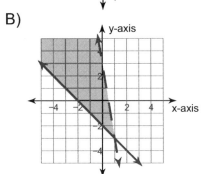

C)
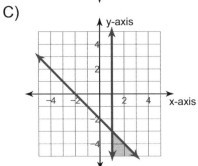

D)
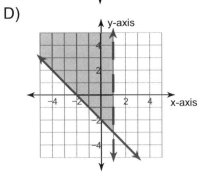

www.math-knots.com | www.a4ace.com

Algebra 1

52) Find the solution for the below system of inequalities by graphing them below

$$x + y \geq 1$$
$$x - y > 3$$

A)
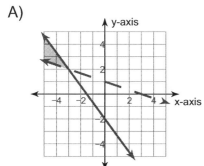

B)
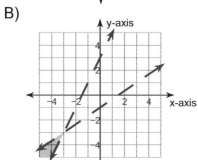

C)
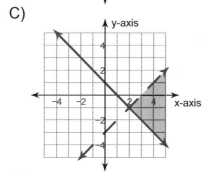

D)
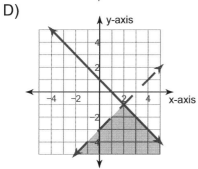

53) Find the solution for the below system of inequalities by graphing them below

$$x - y > -2$$
$$6x - y \geq 3$$

A)
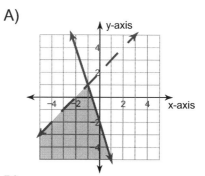

B)
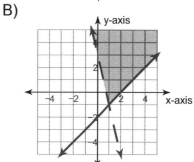

C)
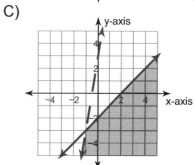

D)
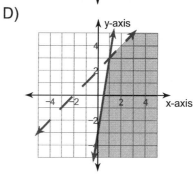

www.math-knots.com | www.a4ace.com

Algebra 1

Vol 2
Week 25
Graphing
Simultaneous
equations

54) Find the solution for the below system of inequalities by graphing them below

$$x - y > 2$$
$$6x - y > -3$$

A)

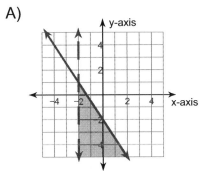

B)

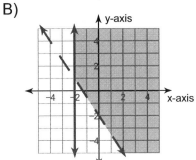

C)

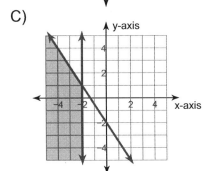

D)

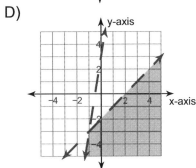

55) Simplify the below

$$2xy^0 \cdot 4xy^{-1}$$

A) $\dfrac{8x^2}{y}$ B) $\dfrac{8}{x^2 y^3}$

C) $\dfrac{9y^7}{x}$ D) $\dfrac{4}{y^5 x^2}$

56) Simplify the below

$$3vu^{-2} \cdot uv$$

A) $\dfrac{3v^2}{u}$ B) $3u^5 v^5$

C) $\dfrac{4v^2}{u^2}$ D) $u^4 v^5$

57) Simplify the below

$$3u^4 v^{-4} \cdot 3u^0 v^{-2} \cdot 4u^4 v^0$$

A) $\dfrac{24u}{v^3}$ B) $12u^2$

C) $u^4 v^2$ D) $\dfrac{36u^8}{v^6}$

58) Simplify the below

$$3u^{-4} v^{-4} \cdot 3u^0 v^3 \cdot 4u^0 v^0$$

A) $\dfrac{36}{u^4 v}$ B) $6u^6 v^2$

C) $\dfrac{2v^6}{u^4}$ D) $\dfrac{3v^6}{u^2}$

Algebra 1

Vol 2
Week 25
Graphing
Simultaneous
equations

59) Simplify the below

$$m^{-1}n^2 \cdot n^{-3}$$

A) $\dfrac{1}{mn}$ B) $\dfrac{8}{m^2}$

C) $\dfrac{3n^3}{m^6}$ D) $\dfrac{3m}{n^4}$

60) Simplify the below

$$xy^{-4} \cdot x^{-1}y^0$$

A) $8x^5y^2$ B) $\dfrac{1}{y^4}$

C) $12x^2y^2$ D) $4x^5$

61) Simplify the below

$$2x^{-4}y^0 \cdot x^2y^{-4}$$

A) $\dfrac{8}{x^3}$ B) $8y^5$

C) $\dfrac{3y^3}{x^2}$ D) $\dfrac{2}{x^2y^4}$

62) Simplify the below

$$4x^{-2}y^0 \cdot 3x^3$$

A) $2yx^2$ B) $\dfrac{2}{x}$

C) $12x$ D) $8y^2$

63) Simplify the below

$$2x^2 \cdot y^2$$

A) $\dfrac{3}{y^5}$ B) $2x^2y^2$

C) $9x^3$ D) $2xy$

64) Simplify the below

$$u \cdot 2u^4$$

A) $2u^2v^3$ B) $4u^5v^6$

C) $2u^5$ D) $4v^4u^6$

65) Simplify the below

$$b^0 \cdot 3ba^3$$

A) $\dfrac{48b^4}{a^2}$ B) $12a^3b^7$

C) $\dfrac{4}{b^2a^4}$ D) $3ba^3$

66) Simplify the below

$$2m^0n^{-4} \cdot 2m^3n^2 \cdot 4m^2n^3$$

A) $16m^5n$ B) $18m^6$

C) $\dfrac{2m^3}{n^5}$ D) $16n^3m^2$

Algebra 1

Vol 2
Week 25
Graphing
Simultaneous
equations

67) Simplify the below

$$2x^2y^4 \cdot 2xy^2$$

A) $4x^3y^6$ B) $\dfrac{3x^7}{y^2}$

C) $3x^7y^4$ D) $12x^5$

68) Simplify the below

$$3x^4 \cdot x^{-3}y^{-2}$$

A) $\dfrac{3x}{y^2}$ B) $\dfrac{9x^3}{y^4}$

C) $\dfrac{8}{x^5y^4}$ D) 4

69) Simplify the below

$$3x^2y^3 \cdot 4x^3y^{-3} \cdot 4x^3$$

A) $48x^8$ B) $\dfrac{9}{y^6}$

C) $\dfrac{4}{xy^3}$ D) $\dfrac{2}{y^2}$

70) Simplify the below

$$2xy^{-1} \cdot 3x^3y^3$$

A) $\dfrac{3x^3}{y^2}$ B) $6x^4y^2$

C) $12x^8y^4$ D) $\dfrac{4y^7}{x^4}$

71) Simplify the below

$$v^{-3} \cdot 2u^2v^0$$

A) $\dfrac{32}{u^5}$ B) $\dfrac{2u^2}{v^3}$

C) $\dfrac{2v^4}{u}$ D) $12u^6$

72) Simplify the below

$$b^{-1} \cdot 2a^{-2}b^0$$

A) $\dfrac{2}{ba^2}$ B) $8b^5a^8$

C) $\dfrac{3a}{b^4}$ D) $\dfrac{32}{ab^4}$

73) Simplify the below

$$3x^2y^3 \cdot 4x^{-4}$$

A) $3x^4y^5$ B) $4xy^5$

C) $\dfrac{12y^3}{x^2}$ D) $\dfrac{4}{x^2y}$

74) Simplify the below

$$2u^{-4}v^{-1} \cdot 2u^3v^4 \cdot 4u^4$$

A) $3u^6v^4$ B) $6u^9v^2$

C) $16v^3u^3$ D) $\dfrac{3u^3}{v^2}$

 **Algebra 1**

Vol 2
Week 25
Graphing
Simultaneous
equations

75) Simplify the below

$$\left(h^{-3}\right)^2 \cdot j^4 k^{-3}$$

A) $8k^9 h^9 j^4$ B) $\dfrac{j^2}{h^4}$

C) $\dfrac{j^4}{h^6 k^3}$ D) $\dfrac{2k}{h^2}$

76) Simplify the below

$$x^{-4} y^2 z^4 \cdot \left(x^0 y^3 z^4 \cdot 2y^{-2} z^{-1}\right)^{-2}$$

A) $16x^4 y^2 z^3$ B) $\dfrac{4z^5}{y^4 x}$

C) $\dfrac{16x^{10}}{y^2 z^{14}}$ D) $\dfrac{1}{4x^4 z^2}$

77) Simplify the below

$$p \cdot \left(2m^4 n^3 p^{-1}\right)^{-2}$$

A) $\dfrac{p^3}{4m^8 n^6}$ B) $m^{20} n^{15} p^9$

C) $\dfrac{16m^{20}}{p^{12}}$ D) $4mn^3 p^3$

78) Simplify the below

$$\left(nm^{-4} p^4\right)^{-4} \cdot p^2$$

A) $\dfrac{8m^9}{p^4}$ B) $m^9 n^{12} p^3$

C) $\dfrac{m^{16}}{n^4 p^{14}}$ D) $16p^2$

79) Simplify the below

$$2hj^3 \cdot \left(2hj^3 k^2\right)^3$$

A) $16h^4 j^{12} k^6$ B) $\dfrac{k^3 j}{h^2}$

C) $\dfrac{32h^7 k^8}{j}$ D) $\dfrac{4k^{14}}{h^6 j^{12}}$

80) Simplify the below

$$\left(x^{-1} y^4 z^3\right)^4 \cdot x^3 z^3$$

A) $\dfrac{y^{16} z^{15}}{x}$ B) $2z^5 x^2 y^3$

C) 1 D) $\dfrac{z}{x^6 y^2}$

81) Simplify the below

$$\left(m^0 p^2\right)^{-2} \cdot 2n^{-3}$$

A) mp^3 B) $\dfrac{2}{p^4 n^3}$

C) $\dfrac{2n^2}{p^2}$ D) $\dfrac{1}{8m^2 p^{10}}$

82) Simplify the below

$$\left(2p^{-2} r^0 \cdot 2q^3 r^4\right)^{-2}$$

A) $\dfrac{p^{16}}{4q^8 r^9}$ B) $\dfrac{p^4}{16q^6 r^8}$

C) $\dfrac{16r^{12}}{p^4}$ D) $\dfrac{q^{13} p^4}{r^2}$

Algebra 1

83) Simplify the below

$$zy^4 \cdot \left(2xy^{-2}z^0\right)^2$$

A) $2y^3z^{15}x$ B) $4x^2z$

C) $\dfrac{x^3y^{16}z^{11}}{16}$ D) $256x^{12}y^{12}z^4$

84) Simplify the below

$$2kh^{-1}j^3 \cdot \left(k^3\right)^3$$

A) $\dfrac{1}{8j^{15}k^4}$ B) $\dfrac{2k^{10}j^3}{h}$

C) $\dfrac{1}{16k^6j^6}$ D) $\dfrac{h^3k}{2j}$

85) Simplify the below

$$\left(2nm^4p^2\right)^{-3} \cdot 2m^{-4}n^3$$

A) $\dfrac{p^{12}}{2n}$ B) $2pn^3m$

C) $\dfrac{m^8n^6p^{10}}{2}$ D) $\dfrac{1}{4m^{16}p^6}$

86) Simplify the below

$$\left(2pm^2\right)^4 \cdot m^2n^{-1}p^0$$

A) $\dfrac{p^2}{n^2m^8}$ B) $\dfrac{16m^{10}p^4}{n}$

C) $\dfrac{64m^2p^{14}}{n^6}$ D) $\dfrac{p^{15}n^3}{m^{10}}$

87) Simplify the below

$$zx^0y^2 \cdot \left(2x^2y^0z^4\right)^2$$

A) $4z^9y^2x^4$ B) $\dfrac{2x^7}{y^{15}z^2}$

C) $2x^8y^4z^7$ D) $\dfrac{y^{19}}{16x^4z^6}$

88) Simplify the below

$$\left(yx^{-1}z^{-3}\right)^2 \cdot 2xzy^4$$

A) $\dfrac{x^{21}z^3}{y^{15}}$ B) $x^4z^{10}y^2$

C) $\dfrac{2y^6}{xz^5}$ D) $\dfrac{x^{10}}{y^2}$

89) Simplify the below

$$h^{-3}j^3k^{-3} \cdot \left(2h^{-3}j^3k^2\right)^{-4}$$

A) $\dfrac{8h}{k}$ B) $\dfrac{h^9}{16k^{11}j^9}$

C) h^6j^8 D) $\dfrac{h^{12}j^{12}}{16}$

90) Simplify the below

$$\left(2x^3y^{-2}z^3 \cdot 2x^{-1}y^{-2}z^{-2}\right)^2$$

A) $\dfrac{1}{16x^4y^7z^3}$ B) $\dfrac{16x^4z^2}{y^8}$

C) $\dfrac{z^4}{y^6}$ D) x^5z^8

Algebra 1

91) Simplify the below

$$\left(x^{-4}z^2\right)^2 \cdot 2xz^{-1}$$

A) $\dfrac{2z^3}{x^7}$
B) $\dfrac{y^6z^6}{4}$

C) $\dfrac{x^4}{4}$
D) $\dfrac{8x^6}{y^6}$

92) Simplify the below

$$2zx^4y^{-3} \cdot \left(x^2y^0\right)^4$$

A) $\dfrac{2x^2z^4}{y^3}$
B) $\dfrac{z^3}{x^7y^{15}}$

C) $\dfrac{2z^6}{y^3x^{10}}$
D) $\dfrac{2zx^{12}}{y^3}$

93) Simplify the below

$$hk^4 \cdot \left(kh^3j^2\right)^0$$

A) hk^4
B) $8j^6k^4$

C) $\dfrac{16j^2k^{16}}{h^9}$
D) $\dfrac{1}{16h^{24}j^4}$

94) Simplify the below

$$\left(2zx^{-3}y^{-4}\right)^2 \cdot 2x^3y^2z^{-2}$$

A) $16x^2y^{12}z^3$
B) $\dfrac{8}{x^3y^6}$

C) $\dfrac{z^3}{y^8}$
D) $2z^4x^4y^8$

95) Simplify the below

$$\left(x^4y^3z^0 \cdot z^{-3}\right)^0$$

A) 1
B) $x^8y^9z^{12}$

C) $\dfrac{16x^{10}z^6}{y^7}$
D) $\dfrac{1}{x^4y^8z^3}$

96) Simplify the below

$$\left(2h^3j^{-2}k^4\right)^3 \cdot 2j$$

A) $\dfrac{32j^{14}h^8}{k^9}$
B) $j^{17}h^8k^{12}$

C) $\dfrac{1}{4jk^6h}$
D) $\dfrac{16h^9k^{12}}{j^5}$

97) Simplify the below

$$\dfrac{7n^0p^{-1}}{3m^{-4}n^2}$$

A) $\dfrac{7m^8n^2}{10}$
B) $\dfrac{6}{5p^8m}$

C) $\dfrac{7m^4}{3pn^2}$
D) $\dfrac{m^3n^{16}}{4p^4}$

98) Simplify the below

$$\dfrac{3nm^8p^{-6}}{8m^3n^6}$$

A) $\dfrac{p^4}{8m^8n^2}$
B) $\dfrac{mp^6}{n^7}$

C) $\dfrac{5m^{15}n^3p}{3}$
D) $\dfrac{3m^5}{8p^6n^5}$

 Algebra 1

Vol 2
Week 25
Graphing
Simultaneous
equations

99) Simplify the below

$$\frac{5p^{-7}q^2r^3}{8p^8q^9r^{-3}}$$

A) $\dfrac{5r^6}{8p^{15}q^7}$　　B) $\dfrac{p^{11}q^9r^5}{9}$

C) $\dfrac{10}{7q^{12}p^3}$　　D) $2p^{13}r^2$

100) Simplify the below

$$\frac{3x^0z^{-2}}{10x^3y^0z^8}$$

A) $\dfrac{5x^{16}y^8}{2z^{15}}$　　B) $\dfrac{3}{10z^{10}x^3}$

C) $\dfrac{10z^6}{3xy^2}$　　D) $\dfrac{x^6}{10y^3z^2}$

101) Simplify the below

$$\frac{p^{10}q^{-5}r^{-1}}{4p^{10}q^8r^9}$$

A) $\dfrac{10p^7r^{10}}{9}$　　B) $q^{10}pr^3$

C) $\dfrac{1}{4q^{13}r^{10}}$　　D) $\dfrac{p^4}{4q^2r^3}$

102) Simplify the below

$$\frac{9m^{10}q^3}{7m^5p^{10}q^{10}}$$

A) $\dfrac{9}{8p^{12}q^9}$　　B) $\dfrac{5q^2}{4m^2p^8}$

C) $\dfrac{9m^5}{7p^{10}q^7}$　　D) $\dfrac{5}{9q^{16}m^4p^2}$

103) Simplify the below

$$\frac{3z^0}{7x^{-9}z^{10}}$$

A) $\dfrac{3x^9}{7z^{10}}$　　B) $\dfrac{7x^3}{z^{19}y^8}$

C) $\dfrac{4}{3z^5xy^2}$　　D) $\dfrac{3z^6}{2}$

104) Simplify the below

$$\frac{4x^{-8}y^{-3}z^{10}}{3x^{-4}y^0z^7}$$

A) $\dfrac{3y^6z^6}{x^{16}}$　　B) $\dfrac{6x^3y^4}{z^{10}}$

C) $\dfrac{4z^3}{3x^4y^3}$　　D) $\dfrac{8x^5y^3}{3z^{15}}$

105) Simplify the below

$$\frac{9x^{-8}y^8z^{-6}}{10y^7z^{-3}}$$

A) $\dfrac{3y^{11}xz^8}{2}$　　B) $\dfrac{9y}{10x^8z^3}$

C) $\dfrac{2z^6y^5}{7x^3}$　　D) $\dfrac{5}{9x^7y^{10}z^5}$

106) Simplify the below

$$\frac{9n^{-4}p^9}{6pn^{-10}}$$

A) $m^{14}p^8$　　B) $\dfrac{9n^6p^4m}{10}$

C) $\dfrac{3p^8n^6}{2}$　　D) $\dfrac{8m^2}{p}$

Algebra 1

107) Simplify the below

$$\frac{7n^2}{7m^{-4}n^{10}p^{-6}}$$

A) $\dfrac{p^{18}m^8n^9}{4}$ B) $\dfrac{m^3p^2}{4}$

C) $\dfrac{m^4p^{10}}{5n^8}$ D) $\dfrac{m^4p^6}{n^8}$

108) Simplify the below

$$\frac{4x^{-7}z^0}{7x^5y^2}$$

A) $\dfrac{x^9y^9z^2}{2}$ B) $\dfrac{2xz^4}{9y^{12}}$

C) $\dfrac{4}{7x^{12}y^2}$ D) $\dfrac{3}{2x^3y^{10}}$

109) Simplify the below

$$\frac{7y^{-4}z^6}{9y^7z^5}$$

A) $\dfrac{8z^2}{9x^6y^3}$ B) $\dfrac{7z}{9y^{11}}$

C) $\dfrac{5x^4z^4}{y^6}$ D) $\dfrac{9x^6z^2}{4}$

110) Simplify the below

$$\frac{p^{-3}q^0r^9}{3p^0q^3r^5}$$

A) $\dfrac{r^4}{3p^3q^3}$ B) $\dfrac{4r^{14}}{7p^6q^9}$

C) $\dfrac{9r^2}{10p^{11}q^8}$ D) $\dfrac{5p^{15}q^8r^2}{4}$

111) Simplify the below

$$\frac{10x^5}{7x^6y^3z^5}$$

A) $\dfrac{2x^2z^{10}y^8}{5}$ B) $\dfrac{y^4}{z^{11}}$

C) $\dfrac{10}{7xy^3z^5}$ D) $\dfrac{9x^9y^{11}}{z^6}$

112) Simplify the below

$$\frac{10p^5q^5r^{-1}}{7pq^{-8}r^{-5}}$$

A) $\dfrac{10p^4q^{13}r^4}{7}$ B) $2q^{15}r^7p$

C) $\dfrac{2}{3p^2q^2r^5}$ D) $\dfrac{r^{10}}{p^6q^9}$

113) Simplify the below

$$\frac{6x^4}{2x^{-5}y^4z^0}$$

A) $\dfrac{4z^{19}y}{5x^3}$ B) $\dfrac{4xz^7}{7y^4}$

C) $\dfrac{3x^9}{y^4}$ D) $\dfrac{3x^{10}y^2}{4z^8}$

114) Simplify the below

$$\frac{4x^0y^{-9}z^{-1}}{3xy^0z^{-1}}$$

A) $\dfrac{4}{3y^9x}$ B) $\dfrac{3x}{y}$

C) $\dfrac{9z^5}{5x^{10}y^2}$ D) $\dfrac{y^{13}}{x^2z^6}$

 Algebra 1

Vol 2
Week 25
Graphing
Simultaneous
equations

115) Simplify the below

$$\frac{3y^{-7}z^{-10}}{7x^8y^6z^{-7}}$$

A) $\dfrac{3}{7y^{13}z^3x^8}$ B) $\dfrac{2z^9}{3y^6x^8}$

C) $\dfrac{x^3z^2}{9}$ D) $\dfrac{2x^6z^5}{3y^9}$

116) Simplify the below

$$\frac{10j^5}{6h^{-1}j^2}$$

A) $\dfrac{5j^9}{4hk^4}$ B) $\dfrac{h^9j^4k^{10}}{2}$

C) $\dfrac{5hj^3}{3}$ D) $\dfrac{k^{10}}{2h^3j^8}$

117) Simplify the below

$$\frac{6m^{-1}}{6m^9n^7p^9}$$

A) $\dfrac{3}{m^3n^3p}$ B) $\dfrac{5m^8n^9p^6}{2}$

C) $\dfrac{10}{p^4mn^{10}}$ D) $\dfrac{1}{m^{10}n^7p^9}$

118) Simplify the below

$$\frac{10x^0y^{-6}z^{-8}}{3x^9y^8z^2}$$

A) $\dfrac{6x^3y^4z^7}{7}$ B) $5y^{11}z^{18}$

C) $\dfrac{2}{3x^2}$ D) $\dfrac{10}{3y^{14}z^{10}x^9}$

119) Simplify the below

$$10x^7y^7z^8 \cdot 3yx^0z^0$$

120) Simplify the below

$$10r^{-2} \cdot 5p^3q^{-4}r^3 \cdot 5p^{-9}q^2$$

121) Simplify the below

$$7ca^0b^{-2} \cdot 5a^4b^4c^4$$

122) Simplify the below

$$10x^{-4}z^2 \cdot 8x^{-1}z^0 \cdot 4x^{10}y^{-5}z^3$$

123) Simplify the below

$$m^8p^{-7}q^{10} \cdot 9m^2p^0q^2$$

124) Simplify the below

$$6pq^9 \cdot 5m^9p^8q^{-7} \cdot 2m^2p^5q^6$$

125) Simplify the below

$$3p^{10} \cdot 4m^{-3}n^{-1}p^{-3}$$

126) Simplify the below

$$10m^7n^{-2} \cdot 9m^{-7}n^{-7}p^{-10}$$

127) Simplify the below

$$7c^0 \cdot a^0b^{-10}c^{-9}$$

 Algebra 1

Vol 2
Week 25
Graphing
Simultaneous
equations

128) Simplify the below

$$2x^4y^{-1} \cdot 5x^6y^6z^2$$

129) Simplify the below

$$7n^0p^8 \cdot 7mn^{-8}p^8$$

130) Simplify the below

$$2x^4z^6 \cdot 3x^{-10}y^{-8}z^{-4}$$

131) Simplify the below

$$8x^2y^9 \cdot 10x^8y^7z^{-9}$$

132) Simplify the below

$$8j^7k^9 \cdot 3h^0j^{-9}k^{10}$$

133) Simplify the below

$$9y^{-6}z^{-3} \cdot 4x^3z^2$$

134) Simplify the below

$$7x^5y^8z^9 \cdot 7x^9y^5z^{-4}$$

135) Simplify the below

$$6x^8y^6 \cdot 6x^4y^{10}z^{10}$$

136) Simplify the below

$$3a^9b^9 \cdot 6ab^0c^{10} \cdot 10a^{-5}c^6$$

137) Simplify the below

$$6x^0y^9z^{-3} \cdot 10x^6y^2z^8$$

138) Simplify the below

$$7pq^{-1} \cdot 3m^3p^{-3}q^{-10}$$

139) Simplify the below

$$6zx^3y^3 \cdot 5x^3y^7z^0$$

140) Simplify the below

$$10p^9 \cdot 2m^2p^4q^4$$

1) Simplify the below :

$$\sqrt{98}$$

A) $7\sqrt{2}$ B) 4

C) $16\sqrt{2}$ D) $3\sqrt{2}$

5) Simplify the below :

$$\sqrt{80}$$

A) $12\sqrt{2}$ B) 10

C) 8   D) $4\sqrt{5}$

2) Simplify the below :

$$\sqrt{147}$$

A) 12   B) $7\sqrt{3}$

C) $5\sqrt{5}$ D) $4\sqrt{6}$

6) Simplify the below :

$$\sqrt{384}$$

A) $14\sqrt{2}$ B) $4\sqrt{7}$

C) $4\sqrt{3}$ D) $8\sqrt{6}$

3) Simplify the below :

$$\sqrt{196}$$

A) $5\sqrt{5}$ B) $4\sqrt{6}$

C) $4\sqrt{7}$ D) 14

7) Simplify the below :

$$\sqrt{128}$$

A) $3\sqrt{6}$ B) $8\sqrt{2}$

C) $5\sqrt{7}$ D) $7\sqrt{2}$

4) Simplify the below :

$$\sqrt{75}$$

A) $6\sqrt{7}$ B) $5\sqrt{3}$

C) $6\sqrt{3}$ D) $7\sqrt{3}$

8) Simplify the below :

$$\sqrt{252}$$

A) $4\sqrt{2}$ B) $6\sqrt{3}$

C) 4   D) $6\sqrt{7}$

www.math-knots.com | www.a4ace.com

9) Simplify the below :

$$\sqrt{12}$$

A) $2\sqrt{3}$ B) $5\sqrt{7}$

C) $2\sqrt{6}$ D) 10

10) Simplify the below :

$$\sqrt{144}$$

A) $5\sqrt{7}$ B) 4

C) 12 D) $2\sqrt{5}$

11) Simplify the below :

$$\sqrt{175}$$

A) 14 B) $4\sqrt{2}$

C) $5\sqrt{7}$ D) $7\sqrt{6}$

12) Simplify the below :

$$\sqrt{512}$$

A) $2\sqrt{3}$ B) $3\sqrt{3}$

C) $6\sqrt{2}$ D) $16\sqrt{2}$

13) Simplify the below :

$$\sqrt{392}$$

A) $3\sqrt{2}$ B) $14\sqrt{2}$

C) $3\sqrt{3}$ D) $5\sqrt{2}$

14) Simplify the below :

$$\sqrt{256}$$

A) $4\sqrt{3}$ B) $4\sqrt{7}$

C) 16 D) $4\sqrt{2}$

15) Simplify the below :

$$\sqrt{20}$$

A) $2\sqrt{5}$ B) $16\sqrt{2}$

C) $4\sqrt{6}$ D) $2\sqrt{6}$

16) Simplify the below :

$$\sqrt{64}$$

A) 10 B) $2\sqrt{7}$

C) $7\sqrt{7}$ D) 8

Algebra 1

17) Simplify the below :

$$\sqrt{28}$$

A) $2\sqrt{7}$ B) $5\sqrt{6}$

C) $8\sqrt{7}$ D) $14\sqrt{2}$

18) Simplify the below :

$$\sqrt{45}$$

A) $4\sqrt{7}$ B) $6\sqrt{2}$

C) $3\sqrt{3}$ D) $3\sqrt{5}$

19) Simplify the below :

$$\sqrt{196x^2}$$

A) $7x^2\sqrt{7}$ B) $14x$

C) $3x^2\sqrt{7}$ D) $16x^2\sqrt{2}$

20) Simplify the below :

$$\sqrt{512p^2}$$

A) $4\sqrt{2p}$ B) $16p\sqrt{2}$

C) $4p^2\sqrt{6}$ D) $6p\sqrt{2}$

21) Simplify the below :

$$\sqrt{50x^4}$$

A) $5x^2\sqrt{2}$ B) $3x^2\sqrt{3}$

C) $14\sqrt{x}$ D) $5\sqrt{6x}$

22) Simplify the below :

$$\sqrt{125x^4}$$

A) $4x^2\sqrt{2}$ B) $3x\sqrt{2x}$

C) $5x^2\sqrt{5}$ D) $5\sqrt{7x}$

23) Simplify the below :

$$\sqrt{112a}$$

A) $5a\sqrt{2}$ B) $4\sqrt{7a}$

C) $8\sqrt{5a}$ D) $14a^2\sqrt{2}$

24) Simplify the below :

$$\sqrt{245k^4}$$

A) $4k\sqrt{k}$ B) $7k\sqrt{2}$

C) $6\sqrt{6k}$ D) $7k^2\sqrt{5}$

 www.math-knots.com | www.a4ace.com

25) Simplify the below :

$$\sqrt{45x^4}$$

A) $8x\sqrt{7x}$ B) $3x\sqrt{2}$

C) $5x^2\sqrt{5}$ D) $3x^2\sqrt{5}$

26) Simplify the below :

$$\sqrt{150r}$$

A) $5r\sqrt{6r}$ B) $5\sqrt{6r}$

C) $4r^2\sqrt{5}$ D) $6r\sqrt{6}$

27) Simplify the below :

$$\sqrt{128x}$$

A) $8x^2$ B) $8\sqrt{2x}$

C) $10x^2$ D) $3\sqrt{2x}$

28) Simplify the below :

$$\sqrt{63r^2}$$

A) $4r\sqrt{3}$ B) $10r^2$

C) $2r\sqrt{3}$ D) $3r\sqrt{7}$

29) Simplify the below :

$$\sqrt{245x^2}$$

A) $7x\sqrt{5}$ B) $8\sqrt{2x}$

C) $14x\sqrt{2}$ D) $5\sqrt{7x}$

30) Simplify the below :

$$\sqrt{144r^4}$$

A) $12r^2$ B) $5r^2\sqrt{6}$

C) $7r^2\sqrt{6}$ D) $4r^2\sqrt{7}$

31) Simplify the below :

$$\sqrt{147p^2}$$

A) $4p\sqrt{6p}$ B) $7p^2\sqrt{5}$

C) $4\sqrt{2p}$ D) $7p\sqrt{3}$

32) Simplify the below :

$$\sqrt{36b}$$

A) $7b\sqrt{7}$ B) $10\sqrt{b}$

C) $5b\sqrt{2b}$ D) $6\sqrt{b}$

33) Simplify the below :

$$\sqrt{18r^2}$$

A) $5r\sqrt{2r}$ B) $3r\sqrt{7}$

C) $7\sqrt{6r}$ D) $3r\sqrt{2}$

34) Simplify the below :

$$\sqrt{45r^3}$$

A) $4r^2$ B) $4r\sqrt{3}$

C) $3r\sqrt{5r}$ D) $16\sqrt{2r}$

35) Simplify the below :

$$\sqrt{8a^2b^3}$$

A) $7ab\sqrt{6ab}$ B) $2ab\sqrt{2b}$

C) $8a^2\sqrt{2b}$ D) $12a^2b\sqrt{2}$

36) Simplify the below :

$$\sqrt{128uv}$$

A) $6v^2u\sqrt{6u}$ B) $8u^2v\sqrt{7}$

C) $8\sqrt{2uv}$ D) $8u^2\sqrt{3v}$

37) Simplify the below :

$$\sqrt{54ab^3}$$

A) $6ab\sqrt{6b}$ B) $3b\sqrt{6ab}$

C) $16a^2\sqrt{2b}$ D) $10a\sqrt{b}$

38) Simplify the below :

$$\sqrt{144m^2n^2}$$

A) $6n^2m\sqrt{7}$ B) $12mn$

C) $6m\sqrt{2n}$ D) $4m^2n^2\sqrt{7}$

39) Simplify the below :

$$\sqrt{252x^3y^4}$$

A) $6xy\sqrt{6x}$ B) $3x\sqrt{3xy}$

C) $2y^2x\sqrt{7x}$ D) $6y^2x\sqrt{7x}$

40) Simplify the below :

$$\sqrt{12m^4n^3}$$

A) $2mn\sqrt{2mn}$ B) $2m^2n\sqrt{3n}$

C) $6n\sqrt{5m}$ D) $6m^2\sqrt{3n}$

       www.math-knots.com | www.a4ace.com

41) Simplify the below :

$$\sqrt{12u^2v}$$

A) $2u\sqrt{3v}$ 　　　B) $2\sqrt{3uv}$

C) $2v\sqrt{5uv}$ 　　　D) $7v^2\sqrt{2u}$

42) Simplify the below :

$$\sqrt{32a^3b}$$

A) $4a\sqrt{2ab}$ 　　　B) $14b\sqrt{ab}$

C) $7ab\sqrt{5ab}$ 　　　D) $8ab\sqrt{6b}$

43) Simplify the below :

$$\sqrt{32xy}$$

A) $12xy$ 　　　B) $4y\sqrt{2x}$

C) $12x^2\sqrt{y}$ 　　　D) $4\sqrt{2xy}$

44) Simplify the below :

$$\sqrt{12x^2y^4}$$

A) $2y^2x\sqrt{3}$ 　　　B) $8xy\sqrt{6xy}$

C) $5x^2y\sqrt{5y}$ 　　　D) $7\sqrt{2xy}$

45) Simplify the below :

$$\sqrt{20x^3y^3}$$

A) $2xy\sqrt{5xy}$ 　　　B) $5xy\sqrt{2xy}$

C) $14\sqrt{2xy}$ 　　　D) $4xy\sqrt{7x}$

46) Simplify the below :

$$\sqrt{45x^2y^4}$$

A) $3y^2x\sqrt{5}$ 　　　B) $4x\sqrt{2xy}$

C) $8y\sqrt{2x}$ 　　　D) $14x^2y^2\sqrt{2}$

47) Simplify the below :

$$\sqrt{80xy}$$

A) $4xy\sqrt{2}$ 　　　B) $16x^2\sqrt{y}$

C) $8x^2\sqrt{3y}$ 　　　D) $4\sqrt{5xy}$

48) Simplify the below :

$$\sqrt{8x^4y^2}$$

A) $16y^2\sqrt{2x}$ 　　　B) $2x^2y\sqrt{2}$

C) $7y^2x\sqrt{5x}$ 　　　D) $5x^2y\sqrt{2}$

www.math-knots.com | www.a4ace.com

49) Simplify the below :

$$\sqrt{384x^4y^4}$$

A) $6x^2y$

B) $8x^2y^2\sqrt{6}$

C) $10y\sqrt{2x}$

D) $6x^2y^2\sqrt{5}$

50) Simplify the below :

$$\sqrt{512m^4n^3}$$

A) $16m^2n\sqrt{2n}$

B) $4m^2\sqrt{6n}$

C) $5m^2n\sqrt{7n}$

D) $10m^2n$

51) Simplify the below :

$$-4\sqrt{128} + 2\sqrt{128}$$

A) $-32\sqrt{2}$

B) $-48\sqrt{2}$

C) $-64\sqrt{2}$

D) $-16\sqrt{2}$

52) Simplify the below :

$$-2\sqrt{24} + 3\sqrt{54}$$

A) $10\sqrt{6}$

B) $5\sqrt{6}$

C) $6\sqrt{6}$

D) $\sqrt{6}$

53) Simplify the below :

$$-4\sqrt{48} + 4\sqrt{3}$$

A) $-8\sqrt{3}$

B) $-12\sqrt{3}$

C) $-4\sqrt{3}$

D) 0

54) Simplify the below :

$$4\sqrt{6} - 4\sqrt{6}$$

A) $8\sqrt{6}$

B) 0

C) $4\sqrt{6}$

D) $-4\sqrt{6}$

55) Simplify the below :

$$-3\sqrt{8} - \sqrt{8}$$

A) $-26\sqrt{2}$

B) $-14\sqrt{2}$

C) $-20\sqrt{2}$

D) $-8\sqrt{2}$

56) Simplify the below :

$$3\sqrt{63} + 3\sqrt{7}$$

A) $33\sqrt{7}$

B) $12\sqrt{7}$

C) $24\sqrt{7}$

D) $15\sqrt{7}$

57) Simplify the below :

$$-4\sqrt{6} - 3\sqrt{96}$$

A) $-44\sqrt{6}$ B) $-32\sqrt{6}$

C) $-20\sqrt{6}$ D) $-16\sqrt{6}$

58) Simplify the below :

$$-4\sqrt{2} - 3\sqrt{32}$$

A) $-20\sqrt{2}$ B) $-16\sqrt{2}$

C) $-32\sqrt{2}$ D) $-36\sqrt{2}$

59) Simplify the below :

$$-4\sqrt{2} - 2\sqrt{18}$$

A) $-16\sqrt{2}$ B) $-10\sqrt{2}$

C) $-22\sqrt{2}$ D) $-28\sqrt{2}$

60) Simplify the below :

$$-3\sqrt{80} + 3\sqrt{5}$$

A) $-33\sqrt{5}$ B) $-45\sqrt{5}$

C) $-9\sqrt{5}$ D) $-21\sqrt{5}$

61) Simplify the below :

$$-4\sqrt{128} - 2\sqrt{8}$$

A) $-44\sqrt{2}$ B) $-36\sqrt{2}$

C) $-76\sqrt{2}$ D) $-40\sqrt{2}$

62) Simplify the below :

$$-4\sqrt{8} + 2\sqrt{8}$$

A) $-4\sqrt{2}$ B) $-8\sqrt{2}$

C) $-12\sqrt{2}$ D) $-16\sqrt{2}$

63) Simplify the below :

$$3\sqrt{128} - 2\sqrt{72}$$

A) $36\sqrt{2}$ B) $60\sqrt{2}$

C) $48\sqrt{2}$ D) $12\sqrt{2}$

64) Simplify the below :

$$-4\sqrt{5} + 3\sqrt{5}$$

A) $-\sqrt{5}$ B) $\sqrt{5}$

C) $-2\sqrt{5}$ D) $-5\sqrt{5}$

65) Simplify the below :

$$2\sqrt{48} - 2\sqrt{48}$$

A) 0   B) $8\sqrt{3}$

C) $16\sqrt{3}$ D) $-8\sqrt{3}$

66) Simplify the below :

$$-4\sqrt{5} - \sqrt{5}$$

A) $-9\sqrt{5}$ B) $-5\sqrt{5}$

C) $-13\sqrt{5}$ D) $-14\sqrt{5}$

67) Simplify the below :

$$-4\sqrt{7} - 3\sqrt{28}$$

A) $-18\sqrt{7}$ B) $-14\sqrt{7}$

C) $-22\sqrt{7}$ D) $-10\sqrt{7}$

68) Simplify the below :

$$-4\sqrt{72} + 3\sqrt{8}$$

A) $-60\sqrt{2}$ B) $-36\sqrt{2}$

C) $-42\sqrt{2}$ D) $-18\sqrt{2}$

69) Simplify the below :

$$2\sqrt{24} \cdot -3\sqrt{18}$$

A) $\sqrt{42}$ B) $-72\sqrt{3}$

C) 432   D) $12\sqrt{3}$

70) Simplify the below :

$$5\sqrt{35} \cdot \sqrt{35}$$

A) 35   B) 1225

C) 175   D) $\sqrt{70}$

71) Simplify the below :

$$\sqrt{10} \cdot 7\sqrt{2}$$

A) $2\sqrt{3}$ B) $14\sqrt{5}$

C) $2\sqrt{5}$ D) 20

72) Simplify the below :

$$3\sqrt{6} \cdot 6\sqrt{12}$$

A) $6\sqrt{2}$ B) $3\sqrt{2}$

C) $108\sqrt{2}$ D) 72

73) Simplify the below :

$$\sqrt{14} \cdot -5\sqrt{12}$$

A) $2\sqrt{42}$ B) $-10\sqrt{42}$

C) 168   D) $\sqrt{26}$

74) Simplify the below :

$$-7\sqrt{12} \cdot \sqrt{15}$$

A) $-42\sqrt{5}$ B) $6\sqrt{5}$

C) $3\sqrt{3}$ D) 180

75) Simplify the below :

$$\sqrt{12} \cdot 4\sqrt{8}$$

A) $16\sqrt{6}$ B) $4\sqrt{6}$

C) 96   D) $2\sqrt{5}$

76) Simplify the below :

$$\sqrt{7} \cdot 2\sqrt{6}$$

A) $2\sqrt{42}$ B) $\sqrt{13}$

C) 42   D) $\sqrt{42}$

77) Simplify the below :

$$2\sqrt{42} \cdot \sqrt{28}$$

A) $\sqrt{70}$ B) $28\sqrt{6}$

C) $14\sqrt{6}$ D) 1176

78) Simplify the below :

$$\sqrt{10} \cdot \sqrt{35}$$

A) 350   B) $3\sqrt{5}$

C) $\sqrt{210}$ D) $5\sqrt{14}$

79) Simplify the below :

$$\sqrt{18} \cdot \sqrt{3}$$

A) $\sqrt{210}$ B) $\sqrt{21}$

C) $3\sqrt{6}$ D) 54

80) Simplify the below :

$$\sqrt{18} \cdot -7\sqrt{6}$$

A) $2\sqrt{6}$ B) $6\sqrt{3}$

C) $-42\sqrt{3}$ D) 108

81) Simplify the below :

$$3\sqrt{28} \cdot \sqrt{28}$$

A) 28　　B) $2\sqrt{14}$

C) 84　　D) 784

82) Simplify the below :

$$\sqrt{6} \cdot \sqrt{42}$$

A) $\sqrt{42}$　　B) 252

C) $4\sqrt{3}$　　D) $6\sqrt{7}$

83) Simplify the below :

$$-4\sqrt{12} \cdot \sqrt{6}$$

A) $3\sqrt{2}$　　B) 72

C) $6\sqrt{2}$　　D) $-24\sqrt{2}$

84) Simplify the below :

$$-7\sqrt{28} \cdot 3\sqrt{14}$$

A) 392　　B) $\sqrt{42}$

C) $-294\sqrt{2}$　　D) $14\sqrt{2}$

85) Simplify the below :

$$\sqrt{7} \cdot \sqrt{5}$$

A) $5\sqrt{210}$　　B) $\sqrt{35}$

C) 35　　D) $2\sqrt{3}$

86) Simplify the below :

$$\sqrt{2} \cdot -5\sqrt{2}$$

A) $-5\sqrt{70}$　　B) 4

C) 2　　D) -10

87) Simplify the below :

$$\sqrt{7}\left(10\sqrt{15} + 2\right)$$

A) $6\sqrt{6} + 56\sqrt{2}$

B) $10\sqrt{105} + 2\sqrt{7}$

C) $-63\sqrt{5} - 30\sqrt{2}$

D) $45\sqrt{6} + 7$

88) Simplify the below :

$$3\sqrt{2}\left(9 + \sqrt{10}\right)$$

A) $27\sqrt{2} + 6\sqrt{5}$　　B) $2\sqrt{5} + 9$

C) $-6\sqrt{210} + 4$　　D) 15

89) Simplify the below :

$$\sqrt{10}\left(\sqrt{10} + 10\sqrt{10}\right)$$

A) 110   B) $2\sqrt{2} + 5\sqrt{3}$

C) 124   D) $3\sqrt{10} + 6$

90) Simplify the below :

$$-3\sqrt{3}\left(-6\sqrt{6} + 6\sqrt{15}\right)$$

A) $60 + \sqrt{30}$

B) $54\sqrt{2} - 54\sqrt{5}$

C) $5\sqrt{2} + 8$

D) 12

91) Simplify the below :

$$\sqrt{14}\left(\sqrt{6} + \sqrt{10}\right)$$

A) $2\sqrt{21} + 2\sqrt{35}$

B) $-7\sqrt{210} + 5$

C) $9\sqrt{3} + 5$

D) $\sqrt{42} + 3$

92) Simplify the below :

$$2\sqrt{42}\left(\sqrt{3} - 9\sqrt{21}\right)$$

A) $2\sqrt{210} - 24\sqrt{2}$

B) $-90\sqrt{3} + 5$

C) $6\sqrt{10} + 9$

D) $6\sqrt{14} - 378\sqrt{2}$

93) Simplify the below :

$$\sqrt{6}\left(5\sqrt{20} + 6\right)$$

A) $6\sqrt{3} + 9\sqrt{30}$

B) -32

C) $35\sqrt{10} + 10$

D) $10\sqrt{30} + 6\sqrt{6}$

94) Simplify the below :

$$-\sqrt{10}\left(2 + \sqrt{10}\right)$$

A) $-2\sqrt{10} - 10$

B) $10\sqrt{3}$

C) $10\sqrt{2} + 7$

D) $28\sqrt{6} + \sqrt{70}$

95) Simplify the below :

$$3\sqrt{70}\left(\sqrt{30} + 6\right)$$

A) $2\sqrt{2} + 10$

B) $10\sqrt{10} + 5$

C) $\sqrt{70} + 7\sqrt{2}$

D) $30\sqrt{21} + 18\sqrt{70}$

96) Simplify the below :

$$\sqrt{6}\left(\sqrt{30} + 10\sqrt{18}\right)$$

A) $\sqrt{30} + 4$

B) $6\sqrt{5} + 60\sqrt{3}$

C) $6\sqrt{3} + 4\sqrt{105}$

D) $7\sqrt{2} + 27$

97)   Simplify the below :

$$\sqrt{6}\left(\sqrt{15}+6\right)$$

A) $-3\sqrt{10}+5\sqrt{3}$

B) $3\sqrt{10}+6\sqrt{6}$

C) $189-28\sqrt{2}$

D) $-5\sqrt{5}+9$

98)   Simplify the below :

$$4\sqrt{70}\left(\sqrt{14}+\sqrt{30}\right)$$

A) $4\sqrt{2}+6$

B) $56\sqrt{5}+40\sqrt{21}$

C) $14\sqrt{2}+7$

D) $-12\sqrt{7}+4$

99)   Simplify the below :

$$8\sqrt{42}\left(10\sqrt{12}+7\sqrt{5}\right)$$

A) $480\sqrt{14}+56\sqrt{210}$

B) $5\sqrt{3}+10$

C) $9\sqrt{2}+5$

D) $-5\sqrt{10}+8$

100) Simplify the below :

$$-9\sqrt{30}\left(\sqrt{6}+6\right)$$

A) $-54\sqrt{5}-54\sqrt{30}$

B) $-144+4\sqrt{5}$

C) $36\sqrt{7}+18$

D) $30-16\sqrt{7}$

101) Simplify the below :

$$-8\sqrt{3}\left(9-7\sqrt{7}\right)$$

A) $-12\sqrt{7}+7$

B) $-45\sqrt{5}+8$

C) $-20\sqrt{2}+7$

D) $-72\sqrt{3}+56\sqrt{21}$

102) Simplify the below :

$$\sqrt{6}\left(4-10\sqrt{5}\right)$$

A) $8\sqrt{6}+2$

B) $4\sqrt{6}-10\sqrt{30}$

C) $28+6\sqrt{2}$

D) $2\sqrt{7}+\sqrt{42}$

103) Simplify the below :

$$\sqrt{10x}\cdot-2\sqrt{5x^3}$$

A) $5\sqrt{2}$ B) 50

C) $-10x^2\sqrt{2}$ D) $\sqrt{15}$

104) Simplify the below :

$$4\sqrt{20x^2}\cdot4\sqrt{15x^2}$$

A) $10\sqrt{3}$ B) $\sqrt{35}$

C) 300 D) $160x^2\sqrt{3}$

105) Simplify the below :

$$\sqrt{3n} \cdot \sqrt{15n^2}$$

A) $3\sqrt{5}$ B) $3\sqrt{2}$

C) $3n\sqrt{5n}$ D) 45

106) Simplify the below :

$$\sqrt{3n^2} \cdot -3\sqrt{15n^2}$$

A) 45   B) $3\sqrt{2}$

C) $3\sqrt{5}$ D) $-9n^2\sqrt{5}$

107) Simplify the below :

$$3\sqrt{6r} \cdot \sqrt{12r^2}$$

A) $18r\sqrt{2r}$ B) $3\sqrt{2}$

C) $6\sqrt{2}$ D) 72

108) Simplify the below :

$$\sqrt{5b^2} \cdot 2\sqrt{20b^2}$$

A) 5   B) $20b^2$

C) 10   D) 100

109) Simplify the below :

$$-5\sqrt{20x^2} \cdot -2\sqrt{10x}$$

A) $10\sqrt{2}$ B) $100x\sqrt{2x}$

C) $\sqrt{30}$ D) 200

110) Simplify the below :

$$\sqrt{12a} \cdot \sqrt{15a^2}$$

A) $6a\sqrt{5a}$ B) 180

C) $6\sqrt{5}$ D) $3\sqrt{3}$

111) Simplify the below :

$$\sqrt{2k} \cdot \sqrt{2k}$$

A) 2   B) $2k$

C) $\sqrt{30}$ D) 4

112) Simplify the below :

$$\sqrt{15r^3} \cdot \sqrt{12r^3}$$

A) 180   B) $6r^3\sqrt{5}$

C) $3\sqrt{3}$ D) $6\sqrt{5}$

113) Simplify the below :

$$\sqrt{8n^3} \cdot -3\sqrt{20n^2}$$

A) 160

B) $2\sqrt{7}$

C) $-12n^2\sqrt{10n}$

D) $4\sqrt{10}$

114) Simplify the below :

$$4\sqrt{3m^3} \cdot \sqrt{3m^2}$$

A) $\sqrt{6}$

B) 3

C) 9

D) $12m^2\sqrt{m}$

115) Simplify the below :

$$\sqrt{12p} \cdot \sqrt{20p}$$

A) $4\sqrt{2}$

B) $4p\sqrt{15}$

C) $4\sqrt{15}$

D) 240

116) Simplify the below :

$$\sqrt{10x} \cdot -\sqrt{10x^2}$$

A) $2\sqrt{5}$

B) 10

C) 100

D) $-10x\sqrt{x}$

117) Simplify the below :

$$5\sqrt{6p^3} \cdot -2\sqrt{6p^3}$$

A) $2\sqrt{3}$

B) $-60p^3$

C) 6

D) 36

118) Simplify the below :

$$\sqrt{20k^2} \cdot \sqrt{5k^2}$$

A) 100

B) $10k^2$

C) 10

D) 5

119) Simplify the below :

$$\sqrt{30r}\left(-4\sqrt{18r} + \sqrt{5}\right)$$

A) $-24r\sqrt{15} + 5\sqrt{6r}$

B) $-20r\sqrt{6r} + 5\sqrt{5r}$

C) $7r\sqrt{3} + 48r\sqrt{r}$

D) $4\sqrt{42r} + \sqrt{105}$

120) Simplify the below :

$$\sqrt{35}\left(\sqrt{21} - 7\sqrt{18}\right)$$

A) $7\sqrt{2n} + 6n\sqrt{5n}$

B) $-3n\sqrt{7n} + 6$

C) $4n\sqrt{3} + 3\sqrt{42}$

D) $7\sqrt{15} - 21\sqrt{70}$

121) Simplify the below :

$$-2\sqrt{3}\left(2 + 5\sqrt{3x}\right)$$

A) $3\sqrt{5x} + 7x$

B) $3x\sqrt{3} - 12x\sqrt{5x}$

C) $-28x\sqrt{3} - 20\sqrt{6x}$

D) $-4\sqrt{3} - 30\sqrt{x}$

122) Simplify the below :

$$-5\sqrt{15r}\left(-5\sqrt{6r} + \sqrt{15}\right)$$

A) $14\sqrt{r} + 4r$

B) $6r\sqrt{7r} + 4$

C) $75r\sqrt{10} - 75\sqrt{r}$

D) $5\sqrt{2r} + 15\sqrt{3r}$

123) Simplify the below :

$$\sqrt{2a}\left(\sqrt{6} - 5\sqrt{2a}\right)$$

A) $2\sqrt{3a} - 10a$

B) $7a\sqrt{7a} + \sqrt{42a}$

C) $-15a\sqrt{6a} + 2a$

D) $4a\sqrt{7} + 4a$

124) Simplify the below :

$$-\sqrt{6b}\left(6b - 5\sqrt{6}\right)$$

A) $5\sqrt{5b} + 2$

B) $3b\sqrt{3} + 3b$

C) $7b\sqrt{2} - 49b\sqrt{5b}$

D) $-6b\sqrt{6b} + 30\sqrt{b}$

125) Simplify the below :

$$\sqrt{7}\left(-2\sqrt{3} + \sqrt{5a}\right)$$

A) $7a\sqrt{2} + 3$

B) $3a\sqrt{3} + 3$

C) $6a\sqrt{7} + 5$

D) $-2\sqrt{21} + \sqrt{35a}$

126) Simplify the below :

$$-3\sqrt{2x}\left(3\sqrt{2x} + 4\right)$$

A) $-28x\sqrt{2} - 42x\sqrt{3}$

B) $-18x - 12\sqrt{2x}$

C) $4x\sqrt{3x} + 3x\sqrt{3}$

D) $2x\sqrt{7} + 2$

127) Simplify the below :

$$\sqrt{3x}\left(\sqrt{2x} + 7\right)$$

A) $15\sqrt{3x} + 4\sqrt{5x}$

B) $x\sqrt{6} + 7\sqrt{3x}$

C) $-7\sqrt{2x} + 4x$

D) $2\sqrt{3x} + 3$

128) Simplify the below :

$$\sqrt{5m}\left(3m + \sqrt{5}\right)$$

A) $6m\sqrt{m} + 30m\sqrt{7}$

B) $-20m\sqrt{7} + 24m\sqrt{5m}$

C) $\sqrt{42} + \sqrt{30m}$

D) $3m\sqrt{5m} + 5\sqrt{m}$

129) Simplify the below :

$$\sqrt{5x}\left(4+\sqrt{18}\right)$$

A) $2x\sqrt{3x}+7$

B) $4\sqrt{5x}+3\sqrt{10x}$

C) $84x\sqrt{x}+6x$

D) $-4\sqrt{2x}+\sqrt{42}$

130) Simplify the below :

$$\sqrt{2}\left(2\sqrt{6p}-3\sqrt{15p}\right)$$

A) $12p\sqrt{p}+2$

B) $4\sqrt{3p}-3\sqrt{30p}$

C) $-28\sqrt{5p}+6p\sqrt{3}$

D) $\sqrt{70}+4p$

131) Simplify the below :

$$\sqrt{30}\left(5x+\sqrt{12}\right)$$

A) $35\sqrt{6x}+3\sqrt{3x}$

B) $3x\sqrt{7x}-36x\sqrt{3}$

C) $14x+5\sqrt{2x}$

D) $5x\sqrt{30}+6\sqrt{10}$

132) Simplify the below :

$$\sqrt{35m}\left(\sqrt{7m}+\sqrt{6m}\right)$$

A) $42m\sqrt{5m}+12m$

B) $7m\sqrt{5}+m\sqrt{210}$

C) $20\sqrt{7m}+6m$

D) $12m+5m\sqrt{5}$

133) Simplify the below :

$$\sqrt{3}\left(4+\sqrt{12n}\right)$$

A) $4\sqrt{3}+6\sqrt{n}$

B) $6n\sqrt{2n}-21n\sqrt{7n}$

C) $25n\sqrt{2n}+3$

D) $2\sqrt{5n}+6$

134) Simplify the below :

$$\sqrt{35}\left(3+\sqrt{14x}\right)$$

A) $6x\sqrt{2}+2$

B) $4\sqrt{7x}+\sqrt{42x}$

C) $30x\sqrt{6}+4\sqrt{2x}$

D) $3\sqrt{35}+7\sqrt{10x}$

135) Simplify the below :

$$-3\sqrt{7n}\left(\sqrt{3}-2\sqrt{2}\right)$$

A) $7n\sqrt{2n}+6$

B) $\sqrt{70n}+3$

C) $-3\sqrt{21n}+6\sqrt{14n}$

D) $4\sqrt{5n}+4n\sqrt{7n}$

136) Simplify the below :

$$\sqrt{35n}\left(\sqrt{10n}+7\sqrt{14n}\right)$$

A) $14n+6n\sqrt{3n}$

B) $6\sqrt{3n}+3n\sqrt{6n}$

C) $2\sqrt{7n}+12n\sqrt{3}$

D) $5n\sqrt{14}+49n\sqrt{10}$

137) Simplify the below :

$$\frac{\sqrt{8}}{\sqrt{20}}$$

A) $\frac{3\sqrt{3}}{2}$ B) $\frac{\sqrt{10}}{5}$

C) $-\frac{\sqrt{5}}{5}$ D) $\frac{\sqrt{10}}{2}$

138) Simplify the below :

$$\frac{4\sqrt{3}}{\sqrt{5}}$$

A) $\frac{2\sqrt{15}}{3}$ B) $\frac{4\sqrt{15}}{5}$

C) $\frac{\sqrt{5}}{5}$ D) $\frac{\sqrt{15}}{12}$

139) Simplify the below :

$$\frac{3\sqrt{8}}{4\sqrt{20}}$$

A) $\sqrt{5}$ B) $\frac{3\sqrt{10}}{20}$

C) $\sqrt{2}$ D) $\frac{2\sqrt{10}}{3}$

140) Simplify the below :

$$\frac{2\sqrt{3}}{\sqrt{5}}$$

A) $\frac{2\sqrt{15}}{5}$ B) $\frac{3\sqrt{2}}{2}$

C) $\frac{\sqrt{15}}{5}$ D) $\frac{\sqrt{10}}{5}$

141) Simplify the below :

$$\frac{4\sqrt{20}}{5\sqrt{15}}$$

A) $\frac{\sqrt{15}}{5}$ B) $\sqrt{6}$

C) $\frac{2\sqrt{10}}{5}$ D) $\frac{8\sqrt{3}}{15}$

142) Simplify the below :

$$\frac{\sqrt{5}}{5\sqrt{3}}$$

A) $\frac{\sqrt{3}}{2}$ B) $\frac{\sqrt{15}}{15}$

C) $\frac{\sqrt{5}}{2}$ D) $\sqrt{15}$

143) Simplify the below :

$$-\frac{5}{\sqrt{3}}$$

A) $-\frac{\sqrt{3}}{5}$ B) $-\frac{5\sqrt{3}}{3}$

C) $\frac{\sqrt{10}}{20}$ D) $\frac{\sqrt{3}}{3}$

144) Simplify the below :

$$\frac{5\sqrt{4}}{\sqrt{3}}$$

A) $\frac{\sqrt{6}}{3}$ B) $\frac{2\sqrt{5}}{5}$

C) $\frac{4\sqrt{3}}{3}$ D) $\frac{10\sqrt{3}}{3}$

145) Simplify the below :

$$\frac{3\sqrt{2}}{5\sqrt{5}}$$

A) $\dfrac{5\sqrt{10}}{6}$ B) $\dfrac{3\sqrt{10}}{25}$

C) $\dfrac{\sqrt{10}}{8}$ D) $-2\sqrt{2}$

146) Simplify the below :

$$\frac{5\sqrt{2}}{\sqrt{5}}$$

A) $\dfrac{2\sqrt{3}}{3}$ B) $\sqrt{10}$

C) $\dfrac{\sqrt{6}}{6}$ D) $\dfrac{4\sqrt{6}}{3}$

147) Simplify the below :

$$\frac{4\sqrt{4}}{\sqrt{3}}$$

A) $\dfrac{\sqrt{5}}{2}$ B) $\dfrac{3\sqrt{10}}{10}$

C) $\dfrac{5\sqrt{2}}{2}$ D) $\dfrac{8\sqrt{3}}{3}$

148) Simplify the below :

$$\frac{3\sqrt{12}}{\sqrt{15}}$$

A) $\dfrac{\sqrt{10}}{5}$ B) $\dfrac{\sqrt{15}}{3}$

C) $\dfrac{6\sqrt{5}}{5}$ D) $\dfrac{4\sqrt{15}}{15}$

149) Simplify the below :

$$\frac{\sqrt{3}}{\sqrt{5}}$$

A) $\dfrac{\sqrt{10}}{6}$ B) $\dfrac{\sqrt{15}}{3}$

C) $\dfrac{\sqrt{15}}{5}$ D) $\dfrac{\sqrt{3}}{3}$

150) Simplify the below :

$$\frac{2\sqrt{5}}{3\sqrt{2}}$$

A) $-\dfrac{\sqrt{2}}{2}$ B) $\dfrac{5\sqrt{6}}{3}$

C) $\dfrac{\sqrt{10}}{3}$ D) $\dfrac{3\sqrt{10}}{10}$

151) Simplify the below :

$$\frac{3}{5\sqrt{3}}$$

A) $\dfrac{5\sqrt{3}}{3}$ B) $\dfrac{\sqrt{10}}{10}$

C) $\dfrac{\sqrt{3}}{5}$ D) $\dfrac{2\sqrt{15}}{5}$

152) Simplify the below :

$$-\frac{1}{\sqrt{3}}$$

A) $\sqrt{5}$ B) $\dfrac{\sqrt{5}}{2}$

C) $-\dfrac{\sqrt{3}}{3}$ D) $\dfrac{\sqrt{15}}{3}$

www.math-knots.com | www.a4ace.com

 **Algebra 1**

153) Simplify the below :

$$\frac{5\sqrt{5}}{3\sqrt{3}}$$

A) $\sqrt{3}$ B) $\frac{2\sqrt{5}}{5}$

C) $\frac{5\sqrt{15}}{9}$ D) $\sqrt{15}$

154) Simplify the below :

$$\frac{\sqrt{12}}{2\sqrt{15}}$$

A) $\frac{\sqrt{5}}{5}$ B) $\sqrt{5}$

C) $\frac{\sqrt{2}}{4}$ D) $\frac{3\sqrt{10}}{5}$

155) Simplify the below :

$$\frac{\sqrt{2x^4}}{4\sqrt{10x^2}}$$

A) $\frac{x\sqrt{5}}{20}$ B) $\frac{16\sqrt{3}}{21}$

C) $\frac{4\sqrt{5}}{|x|}$ D) $\frac{\sqrt{15}}{15}$

156) Simplify the below :

$$\frac{4\sqrt{4a^4b^4}}{\sqrt{3a^2b}}$$

A) $\frac{8ab\sqrt{3b}}{3}$ B) $\frac{\sqrt{30}}{6}$

C) $\frac{7\sqrt{14}}{12}$ D) $\frac{\sqrt{3b}}{8b^2|a|}$

157) Simplify the below :

$$\frac{4\sqrt{8m^2}}{\sqrt{6m^2}}$$

A) $\frac{\sqrt{5}}{10}$ B) $\frac{8\sqrt{3}}{3}$

C) $\frac{\sqrt{3}}{8}$ D) $\frac{\sqrt{6}}{3}$

158) Simplify the below :

$$\frac{\sqrt{32a}}{7\sqrt{20a}}$$

A) $\frac{2\sqrt{14}}{49}$ B) $\frac{2\sqrt{10}}{35}$

C) $\frac{4\sqrt{3}}{9}$ D) $\frac{\sqrt{30}}{5}$

159) Simplify the below :

$$\frac{\sqrt{7xy^2}}{\sqrt{6x^2y}}$$

A) $\frac{\sqrt{42}}{7}$ B) $\frac{\sqrt{42xy}}{6x}$

C) $\frac{\sqrt{14}}{2}$ D) $\frac{\sqrt{30}}{48}$

160) Simplify the below :

$$\frac{\sqrt{3xy^4}}{\sqrt{6x^2y^3}}$$

A) $\frac{\sqrt{5}}{14}$ B) $\frac{2\sqrt{5}}{25}$

C) $\frac{|y|\sqrt{2yx}}{y^2}$ D) $\frac{\sqrt{2xy}}{2x}$

161) Simplify the below :

$$\frac{4\sqrt{7m^4n^3}}{\sqrt{6m^4n^2}}$$

A) $\dfrac{\sqrt{3}}{21}$　　B) $\dfrac{2\sqrt{42n}}{3}$

C) $\dfrac{\sqrt{42n}}{28|n|}$　　D) $\dfrac{8\sqrt{5}}{5}$

162) Simplify the below :

$$\frac{8\sqrt{3n^2}}{\sqrt{18n^4}}$$

A) $\dfrac{4\sqrt{6}}{3n}$　　B) $\dfrac{|n|\sqrt{6}}{8}$

C) $\dfrac{2\sqrt{5}}{5}$　　D) $\dfrac{\sqrt{30}}{10}$

163) Simplify the below :

$$\frac{\sqrt{8r^3}}{6\sqrt{28r^3}}$$

A) $\dfrac{\sqrt{14}}{42}$　　B) $\dfrac{7\sqrt{3}}{6}$

C) $\dfrac{\sqrt{3}}{12}$　　D) $\dfrac{\sqrt{14}}{7}$

164) Simplify the below :

$$\frac{\sqrt{4k^4}}{\sqrt{7k}}$$

A) $\dfrac{3\sqrt{7}}{56}$　　B) $\dfrac{8\sqrt{15}}{3}$

C) $\dfrac{\sqrt{7k}}{2k^2}$　　D) $\dfrac{2k\sqrt{7k}}{7}$

165) Simplify the below :

$$\frac{6\sqrt{4r}}{2\sqrt{7r}}$$

A) $-\dfrac{7\sqrt{3}}{8}$　　B) $\dfrac{4\sqrt{6}}{5}$

C) $\dfrac{2\sqrt{6}}{7}$　　D) $\dfrac{6\sqrt{7}}{7}$

166) Simplify the below :

$$\frac{6\sqrt{6x^4}}{\sqrt{5x^3}}$$

A) $\sqrt{3}$　　B) $\dfrac{6\sqrt{30x}}{5}$

C) $-\dfrac{\sqrt{3}}{2}$　　D) $\dfrac{\sqrt{30}}{6}$

167) Simplify the below :

$$\frac{\sqrt{5x^4y^4}}{5\sqrt{2x^3y^4}}$$

A) $\dfrac{\sqrt{10x}}{10}$　　B) $\dfrac{\sqrt{42}}{30}$

C) $\dfrac{\sqrt{6}}{4}$　　D) $\dfrac{\sqrt{10x}}{|x|}$

168) Simplify the below :

$$\frac{\sqrt{9u^2v}}{3\sqrt{24u^3v^2}}$$

A) $\dfrac{4\sqrt{6}}{3}$　　B) $2\sqrt{6uv}$

C) $\dfrac{2\sqrt{10}}{5}$　　D) $\dfrac{\sqrt{6uv}}{12uv}$

169) Simplify the below :

$$\frac{6\sqrt{2x^3}}{3\sqrt{5x^3}}$$

A) $-\dfrac{2\sqrt{7}}{21}$

B) $\dfrac{\sqrt{10}}{4}$

C) $\dfrac{2\sqrt{10}}{5}$

D) $\dfrac{\sqrt{3}}{3}$

170) Simplify the below :

$$\frac{\sqrt{8u^2v^3}}{\sqrt{5u^4v^3}}$$

A) $\dfrac{u^2\sqrt{10}}{4|u|}$

B) $\dfrac{\sqrt{14}}{2}$

C) $\dfrac{2\sqrt{21}}{7}$

D) $\dfrac{2\sqrt{10}}{5u}$

171) Simplify the below :

$$\frac{\sqrt{8n^3}}{3\sqrt{6n^4}}$$

A) $\dfrac{7\sqrt{15}}{3}$

B) $\dfrac{2\sqrt{3n}}{9n}$

C) $\dfrac{\sqrt{7}}{14}$

D) $\dfrac{\sqrt{14}}{56}$

172) Simplify the below :

$$\frac{\sqrt{16k}}{8\sqrt{6k^2}}$$

A) $2\sqrt{6k}$

B) $\dfrac{\sqrt{6k}}{12k}$

C) $\dfrac{\sqrt{6}}{20}$

D) $\dfrac{\sqrt{6}}{21}$

173) Simplify the below :

$$\frac{5\sqrt{5}+\sqrt{7}}{7\sqrt{14}}$$

A) $\dfrac{5\sqrt{14}+3\sqrt{10}}{4}$

B) $\dfrac{5\sqrt{70}+7\sqrt{2}}{98}$

C) $\dfrac{105-7\sqrt{10}}{43}$

D) $\dfrac{-4\sqrt{31}-3\sqrt{62}}{217}$

174) Simplify the below :

$$\frac{3-6\sqrt{5}}{4\sqrt{3}}$$

A) $\dfrac{\sqrt{3}-2\sqrt{15}}{4}$

B) $\dfrac{24-4\sqrt{6}}{9}$

C) $\dfrac{4\sqrt{3}-9\sqrt{2}}{9}$

D) $\dfrac{\sqrt{102}-8\sqrt{119}}{17}$

175) Simplify the below :

$$\frac{2\sqrt{7} - 4\sqrt{2}}{3\sqrt{27}}$$

A) $\dfrac{2\sqrt{21} - 4\sqrt{6}}{27}$

B) $\dfrac{-\sqrt{115} + 7\sqrt{23}}{44}$

C) $\dfrac{3\sqrt{6} - 4\sqrt{42}}{6}$

D) $\dfrac{-15\sqrt{2} - 48\sqrt{3}}{359}$

176) Simplify the below :

$$\frac{3 + 6\sqrt{3}}{8\sqrt{21}}$$

A) $\dfrac{8\sqrt{2} - \sqrt{14}}{16}$

B) $\dfrac{-8\sqrt{21} + 48\sqrt{7}}{33}$

C) $\dfrac{-6\sqrt{3} + \sqrt{6}}{9}$

D) $\dfrac{\sqrt{21} + 6\sqrt{7}}{56}$

177) Simplify the below :

$$\frac{-4 - \sqrt{6}}{3\sqrt{13}}$$

A) $\dfrac{2\sqrt{15} + 2\sqrt{35}}{15}$

B) $\dfrac{-4\sqrt{13} - \sqrt{78}}{39}$

C) $\dfrac{-4\sqrt{3} - \sqrt{21}}{48}$

D) $\dfrac{7\sqrt{6} - 6\sqrt{42}}{36}$

178) Simplify the below :

$$\frac{3 + 7\sqrt{7}}{\sqrt{2}}$$

A) $\dfrac{-3\sqrt{2} + 7\sqrt{14}}{334}$

B) $\dfrac{3\sqrt{2} + 7\sqrt{14}}{2}$

C) $\dfrac{3\sqrt{7} - 14}{112}$

D) $\dfrac{\sqrt{10} + 3\sqrt{5}}{25}$

179) Simplify the below :

$$\frac{-1 - 3\sqrt{2}}{\sqrt{26}}$$

A) $\dfrac{-\sqrt{26} - 6\sqrt{13}}{26}$

B) $\dfrac{\sqrt{26} - 6\sqrt{13}}{17}$

C) $\dfrac{6\sqrt{7} + \sqrt{14}}{28}$

D) $\dfrac{-5\sqrt{7} + 2\sqrt{14}}{7}$

180) Simplify the below :

$$\frac{4\sqrt{5} + 6}{4\sqrt{18}}$$

A) $\dfrac{12\sqrt{10} - 18\sqrt{2}}{11}$

B) $\dfrac{2\sqrt{10} + 3\sqrt{2}}{12}$

C) $\dfrac{18\sqrt{30} + 3\sqrt{210}}{29}$

D) $\dfrac{4\sqrt{22} - 3\sqrt{33}}{11}$

181) Simplify the below :

$$\frac{5 - 5\sqrt{7}}{\sqrt{10}}$$

A) $\dfrac{-\sqrt{15} - 3\sqrt{5}}{60}$

B) $\dfrac{-\sqrt{10} - \sqrt{70}}{30}$

C) $\dfrac{\sqrt{10} - \sqrt{70}}{2}$

D) $\dfrac{6\sqrt{5} - \sqrt{30}}{40}$

182) Simplify the below :

$$\frac{8 + \sqrt{5}}{8\sqrt{15}}$$

A) $\dfrac{8\sqrt{15} + 5\sqrt{3}}{120}$

B) $\dfrac{64\sqrt{15} - 40\sqrt{3}}{59}$

C) $\dfrac{-4\sqrt{2} - 3\sqrt{14}}{10}$

D) $\dfrac{7\sqrt{3} - \sqrt{21}}{42}$

183)   Simplify the below :

$$\frac{8 - \sqrt{2}}{3\sqrt{24}}$$

A) $\dfrac{8\sqrt{2} - 5\sqrt{14}}{24}$

B) $\dfrac{\sqrt{22} - 2\sqrt{33}}{11}$

C) $\dfrac{4\sqrt{6} - \sqrt{3}}{18}$

D) $\dfrac{-3\sqrt{22} - \sqrt{66}}{6}$

184)   Simplify the below :

$$\frac{4 - \sqrt{2}}{\sqrt{29}}$$

A) $\dfrac{4\sqrt{2} - 3\sqrt{10}}{8}$

B) $\dfrac{4\sqrt{29} + \sqrt{58}}{14}$

C) $\dfrac{6\sqrt{2} + \sqrt{10}}{36}$

D) $\dfrac{4\sqrt{29} - \sqrt{58}}{29}$

185)   Simplify the below :

$$\frac{-1 + \sqrt{5}}{\sqrt{32}}$$

A) $\dfrac{6\sqrt{3} - \sqrt{21}}{3}$

B) $\dfrac{-\sqrt{2} + \sqrt{10}}{8}$

C) $\dfrac{\sqrt{70} + 16\sqrt{7}}{14}$

D) $\sqrt{2} + \sqrt{10}$

186)   Simplify the below :

$$\frac{2 - 8\sqrt{3}}{8\sqrt{15}}$$

A) $\dfrac{2\sqrt{26} - \sqrt{130}}{156}$

B) $\dfrac{\sqrt{15} - 12\sqrt{5}}{60}$

C) $2 + 6\sqrt{3}$

D) $\dfrac{4\sqrt{114} - 5\sqrt{95}}{19}$

187) Simplify the below :

$$\frac{6 + 8\sqrt{3}}{5\sqrt{14}}$$

A) $\dfrac{3\sqrt{14} + 4\sqrt{42}}{35}$

B) $\dfrac{5\sqrt{3} - \sqrt{15}}{18}$

C) $\dfrac{4\sqrt{13} + \sqrt{26}}{13}$

D) $\dfrac{-15\sqrt{14} + 20\sqrt{42}}{78}$

188) Simplify the below :

$$\frac{\sqrt{7} + 5\sqrt{3}}{6\sqrt{32}}$$

A) $\dfrac{6\sqrt{22} - \sqrt{154}}{22}$

B) $\dfrac{\sqrt{14} + 5\sqrt{6}}{48}$

C) $\dfrac{4\sqrt{2} - 6\sqrt{6}}{23}$

D) $\dfrac{7\sqrt{13} + 4\sqrt{26}}{104}$

189) Simplify the below :

$$\frac{3\sqrt{3} + 6\sqrt{2}}{2\sqrt{21}}$$

A) $\dfrac{-4\sqrt{2} + 8}{3}$

B) $\dfrac{-9\sqrt{2} + 8\sqrt{3}}{5}$

C) $\dfrac{8\sqrt{42} + \sqrt{21}}{14}$

D) $\dfrac{3\sqrt{7} + 2\sqrt{42}}{14}$

190) Simplify the below :

$$\frac{\sqrt{5} + 2\sqrt{3}}{7\sqrt{20}}$$

A) $\dfrac{\sqrt{42} + \sqrt{21}}{49}$

B) $-12\sqrt{7} - 6\sqrt{42}$

C) $\dfrac{-5\sqrt{29} - 6\sqrt{58}}{47}$

D) $\dfrac{5 + 2\sqrt{15}}{70}$

www.math-knots.com | www.a4ace.com

191) Simplify the below :

$$\frac{-1 - \sqrt{m^2}}{\sqrt{23m^4}}$$

A) $\dfrac{30\sqrt{5} - 5\sqrt{10}}{34}$

B) $\dfrac{-5\sqrt{26} + \sqrt{182}}{18}$

C) $\dfrac{2\sqrt{14} - 3\sqrt{21}}{21}$

D) $\dfrac{-\sqrt{23} - m\sqrt{23}}{23m^2}$

192) Simplify the below :

$$\frac{5\sqrt{6v^3} + \sqrt{7v^4}}{\sqrt{5v^2}}$$

A) $\dfrac{7\sqrt{22} - 2\sqrt{11}}{176}$

B) $\dfrac{5\sqrt{30v} + v\sqrt{35}}{5}$

C) $\dfrac{2\sqrt{2} + 5\sqrt{10}}{8}$

D) $\dfrac{5\sqrt{30v} - |v|\sqrt{35}}{150|v| - 7v^2}$

193) Simplify the below :

$$\frac{2 + 5\sqrt{3x^4y^2}}{7\sqrt{6xy^2}}$$

A) $\dfrac{2\sqrt{6x} + 15x^2y\sqrt{2x}}{42yx}$

B) $\dfrac{-6\sqrt{2} + \sqrt{10}}{8}$

C) $\dfrac{35\sqrt{17} - 7\sqrt{102}}{19}$

D) $-\dfrac{24\sqrt{3}}{7}$

194) Simplify the below :

$$\frac{\sqrt{b} + \sqrt{2b^2}}{6\sqrt{23b^4}}$$

A) $\dfrac{4\sqrt{14} + 28\sqrt{2}}{7}$

B) $\dfrac{5\sqrt{6} - 3\sqrt{2}}{2}$

C) $\dfrac{42\sqrt{2} - 10\sqrt{3}}{269}$

D) $\dfrac{\sqrt{23b} + b\sqrt{46}}{138b^2}$

195) Simplify the below :

$$\frac{6 + 7\sqrt{6x^2}}{\sqrt{14x^2}}$$

A) $\dfrac{-3\sqrt{10} + 2\sqrt{5}}{40}$

B) $\dfrac{3\sqrt{14} + 7x\sqrt{21}}{7x}$

C) $\dfrac{7\sqrt{11}}{88}$

D) $\dfrac{8\sqrt{5} + \sqrt{35}}{35}$

196) Simplify the below :

$$\frac{4\sqrt{7n^3} + \sqrt{3n}}{\sqrt{2n^3}}$$

A) $\dfrac{4n\sqrt{14} + \sqrt{6}}{2n}$

B) $\dfrac{\sqrt{57} - 3\sqrt{19}}{152}$

C) $\dfrac{3\sqrt{23} + \sqrt{138}}{23}$

D) $\dfrac{2\sqrt{5} - 2\sqrt{15}}{5}$

197) Simplify the below :

$$\frac{3 - 2\sqrt{6b}}{\sqrt{8b^3}}$$

A) $\dfrac{2\sqrt{29} - 4\sqrt{87}}{29}$

B) $\dfrac{16\sqrt{6} + 2\sqrt{42}}{57}$

C) $\dfrac{4\sqrt{26} - 4\sqrt{182}}{13}$

D) $\dfrac{3\sqrt{2b} - 4b\sqrt{3}}{4b^2}$

198) Simplify the below :

$$\frac{5\sqrt{2n^2} - 6}{\sqrt{2n}}$$

A) $-\dfrac{\sqrt{14}}{3}$

B) $\dfrac{\sqrt{34} + 2\sqrt{17}}{34}$

C) $\dfrac{5n\sqrt{n} - 3\sqrt{2n}}{n}$

D) $\dfrac{5|n|\sqrt{n} + 3\sqrt{2n}}{25n^2 - 18}$

Algebra 1

199) Simplify the below :

$$\frac{8uv - \sqrt{5u^2v^2}}{3\sqrt{20u^4v^4}}$$

A) $\dfrac{8\sqrt{5} - 5}{30uv}$

B) $\dfrac{6\sqrt{2} - \sqrt{14}}{12}$

C) $\dfrac{-2\sqrt{11} + \sqrt{77}}{2}$

D) $\dfrac{48uv\sqrt{5} + 30|u| \cdot |v|}{59}$

200) Simplify the below :

$$\frac{2 + 3\sqrt{2x^2}}{8\sqrt{6x^3}}$$

A) $\dfrac{13\sqrt{29}}{203}$

B) $\dfrac{-\sqrt{15} - 4\sqrt{5}}{26}$

C) $\dfrac{6\sqrt{6} - \sqrt{30}}{6}$

D) $\dfrac{\sqrt{6x} + 3x\sqrt{3x}}{24x^2}$

201) Simplify the below :

$$\frac{2 + 2\sqrt{6p^3}}{5\sqrt{17p}}$$

A) $\dfrac{-3\sqrt{21} - 28\sqrt{3}}{42}$

B) $\dfrac{5\sqrt{21} - 3\sqrt{14}}{85p}$

C) $\dfrac{2\sqrt{17p} + 2p^2\sqrt{102}}{85p}$

D) $\dfrac{2\sqrt{22} + 7\sqrt{33}}{11}$

202) Simplify the below :

$$\frac{-7 - \sqrt{2b^4}}{\sqrt{2b^4}}$$

A) $\dfrac{-7b^2\sqrt{2} + 2b^4}{49 - 2b^4}$

B) $\dfrac{-7\sqrt{2} - 2b^2}{2b^2}$

C) $\dfrac{70\sqrt{10} + 80\sqrt{2}}{181}$

D) $\dfrac{\sqrt{3} + \sqrt{6}}{9}$

203) Simplify the below :

$$\frac{4n - 6\sqrt{7n^4}}{\sqrt{18n}}$$

A) $\dfrac{2\sqrt{2n} - 3n\sqrt{14n}}{3}$

B) $\dfrac{\sqrt{30}}{5}$

C) $\dfrac{2\sqrt{11} - \sqrt{110}}{110}$

D) $\dfrac{-4\sqrt{7} - \sqrt{14}}{98}$

204) Simplify the below :

$$\frac{5a^2 - \sqrt{3a^4b^2}}{2\sqrt{29a^3b}}$$

A) $\dfrac{10|a|\sqrt{29ab} + 2|a| \cdot |b|\sqrt{87ab}}{25a^2 - 3a^2b^2}$

B) $\dfrac{5\sqrt{29ab} - b\sqrt{87ab}}{58b}$

C) $\dfrac{3\sqrt{30} + 2\sqrt{15}}{120}$

D) $\dfrac{4\sqrt{26} - 6\sqrt{13}}{13}$

205) Simplify the below :

$$\frac{5 + 4\sqrt{7b^4}}{\sqrt{14b}}$$

A) $\dfrac{5\sqrt{14b} + 28b^2\sqrt{2b}}{14b}$

B) $\dfrac{\sqrt{11} + \sqrt{33}}{66}$

C) $\dfrac{-8\sqrt{5} + 12\sqrt{30}}{25}$

D) $\dfrac{7\sqrt{29} - 8\sqrt{203}}{29}$

206) Simplify the below :

$$\frac{-8x + \sqrt{6x^3}}{7\sqrt{27x}}$$

A) $\dfrac{8\sqrt{19} + \sqrt{133}}{133}$

B) $\dfrac{2\sqrt{29}}{29}$

C) $\dfrac{-8\sqrt{3x} + 3x\sqrt{2}}{63}$

D) $\dfrac{5\sqrt{31} - 7\sqrt{217}}{31}$

www.math-knots.com | www.a4ace.com

1) Solve the below equation

$$k^{\frac{5}{3}} = 243$$

A) $\{8, 5\}$ B) $\{8\}$

C) $\{27\}$ D) $\{8, -10\}$

2) Solve the below equation

$$\frac{1}{729} = p^{-\frac{3}{2}}$$

A) $\{-31, 4\}$ B) $\{81\}$

C) $\{81, 5\}$ D) $\{81, 4\}$

3) Solve the below equation

$$v^{\frac{3}{2}} = 729$$

A) $\{18, 4\}$ B) $\{4\}$

C) $\{81\}$ D) $\{18\}$

4) Solve the below equation

$$2 = x^{\frac{1}{6}}$$

A) $\{-64, 12\}$ B) $\{-64, -12\}$

C) $\{64\}$ D) $\{-64, 11\}$

5) Solve the below equation

$$v^{\frac{5}{4}} = 243$$

A) $\{81, -1\}$ B) $\{-1\}$

C) $\{-81, -1\}$ D) $\{81\}$

6) Solve the below equation

$$32 = n^{\frac{5}{4}}$$

A) $\{-10, 0\}$ B) $\{-6, 0\}$

C) $\{0\}$ D) $\{16\}$

7) Solve the below equation

$$a^{\frac{3}{2}} = 216$$

A) $\{-6\}$ B) $\{6, -6\}$

C) $\{36\}$ D) $\{6, 3\}$

8) Solve the below equation

$$n^{\frac{4}{3}} = 625$$

A) $\{125, -12\}$ B) $\{125\}$

C) $\{125, -125\}$ D) $\{125, -11\}$

9) Find the value of r that satisfies
the below equation

$$8 = r^{\frac{3}{4}}$$

A) $\{0\}$ B) $\{-16, 0\}$

C) $\{16\}$ D) $\{-12, 0\}$

10) Find the value of k that satisfies
the below equation

$$32 = k^{\frac{5}{3}}$$

A) $\{6, -10\}$ B) $\{6, -8\}$

C) $\{8\}$ D) $\{8, -8\}$

11) Find the value of p that satisfies
the below equation

$$p^{\frac{1}{3}} = 2$$

A) $\{2, -8\}$ B) $\{8\}$

C) $\{8, -8\}$ D) $\{8, 1\}$

12) Find the value of x that satisfies
the below equation

$$x^{\frac{1}{3}} = 5$$

A) $\{-125, 10\}$ B) $\{125, 10\}$

C) $\{125\}$ D) $\{125, 32\}$

13) Find the value of a that satisfies
the below equation

$$256 = a^{\frac{4}{3}}$$

A) $\{-64\}$ B) $\{64, -64\}$

C) $\{-64, -4\}$ D) $\{-5, -64\}$

14) Find the value of m that satisfies
the below equation

$$64 = m^{\frac{3}{2}}$$

A) $\{-8\}$ B) $\{18, 8\}$

C) $\{-8, 8\}$ D) $\{16\}$

15) Find the value of b that satisfies
the below equation

$$1024 = b^{\frac{5}{3}}$$

A) $\{-4, -12\}$ B) $\{-4\}$

C) $\{64\}$ D) $\{-4, -81\}$

16) Find the value of p that satisfies
the below equation

$$16 = p^{\frac{2}{3}}$$

A) $\{-1, -64\}$ B) $\{1, -64\}$

C) $\{7, -64\}$ D) $\{64, -64\}$

 www.math-knots.com | www.a4ace.com

17) Solve the below equation

$$\frac{1}{32} = x^{-\frac{5}{3}}$$

A) $\{8\}$ B) $\{-8, 11\}$

C) $\{-8, -12\}$ D) $\{8, 11\}$

18) Solve the below equation

$$1000 = v^{\frac{3}{2}}$$

A) $\{-100, 0\}$ B) $\{0\}$

C) $\{100\}$ D) $\{0, 5\}$

19) Solve the below equation

$$b^{\frac{7}{5}} = 128$$

A) $\{-32, 3\}$ B) $\{3\}$

C) $\{32\}$ D) $\{4, 3\}$

20) Solve the below equation

$$v^{\frac{5}{6}} = 32$$

A) $\{1\}$ B) $\{1, -1\}$

C) $\{64\}$ D) $\{64, 1\}$

21) Solve the below equation

$$v^{-\frac{3}{2}} = \frac{1}{512}$$

A) $\{64, 29\}$ B) $\{64\}$

C) $\{-11, 64\}$ D) $\{64, -2\}$

22) Solve the below equation

$$p^{\frac{4}{5}} = 16$$

A) $\{-9, -32\}$ B) $\{32, -32\}$

C) $\{-32\}$ D) $\{-14, -32\}$

23) Solve the below equation

$$x^{\frac{3}{2}} = 343$$

A) $\{49\}$ B) $\{-10, 11\}$

C) $\{-10, -5\}$ D) $\{-10, 12\}$

24) Solve the below equation

$$n^{\frac{3}{2}} = 512$$

A) $\{1, -2\}$ B) $\{1, 2\}$

C) $\{64, 2\}$ D) $\{64\}$

www.math-knots.com | www.a4ace.com

25) Find the value of x that satisfies
the below equation

$$x^{\frac{3}{4}} = 27$$

A) $\{81\}$

B) $\{-8, 1\}$

C) $\{-8, 7\}$

D) $\{-8\}$

26) Find the value of k that satisfies
the below equation

$$3k^{\frac{1}{4}} - 8 = 1$$

A) $\{81\}$

B) $\{1, -4\}$

C) $\{-4\}$

D) $\{81, -4\}$

27) Find the value of a that satisfies
the below equation

$$(4a)^{\frac{7}{6}} = 128$$

A) $\{-16, 8\}$

B) $\{-16, -10\}$

C) $\{16\}$

D) $\{16, 8\}$

28) Find the value of m that satisfies
the below equation

$$1722 = 7 + 5\,m^{\frac{3}{2}}$$

A) $\{49\}$

B) $\{-4, 7\}$

C) $\{-4\}$

D) $\{-4, -7\}$

29) Find the value of k that satisfies
the below equation

$$512 = (14 - 2k)^{\frac{3}{2}}$$

A) $\{-10, 9\}$

B) $\{-10, 10\}$

C) $\{-10, -1\}$

D) $\{-25\}$

30) Find the value of p that satisfies
the below equation

$$p^{-\frac{3}{2}} + 4 = \frac{2049}{512}$$

A) $\{-80\}$

B) $\{64\}$

C) $\{-80, 9\}$

D) $\{37, -80\}$

31) Find the value of r that satisfies
the below equation

$$-12 = -3\,r^{\frac{2}{3}}$$

A) $\{8, -8\}$

B) $\{8, 11\}$

C) $\{8, -11\}$

D) $\{8, -1\}$

32) Find the value of x that satisfies
the below equation

$$-6 + (4x)^{\frac{3}{2}} = 506$$

A) $\{16\}$

B) $\{12, 6\}$

C) $\{16, -6\}$

D) $\{16, 6\}$

www.math-knots.com | www.a4ace.com

33) Solve the below equation

$$(6 - 2m)^{\frac{3}{2}} = 512$$

A) $\{-29\}$ B) $\{-7, -1\}$

C) $\{-1\}$ D) $\{-29, -1\}$

34) Solve the below equation

$$\frac{2}{3} = 2k^{-\frac{1}{2}}$$

A) $\{9\}$ B) $\{-6\}$

C) $\{6, -6\}$ D) $\{9, -6\}$

35) Solve the below equation

$$\frac{1}{1024} = (24 - 2m)^{-\frac{5}{3}}$$

A) $\{6, 0\}$ B) $\{-20\}$

C) $\{-9, 0\}$ D) $\{-2, 0\}$

36) Solve the below equation

$$27 = (x + 14)^{\frac{3}{4}}$$

A) $\{67\}$ B) $\{-1, -11\}$

C) $\{-1, 12\}$ D) $\{-1, -4\}$

37) Solve the below equation

$$(n - 4)^{-\frac{1}{2}} = \frac{1}{10}$$

A) $\{3\}$ B) $\{-2\}$

C) $\{104\}$ D) $\{3, -2\}$

38) Solve the below equation

$$10 - 5r^{\frac{1}{6}} = 0$$

A) $\{64\}$ B) $\{9, 10\}$

C) $\{9\}$ D) $\{-9, 10\}$

39) Solve the below equation

$$2(v + 1)^{\frac{3}{2}} = 1458$$

A) $\{5\}$ B) $\{80\}$

C) $\{-5, 0\}$ D) $\{5, 0\}$

40) Solve the below equation

$$-5(r + 22)^{\frac{3}{2}} = -135$$

A) $\{-4\}$ B) $\{-13\}$

C) $\{-13, -26\}$ D) $\{-4, -26\}$

41) Find the value of p that satisfies the below equation

$$(8p)^{\frac{3}{4}} = 8$$

A) $\{-2, -6\}$ B) $\{-2, -15\}$

C) $\{2\}$ D) $\{-6, -15\}$

42) Find the value of a that satisfies the below equation

$$\frac{191}{32} = -2a^{-\frac{3}{2}} + 6$$

A) $\{-4\}$ B) $\{16\}$

C) $\{-10, -4\}$ D) $\{10, -4\}$

43) Find the value of a that satisfies the below equation

$$(x+1)^{\frac{5}{4}} = 243$$

A) $\{-12, -80\}$ B) $\{-12, 80\}$

C) $\{80\}$ D) $\{80, 8\}$

44) Find the value of x that satisfies the below equation

$$512 = (x+8)^{\frac{3}{2}}$$

A) $\{56\}$ B) $\{12, -6\}$

C) $\{12, 6\}$ D) $\{-10, 6\}$

45) Find the value of a that satisfies the below equation

$$7 + (4a)^{-\frac{1}{3}} = \frac{29}{4}$$

A) $\{16\}$ B) $\{16, 4\}$

C) $\{1, -8\}$ D) $\{16, -8\}$

46) Find the value of a that satisfies the below equation

$$4 = (a-2)^{\frac{2}{3}}$$

A) $\{10, -6\}$ B) $\{-10, 10\}$

C) $\{10\}$ D) $\{-3, 10\}$

47) Find the value of k that satisfies the below equation

$$-6 + k^{\frac{5}{3}} = 3119$$

A) $\{1\}$ B) $\{-1\}$

C) $\{125\}$ D) $\{1, -1\}$

48) Find the value of n that satisfies the below equation

$$(n+26)^{\frac{3}{2}} = 512$$

A) $\{-10, 9\}$ B) $\{38, 9\}$

C) $\{-10\}$ D) $\{38\}$

www.math-knots.com | www.a4ace.com

Algebra 1

49) Solve the below equation

$$9 = (b + 18)^{\frac{2}{3}}$$

A) $\{9, -45\}$ B) $\{-45\}$

C) $\{7, -45\}$ D) $\{-7, -45\}$

50) Solve the below equation

$$(3x - 32)^{\frac{3}{2}} = 343$$

A) $\{27\}$ B) $\{0, 3\}$

C) $\{0\}$ D) $\{-27, 0\}$

51) Solve the below equation

$$(-12 - k)^{\frac{3}{4}} - 1 = 7$$

A) $\{-28, 11\}$ B) $\{-8, 11\}$

C) $\{-28\}$ D) $\{8, 11\}$

52) Solve the below equation

$$-10 + 5 \cdot (125x)^{\frac{2}{3}} = 115$$

A) $\{39, -1\}$ B) $\{-3, -1\}$

C) $\{-1\}$ D) $\{1, -1\}$

53) Solve the below equation

$$1457 = 2(2r + 21)^{\frac{3}{2}} - 1$$

A) $\{30\}$ B) $\{30, 3\}$

C) $\{30, -30\}$ D) $\{-4, 3\}$

54) Solve the below equation

$$15 = (37 - 4p)^{\frac{2}{3}} - 10$$

A) $\{-22, 4\}$ B) $\{-22\}$

C) $\{5, 4\}$ D) $\left\{-22, \frac{81}{2}\right\}$

55) Solve the below equation

$$5(2r - 4)^{\frac{3}{2}} + 4 = 324$$

A) $\{10, -1\}$ B) $\{10\}$

C) $\{12, -1\}$ D) $\{-10, -1\}$

56) Solve the below equation

$$-7 + (7n + 1)^{\frac{4}{3}} = 249$$

A) $\left\{9, -\frac{65}{7}\right\}$ B) $\left\{-9, -\frac{65}{7}\right\}$

C) $\{-9\}$ D) $\{-9, 9\}$

57) Find the value of m that satisfies the below equation

$$-623 = -5 \cdot (5m)^{\frac{3}{2}} + 2$$

A) $\{5, -2\}$ B) $\{-3, -2\}$

C) $\{5\}$ D) $\{-2\}$

58) Find the value of x that satisfies the below equation

$$-1 - (x - 9)^{\frac{3}{2}} = -126$$

A) $\{-34, 3\}$ B) $\{-34, 1\}$

C) $\{34\}$ D) $\{-34, 20\}$

59) Find the value of n that satisfies the below equation

$$-3(10 - 6n)^{-\frac{4}{3}} = -\frac{3}{256}$$

A) $\left\{-10, -\frac{37}{3}\right\}$ B) $\left\{-9, \frac{37}{3}\right\}$

C) $\left\{-9, -\frac{37}{3}\right\}$ D) $\left\{9, -\frac{37}{3}\right\}$

60) Find the value of x that satisfies the below equation

$$2(-2 - 22x)^{\frac{3}{2}} + 1 = 1025$$

A) $\{36, -24\}$ B) $\{-3\}$

C) $\{-24\}$ D) $\{-24, -8\}$

61) Find the value of x that satisfies the below equation

$$6 - 3(x + 6)^{\frac{3}{2}} = -2181$$

A) $\{-75, -9\}$ B) $\{-75\}$

C) $\{-75, 75\}$ D) $\{75\}$

62) Find the value of n that satisfies the below equation

$$-25 = -5(n - 16)^{\frac{2}{3}} - 5$$

A) $\{24, 8\}$ B) $\{24\}$

C) $\{-12, 24\}$ D) $\{-11, 24\}$

63) Find the value of x that satisfies the below equation

$$-870 = -6 - 4(x + 18)^{\frac{3}{2}}$$

A) $\{18\}$ B) $\{6, 0\}$

C) $\{0\}$ D) $\{-18, 0\}$

64) Find the value of n that satisfies the below equation

$$-8 = -(42 - n)^{\frac{3}{4}}$$

A) $\{9, -9\}$ B) $\{9\}$

C) $\{26\}$ D) $\{9, 37\}$

www.math-knots.com | www.a4ace.com

65) Solve the below equation

$$2561 = 5(1 - 9k)^{\frac{3}{2}} + 1$$

A) $\{-34, 34\}$ B) $\{-34, 10\}$

C) $\{-7\}$ D) $\{-34\}$

66) Solve the below equation

$$-117 = -5(4 - 11x)^{\frac{2}{3}} + 8$$

A) $\{11\}$ B) $\left\{11, \dfrac{129}{11}\right\}$

C) $\left\{-11, \dfrac{129}{11}\right\}$ D) $\{11, 6\}$

67) Solve the below equation

$$64 = 4(-16 - x)^{\frac{4}{3}}$$

A) $\{8, 12\}$ B) $\{20, -8\}$

C) $\{20, 12\}$ D) $\{-24, -8\}$

68) Solve the below equation

$$-12509 = -4(30 - 5x)^{\frac{5}{3}} - 9$$

A) $\{-12, -34\}$ B) $\{-19\}$

C) $\{7, -34\}$ D) $\{7, -12\}$

69) Solve the below equation

$$-2 + 4(n + 9)^{\frac{3}{2}} = 2914$$

A) $\{-8, 12\}$ B) $\{72\}$

C) $\{-8, -12\}$ D) $\{-8\}$

70) Solve the below equation

$$-687 = -1 - 2(n - 27)^{\frac{3}{2}}$$

A) $\{7, -12\}$ B) $\{76\}$

C) $\{76, -12\}$ D) $\{-7, -12\}$

71) Solve the below equation

$$\frac{59}{6} = -(3n + 3)^{-\frac{1}{2}} + 10$$

A) $\{11, 3\}$ B) $\{11\}$

C) $\{-8, -5\}$ D) $\{11, -5\}$

72) Solve the below equation

$$-4 = -2(23 - n)^{\frac{1}{3}}$$

A) $\{15\}$ B) $\{-15, -24\}$

C) $\{-17, -12\}$ D) $\{-15, -12\}$

73) Find the value of b that satisfies the below equation

$$-158 = 4 - 2(85 - 2b)^{\frac{4}{3}}$$

A) $\{56\}$

B) $\{29, 56\}$

C) $\{-12, 56\}$

D) $\{-10, 56\}$

74) Find the value of x that satisfies the below equation

$$-3 \cdot (4x)^{-\frac{5}{6}} + 5 = \frac{157}{32}$$

A) $\{30, -7\}$

B) $\{16\}$

C) $\{-6, -12\}$

D) $\{-6, -7\}$

75) Find the value of b that satisfies the below equation

$$4 = 2 + (-1 - 33b)^{\frac{1}{5}}$$

A) $\{-1, 4\}$

B) $\{-1, -4\}$

C) $\{-1, 1\}$

D) $\{-1\}$

76) Find the value of r that satisfies the below equation

$$25 = -7 + 2(1 - 9r)^{\frac{2}{3}}$$

A) $\left\{-9, \frac{65}{9}\right\}$

B) $\left\{-7, \frac{65}{9}\right\}$

C) $\left\{\frac{65}{9}\right\}$

D) $\left\{11, \frac{65}{9}\right\}$

77) Find the value of p that satisfies the below equation

$$p^{\frac{3}{2}} = 729$$

A) $\{0\}$

B) $\{-5, 0\}$

C) $\{81\}$

D) $\{11, 0\}$

78) Find the value of x that satisfies the below equation

$$-5x^{\frac{3}{2}} = -1715$$

A) $\{-49\}$

B) $\{49\}$

C) $\{49, -8\}$

D) $\{-49, 49\}$

79) Find the value of b that satisfies the below equation

$$(64b)^{\frac{1}{3}} = 4$$

A) $\{2, -8\}$

B) $\{2\}$

C) $\{1\}$

D) $\{-2, -8\}$

80) Find the value of v that satisfies the below equation

$$8 - v^{\frac{3}{2}} = -504$$

A) $\{64, -11\}$

B) $\{-2, -11\}$

C) $\{-11\}$

D) $\{64\}$

www.math-knots.com | www.a4ace.com

81) Solve the below equation

$$-5 + \left(4v + 37\right)^{\frac{5}{3}} = 3120$$

A) $\{-6, 9\}$ B) $\{6, 9\}$

C) $\{6\}$ D) $\{22\}$

82) Solve the below equation

$$64 = k^{\frac{3}{2}}$$

A) $\{1, 10\}$ B) $\{16\}$

C) $\{1, 8\}$ D) $\{1\}$

83) Solve the below equation

$$3x^{\frac{3}{2}} - 4 = 2183$$

A) $\{4\}$ B) $\{4, -4\}$

C) $\{81\}$ D) $\{4, 6\}$

84) Solve the below equation

$$2 = \left(68 - 2k\right)^{\frac{1}{6}}$$

A) $\{2, -10\}$ B) $\{2, -9\}$

C) $\{2\}$ D) $\{2, 10\}$

85) Solve the below equation

$$\left(m + 12\right)^{\frac{3}{2}} - 7 = 20$$

A) $\{7, 10\}$ B) $\{-3\}$

C) $\{-27, 10\}$ D) $\{-27\}$

86) Solve the below equation

$$-9 = 6 - 5 \cdot \left(27n\right)^{\frac{1}{4}}$$

A) $\{-8\}$ B) $\{3, -8\}$

C) $\{3\}$ D) $\{-2, -8\}$

87) Solve the below equation

$$k^{\frac{3}{2}} - 8 = 504$$

A) $\{6\}$ B) $\{3, 6\}$

C) $\{64\}$ D) $\{6, 25\}$

88) Solve the below equation

$$512 = n^{\frac{3}{2}}$$

A) $\{64\}$ B) $\{-10, -35\}$

C) $\{64, 37\}$ D) $\{64, -35\}$

89) Find the value of x that satisfies the below equation

$$\frac{1}{5} = \left(125p\right)^{-\frac{1}{3}}$$

A) $\{1\}$ B) $\{1,-8\}$

C) $\{-8\}$ D) $\{1,8\}$

90) Find the value of x that satisfies the below equation

$$14 = x^{\frac{3}{2}} + 6$$

A) $\{4,-4\}$ B) $\{-11,-4\}$

C) $\{4\}$ D) $\{-11\}$

91) Find the value of n that satisfies the below equation

$$10 = \sqrt{n-6} + 9$$

A) 10   B) -2

C) 7   D) -10

92) Find the value of n that satisfies the below equation

$$\sqrt{19r-2} - 6 = 0$$

A) -6 B) 7

C) 2   D) -3

93) Find the value of n that satisfies the below equation

$$\sqrt{20-2n} - 6 = -2$$

A) 2   B) $2,-3$

C) -2 D) $-2,2$

94) Find the value of v that satisfies the below equation

$$2 = -8 + \sqrt{v-5}$$

A) $-105, 3$ B) -105

C) 105   D) $-105, 105$

95) Find the value of x that satisfies the below equation

$$-7\sqrt{x-8} = -14$$

A) 12   B) $12,-10$

C) 6   D) $12,-12$

96) Find the value of n that satisfies the below equation

$$\sqrt{n-6} + 3 = 4$$

A) -8 B) $-7,9$

C) 7   D) -7

97) Find the value of n that satisfies the below equation

$$3 = \sqrt{-1-10n}$$

A) $-1,6$ B) -8

C) -1 D) -9

98) Find the value of v that satisfies the below equation

$$24 = 6\sqrt{8v}$$

A) $9,-2$ B) 9

C) 2   D) $2,-5$

99) Find the value of k that satisfies the below equation

$$-7\sqrt{k-6} = -63$$

A) -5 B) 8

C) $-8, -87$ D) 87

100) Find the value of m that satisfies the below equation

$$-7 + \sqrt{m+4} = -3$$

A) $-4, 4$ B) 12

C) $-4, 10$ D) 12, 4

101) Find the value of x that satisfies the below equation

$$\sqrt{9x} = 9$$

A) -9 B) 9

C) -3 D) 7

102) Find the value of n that satisfies the below equation

$$7\sqrt{13n-3} = 49$$

A) 1, 3   B) 4

C) -3 D) 1

103) Find the value of n that satisfies the below equation

$$7\sqrt{n+6} = 63$$

A) -4 B) -75

C) 75   D) -6

104) Find the value of n that satisfies the below equation

$$9 = 3\sqrt{2n+27}$$

A) -9 B) 10

C) -3 D) 9, -9

105) Find the value of x that satisfies the below equation

$$4 = \sqrt{\frac{x}{7}}$$

A) 112   B) 1

C) -1 D) 8

106) Find the value of x that satisfies the below equation

$$\sqrt{4-8x} = 6$$

A) 6, -4 B) 6, 4

C) -4 D) 4, -4

107) Find the value of b that satisfies the below equation

$$8 = \sqrt{-6 - 10b}$$

A) – 7

B) 7

C) – 3

D) 1

108) Find the value of r that satisfies the below equation

$$4 = -5 + \sqrt{81r}$$

A) 1

B) – 9 , – 1

C) 1 , 9

D) – 9

109) Find the value of x that satisfies the below equation

$$16 = 4\sqrt{2x + 34}$$

A) – 9

B) 6 , 1

C) 5

D) – 1

110) Find the value of r that satisfies the below equation

$$3 = \sqrt{r + 3}$$

A) 6

B) – 6 , 6

C) – 8 , – 9

D) – 9

111) Find the value of k that satisfies the below equation

$$\sqrt{15k + 4} + 3 = 11$$

A) 4

B) 8

C) 6

D) – 4

112) Find the value of m that satisfies the below equation

$$\sqrt{3m - 37} = \sqrt{m - 9}$$

A) 14

B) 8

C) – 8

D) 10

113) Find the value of x that satisfies the below equation

$$\sqrt{18 - 2x} = \sqrt{3x - 22}$$

A) 10

B) – 5 , 10

C) – 5

D) 8

114) Find the value of n that satisfies the below equation

$$\sqrt{5n} = \sqrt{12 - n}$$

A) – 9 , – 7

B) – 6

C) – 7

D) 2

www.math-knots.com | www.a4ace.com

115) Find the value of n that satisfies the below equation

$$\sqrt{4 - 2n} = \sqrt{1 - 5n}$$

A) 1   B) – 6 , – 1

C) – 6 , – 7   D) – 1

116) Find the value of b that satisfies the below equation

$$\sqrt{8 - b} = \sqrt{3b}$$

A) 2 , – 2   B) – 9

C) 2   D) – 2

117) Find the value of k that satisfies the below equation

$$\sqrt{-3 - k} = \sqrt{-9 - 2k}$$

A) – 6   B) – 6 , 7

C) – 7 , – 6   D) 6

118) Find the value of x that satisfies the below equation

$$\sqrt{63 - x} = \sqrt{\dfrac{x}{6}}$$

A) 54 , 6   B) 54

C) – 9   D) – 3 , 6

119) Find the value of x that satisfies the below equation

$$\sqrt{26 - 2x} = \sqrt{16 - x}$$

A) – 8   B) 10

C) – 7   D) – 10

120) Find the value of k that satisfies the below equation

$$\sqrt{k - 7} = \sqrt{13 - k}$$

A) – 2   B) 10

C) 10 , – 2   D) – 10

121) Find the value of x that satisfies the below equation

$$\sqrt{3x - 9} = \sqrt{2x - 3}$$

A) 6   B) – 10

C) – 6   D) – 5 , 6

122) Find the value of k that satisfies the below equation

$$\sqrt{9k} = \sqrt{2k + 7}$$

A) 9   B) 1

C) 9 , 1   D) – 8

123) Find the value of x that satisfies the below equation

$$\sqrt{x+1} = \sqrt{3x-1}$$

A) 3   B) 1

C) − 1   D) 1 , 3

124) Find the value of x that satisfies the below equation

$$\sqrt{10x-1} = \sqrt{11-2x}$$

A) − 9   B) 1

C) − 4   D) 8

125) Find the value of x that satisfies the below equation

$$\sqrt{3x} = \sqrt{2x+3}$$

A) − 3   B) − 8

C) 3   D) − 5 , − 3

126) Find the value of n that satisfies the below equation

$$\sqrt{n-8} = \sqrt{22-2n}$$

A) − 10   B) 10

C) − 4   D) 10 , − 10

127) Find the value of n that satisfies the below equation

$$\sqrt{2n+1} = \sqrt{3n}$$

A) 7   B) 3

C) 1   D) − 7 , 7

128) Find the value of b that satisfies the below equation

$$\sqrt{3b-26} = \sqrt{14-b}$$

A) − 10 , 10   B) − 10

C) 10   D) 6 , 10

129) Find the value of k that satisfies the below equation

$$\sqrt{9-k} = \sqrt{\frac{k}{8}}$$

A) − 8 , − 7   B) 8

C) − 1 , 8   D) − 7

130) Find the value of n that satisfies the below equation

$$\sqrt{10-n} = \sqrt{2n-11}$$

A) − 3   B) − 7 , − 3

C) 7   D) − 7

www.math-knots.com | www.a4ace.com

 **Algebra 1**

131) Find the value of x that satisfies the below equation

$$\sqrt{2x-7} = \sqrt{8-x}$$

A) 5

B) – 2

C) 5 , 2

D) 3

132) Find the value of x that satisfies the below equation

$$\sqrt{-2-x} = \sqrt{2x+16}$$

A) – 6

B) – 6 , 6

C) – 4

D) 6

133) Find the value of b that satisfies the below equation

$$\sqrt{2b} = \sqrt{3b-1}$$

A) – 1

B) 3

C) 9 , – 1

D) 1

134) Find the value of a that satisfies the below equation

$$\sqrt{2a+7} = \sqrt{3a+9}$$

A) 2

B) – 3

C) – 2 , 2

D) – 2

135) Find the value of a that satisfies the below equation

$$\sqrt{\frac{a}{3}} = \sqrt{8-a}$$

A) 6

B) – 8 , – 1

C) 8

D) – 8

136) Find the value of p that satisfies the below equation

$$\sqrt{5p} = \sqrt{7-2p}$$

A) – 10

B) 10

C) 1

D) – 1

137) Find the value of n that satisfies the below equation

$$\sqrt{-3-2n} = \sqrt{n+9}$$

A) – 4

B) – 3

C) 4

D) – 9

www.math-knots.com | www.a4ace.com

1) Find the solution for the below system of equations using method of substitution

$$8x + 2y = 20$$
$$-24x - 6y = -60$$

A) $(-3, 10)$

B) Infinite number of solutions

C) $(-3, -3)$

D) $(-3, -6)$

2) The difference of two numbers is 3. Their sum is 25. What are the numbers?

A) 11 and 14

B) 12 and 19

C) 14 and 13

D) 18 and 19

3) The senior classes at High School A and High School B planned separate trips to the local amusement park. The senior class at High School A rented and filled 9 vans and 5 buses with 264 students. High School B rented and filled 4 vans and 5 buses with 209 students. Each van and each bus carried the same number of students. Find the number of students in each van and in each bus.

A) Van : 9 , Bus : 31

B) Van : 33 , Bus : 11

C) Van : 9 , Bus : 30

D) Van : 11 , Bus : 33

4) Find the solution for the below system of equations using method of substitution

$$12x - y = 19$$
$$-2x - 4y = 26$$

A) $(-1, 7)$ B) $(1, 7)$

C) $(1, -7)$ D) $(-1, -7)$

5) Find the solution for the below system of equations using method of elimination.

$$5x - 2y = 16$$
$$11x + 13y = -17$$

A) $(-3, 3)$ B) $(2, -3)$

C) $(9, 3)$ D) $(8, -3)$

6) The sum of two numbers is 26. Their difference is 2. Find the numbers?

A) 14 and 10

B) 16 and 16

C) 16 and 22

D) 12 and 14

7) Simplify the below

$$2m^4n^2 \cdot m^3n^3$$

A) $\dfrac{4m^5}{n^2}$ B) $\dfrac{9n^3}{m}$

C) $2m^7n^5$ D) $\dfrac{3}{m^3n^3}$

8) Rob's school is selling tickets to a choral performance. On the first day of ticket sales the school sold 2 adult tickets and 6 student tickets for a total of $78. The school took in $144 on the second day by selling 3 adult tickets and 12 student tickets. What is the price each of one adult ticket and one student ticket?

9) Find the solution for the below system of equations using method of substitution

$$12x + 6y = -30$$
$$-6x + 8y = -18$$

A) $(-1, -3)$ 　　　　 B) $(-1, 3)$

C) $(1, 3)$ 　　　　 D) $(1, -3)$

10) The grade 10 class of MK Academy used cars and vans to go on a field trip.
Class A used 4 vans and 1 bus transported 100 students. Class B used 14 vans and 1 bus to transport 260 students. How many cars and vans are used for transportation.

A) Van : 15 , Bus : 24

B) Van : 22 , Bus : 58

C) Van : 16 , Bus : 36

D) Van : 11 , Bus : 39

11) Find the solution for the below system of equations by graphing them below

$$1 + \frac{1}{2}y = \frac{1}{8}x$$

$$2x = -\frac{32}{7} + \frac{8}{7}y$$

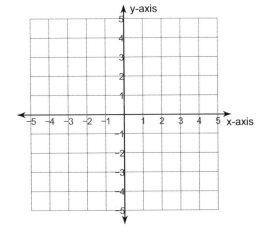

A) $(3, 5)$

B) $(1, -3)$

C) $(-3, 1)$

D) $(-4, -3)$

12) Simplify the below :

$$4\sqrt{32} - \sqrt{8}$$

A) $46\sqrt{2}$ 　　　 B) $30\sqrt{2}$

C) $14\sqrt{2}$ 　　　 D) $44\sqrt{2}$

Algebra 1

Vol 2
Week 28
Assessment 4

13) The grade 9 class of MK Academy is selling tickets for their school annual talent show. On Thursday they sold 5 senior tickets and 12 student tickets and raised $97. On Friday they sold 13 senior tickets and 9 student tickets for a total of $119.Find the price of senior and student ticket?

14) The sum of the digits of a certain two-digit number is 14. Reversing its digits decreases the number by 18. What is the number?

15) Solve the below equation

$$\frac{1}{3125} = m^{-\frac{5}{3}}$$

A) $\{-9, 9\}$

B) $\{-9, -8\}$

C) $\{-9\}$

D) $\{125\}$

16) The grade 7 class of MK Academy is selling tickets for their school annual talent show. On Thursday they sold 5 adult tickets and 11 child tickets and raised $194. On Friday they sold 10 adult tickets and 7 child tickets for a total of $178.Find the price of adult and child ticket?

17) Find the solution for the below system of inequalities by graphing them below

$$y > -2x - 2$$
$$y \leq -\frac{1}{2}x + 1$$

A)

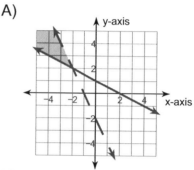

B)

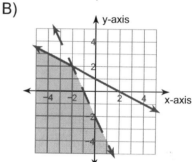

C)

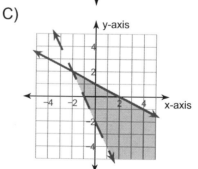

D)

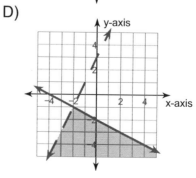

279

18) Find the solution for the below system of equations using method of substitution

$$8x - 10y = -34$$
$$-3x + 8y = 17$$

A) (1 , 10)   B) (-3 , 1)

C) (1 , -3)   D) (10 , 1)

19) The grade 8 class of MK Academy is selling tickets for their school annual talent show. On Thursday they sold 7 adult tickets and 6 child tickets and raised $95. On Friday they sold 3 adult tickets and 1 child tickets for a total of $36.Find the price of adult and child ticket?

20) Simplify the below :

$$\frac{5\sqrt{2}}{5\sqrt{3}}$$

A) $\frac{2\sqrt{3}}{3}$ B) $\frac{\sqrt{6}}{2}$

C) $\frac{\sqrt{15}}{4}$ D) $\frac{\sqrt{6}}{3}$

21) Find the value of r that satisfies the below equation

$$-2\sqrt{r+1} = -10$$

A) { -24 }   B) { 5 , 24 }

C) { 24 }   D) { 4 }

22) Find the value of k that satisfies the below equation

$$-8\sqrt{3k} = -24$$

A) { 6 , -3 }   B) { 5 }

C) { 3 }   D) { 6 , 3 }

23) Find the solution for the below system of equations by graphing them below

$$-2y = -3x - 8$$
$$0 = -8 - 3x + 2y$$

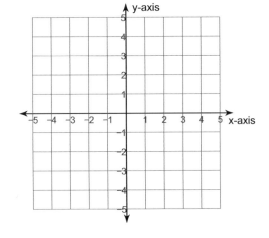

A) (3 , 2)

B) (1 , 2)

C) Infinite number of solutions

D) (2 , -2)

www.math-knots.com | www.a4ace.com

24) The grade 8 class of MK Academy used cars and vans to go on a field trip. Class A used 13 vans and 1 bus transported 215 students. Class B used 14 vans and 1 bus to transport 228 students. How many cars and vans are used for transportation.

A) Van : 7 , Bus : 60

B) Van : 13 , Bus : 46

C) Van : 46 , Bus : 13

D) Van : 7 , Bus : 42

25) Simplify the below :

$$-3\sqrt{6} - 4\sqrt{6}$$

A) $-7\sqrt{6}$

B) $-10\sqrt{6}$

C) $-17\sqrt{6}$

D) $-13\sqrt{6}$

26) The grade 9 class of MK Academy is selling tickets for their school annual talent show. On Thursday they sold 11 adult tickets and 12 child tickets and raised $262. On Friday they sold 1 adult tickets and 2 child tickets for a total of $32. Find the price of adult and child ticket?

27) Simplify the below :

$$\sqrt{42}\left(3n - 7\sqrt{7n}\right)$$

A) $6\sqrt{3n} + 3$

B) $6\sqrt{105n} + 7n$

C) $3n\sqrt{42} - 49\sqrt{6n}$

D) $4n\sqrt{5} + 7$

28) Find the value of n that satisfies the below equation

$$\sqrt{2n + 23} = \sqrt{-1 - n}$$

A) $\{-3\}$ B) $\{8\}$

C) $\{-8\}$ D) $\{-4\}$

29) Solve the below equation

$$8 - 3 \cdot (2n)^{\frac{1}{6}} = 2$$

A) $\{-10, -32\}$ B) $\{-32\}$

C) $\{32\}$ D) $\{-32, 32\}$

30) Simplify the below :

$$7\sqrt{6} \cdot \sqrt{12}$$

A) $6\sqrt{2}$ B) 72

C) $42\sqrt{2}$ D) $3\sqrt{2}$

31) The grade 12 class of MK Academy used cars and vans to go on a field trip. Class A used 12 vans and 12 buses transported 588 students. Class B used 10 vans and 12 buses to transport 544 students. How many buses and vans are used for transportation.

    A) Van : 17 , Bus : 32

    B) Van : 10 , Bus : 36

    C) Van : 10 , Bus : 30

    D) Van : 32 , Bus : 17

32) Find the solution for the below system of equations using method of elimination.

$$10 x + 6 y = 12$$
$$- 6 x + 13 y = 26$$

    A) $(0 , 1)$

    B) $(0 , 2)$

    C) No solution

    D) $(0 , - 2)$

33) The sum of two numbers is 24. Their difference is 2. What are the numbers?

    A) 18 and 19

    B) 11 and 13

    C) 17 and 8

    D) 14 and 7

34) Find the value of two numbers if their sum is 20 and their difference is 4.

    A) 7 and 14

    B) 12 and 5

    C) 5 and 18

    D) 8 and 12

35) Find the solution for the below system of equations by graphing them below

$$y - 4 = 3 x$$
$$x - 2 = 2 y$$

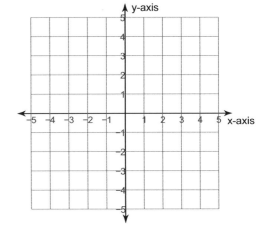

    A) $(- 2 , 2)$

    B) $(2 , - 2)$

    C) $(- 1 , 2)$

    D) $(- 2 , - 2)$

36) The grade 10 class of MK Academy is selling fruits at county fair for a fundraiser. On Saturday they sold 8 small of mangoes and 8 large boxes of mangoes and raised $224. On Sunday they sold 7 small boxes and 6 large boxes and raised $181. Find the cost of a small box and a large box of mangoes.

37) Find the solution for the below system of equations using method of elimination.

$$6x + 3y = 39$$
$$-5x + 7y = 15$$

A) $(4, 5)$

B) $(-8, 5)$

C) $(-8, 7)$

D) Infinite number of solutions

38) The sum of the digits of a certain two-digit number is 5. When you reverse its digits you decrease the number by 9. What is the number?

39) Simplify the below :

$$6\sqrt{35}\left(4\sqrt{15} + 9\right)$$

A) $6\sqrt{2} + 4$

B) $2\sqrt{70} + 2$

C) $120\sqrt{21} + 54\sqrt{35}$

D) $9\sqrt{5} + 6\sqrt{3}$

40) Simplify the below

$$2m^{-2} \cdot mn^2$$

A) $\dfrac{4}{n}$

B) $\dfrac{2n^2}{m}$

C) $\dfrac{3m^8}{n^3}$

D) $4nm^2$

41) Simplify the below :

$$\sqrt{343x^3}$$

A) $3x^2\sqrt{5}$

B) $7x\sqrt{7x}$

C) $2x^2\sqrt{5}$

D) $7x\sqrt{2}$

42) Find the value of x that satisfies the below equation

$$\sqrt{-3-x} = 2$$

A) $\{-7, 7\}$

B) $\{-1\}$

C) $\{3, 1\}$

D) $\{-7\}$

43) Simplify the below :

$$\sqrt{63x^2y^4}$$

A) $14xy$

B) $4x^2y\sqrt{5}$

C) $6x\sqrt{5xy}$

D) $3y^2x\sqrt{7}$

44) Find the solution for the below system of inequalities by graphing them below

$$y < -\frac{1}{2}x - 2$$

$$y \le -\frac{5}{2}x + 2$$

A)
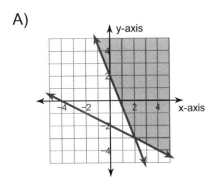

B)
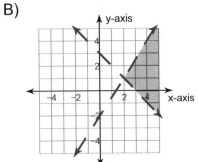

C)
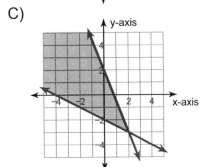

D)
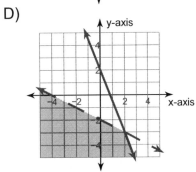

45) Simplify the below :

$$\frac{8\sqrt{2n} + \sqrt{5n}}{8\sqrt{22n}}$$

A) $\dfrac{4\sqrt{22} + \sqrt{33}}{11}$

B) $\dfrac{16\sqrt{11} + \sqrt{110}}{176}$

C) $\dfrac{3\sqrt{2}}{32}$

D) $\dfrac{-5\sqrt{3} - \sqrt{15}}{9}$

46) Simplify the below :

$$\sqrt{14a}\left(\sqrt{14} + 4\sqrt{2a}\right)$$

A) $\sqrt{70a} + 14a$

B) $6a\sqrt{6} + \sqrt{30}$

C) $14\sqrt{a} + 8a\sqrt{7}$

D) $-35\sqrt{5a} + 4\sqrt{3a}$

47) Solve the below equation

$$729 = (-9 - 3n)^{\frac{3}{2}}$$

A) $\{-30\}$ B) $\{5, -30\}$

C) $\{-25\}$ D) $\{-25, -30\}$

48) The grade 10 class of MK Academy used cars and vans to go on a field trip. Class A used 1 van and 14 buses transported 471 students. Class B used 10 vans and 14 buses to transport 552 students. How many cars and vans are used for transportation.

A) Van : 9 , Bus : 33

B) Van : 13 , Bus : 31

C) Van : 14 , Bus : 18

D) Van : 6 , Bus : 34

49) Simplify the below :

$$\sqrt{200x^2y^3}$$

A) $3x^2y^2\sqrt{3}$ B) $3xy\sqrt{2y}$

C) $7x\sqrt{2y}$ D) $10xy\sqrt{2y}$

50) Simplify the below :

$$5\sqrt{4r} \cdot -5\sqrt{8r^2}$$

A) $4\sqrt{2}$ B) $2\sqrt{3}$

C) $-100r\sqrt{2r}$ D) 32

51) Simplify the below :

$$\frac{6m^2 + \sqrt{2m^3}}{\sqrt{6m^2}}$$

A) $\dfrac{3m\sqrt{6} + \sqrt{3m}}{3}$

B) $\dfrac{3\sqrt{31} + 5\sqrt{62}}{31}$

C) $\dfrac{14\sqrt{15} + 7\sqrt{105}}{3}$

D) $\dfrac{6 + \sqrt{42}}{12}$

52) Simplify the below :

$$\frac{\sqrt{3b^2}}{\sqrt{6b^3}}$$

A) $\dfrac{\sqrt{14}}{35}$ B) $\sqrt{2b}$

C) $\dfrac{\sqrt{2b}}{2b}$ D) $\dfrac{\sqrt{21}}{3}$

53) Simplify the below

$$3a^4b^3 \cdot 3a^{-4}b^2$$

A) $9b^5$ B) $6a^4$

C) $\dfrac{12}{a^8}$ D) $\dfrac{b^2}{a^6}$

54) The grade 8 class of MK Academy used cars and vans to go on a fieldtrip. Class A used 2 vans and 9 buses transported 283 students. Class B used 3 vans and 11 buses to transport 352 students. How many cars and vans are used for transportation.

55) The sum of the digits of a certain two-digit number is 9. When you reverse its digits you increase the number by 63. What is the number?

56) Simplify the below :

$$\sqrt{14}\left(\sqrt{21} + 6\sqrt{30}\right)$$

A) $7\sqrt{6} + 12\sqrt{105}$

B) $\sqrt{30} + 8$

C) 47

D) $8\sqrt{7} + 8$

57) Simplify the below :

$$\sqrt{128x^4}$$

A) $2x^2\sqrt{3}$ B) $12x\sqrt{2}$

C) $4x^2\sqrt{3}$ D) $8x^2\sqrt{2}$

58) Simplify the below :

$$\frac{2\sqrt{8}}{2\sqrt{6}}$$

A) $\frac{2\sqrt{6}}{3}$ B) $\frac{2\sqrt{3}}{3}$

C) $\frac{\sqrt{6}}{10}$ D) $\frac{\sqrt{3}}{2}$

59) Simplify the below

$$\left(y^{-1}z^2 \cdot 2zx^4y^{-1}\right)^3$$

A) $16y^3z^2x^8$ B) $\frac{2}{xz^3}$

C) $2y^3z^{12}x^2$ D) $\frac{8z^9x^{12}}{y^6}$

60) Simplify the below :

$$\sqrt{72r^3}$$

A) $3r\sqrt{5r}$ B) $2r^2\sqrt{5}$

C) $14\sqrt{2r}$ D) $6r\sqrt{2r}$

61) Simplify the below :

$$\sqrt{5}\left(-3\sqrt{10} + \sqrt{35}\right)$$

A) $-15\sqrt{2} + 5\sqrt{7}$

B) $144\sqrt{2} + 9$

C) $7\sqrt{5} + 9$

D) $42\sqrt{10} + 2\sqrt{2}$

62) Find the solution for the below system of equations by graphing them below

$$y = -2 - x$$
$$-2 = x + y$$

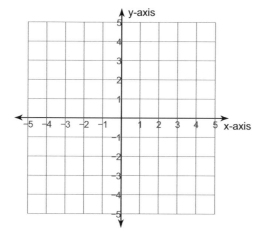

A) $(-5, 2)$

B) $(2, -2)$

C) $(2, 2)$

D) Infinite number of solutions

63) Find the solution for the below system of equations using method of substitution
$$6x - 9y = -30$$
$$-x - 5y = 5$$

A) $(4, 0)$

B) $(-4, 0)$

C) $(-5, 0)$

D) $(11, 0)$

64) Simplify the below :

$$\sqrt{48x^4 y^2}$$

A) $4x^2 y\sqrt{3}$

B) $6x^2\sqrt{7y}$

C) $8y^2 x\sqrt{2x}$

D) $7xy\sqrt{6}$

65) Simplify the below :

$$\sqrt{5}\left(\sqrt{50} + \sqrt{12}\right)$$

A) $10\sqrt{5} + \sqrt{210}$

B) $3\sqrt{10} + 5$

C) $-12 + 4\sqrt{2}$

D) $5\sqrt{10} + 2\sqrt{15}$

66) Simplify the below

$$\left(2x^4 y^{-1}\right)^3 \cdot \left(xy^4 z^4\right)^3$$

A) $\dfrac{x^4}{z^8}$

B) $\dfrac{2z^4}{y^3 x}$

C) $8x^{15} y^9 z^{12}$

D) $\dfrac{4x^{10} y^{10}}{z^2}$

67) Find the value of b that satisfies the below equation

$$\sqrt{2b} = 8$$

A) $\{32\}$

B) $\{32, -1\}$

C) $\{-5, -1\}$

D) $\{-1\}$

Algebra 1

68) When you reverse the digits in a certain two-digit number you decrease its value by 9. Find the number if the sum of its digits is 7.

69) Simplify the below

$$\frac{6x^7 y^5 z^4}{9x^2 y^4 z^{-10}}$$

A) $\dfrac{2z^{14} x^5 y}{3}$ B) $\dfrac{7z^6 y^9}{6}$

C) $\dfrac{7z^9}{2x^{10} y^8}$ D) $\dfrac{5y^{14} z^7}{7x}$

70) Find the solution for the below system of equations using method of elimination.

$$12x + 11y = 34$$
$$-7x + 5y = 3$$

A) $(-4, 1)$ B) $(-4, -1)$

C) $(1, -4)$ D) $(1, 2)$

71) Find the value of x that satisfies the below equation

$$12 = \sqrt{\frac{x}{3}} + 3$$

A) $\{10, -243\}$ B) $\{-243\}$

C) $\{243\}$ D) $\{10, 243\}$

72) Simplify the below :

$$-\sqrt{5v^2} \cdot \sqrt{2v}$$

A) 10   B) $\sqrt{10}$

C) $-v\sqrt{10v}$ D) $\sqrt{7}$

73) Solve the below equation

$$(2x - 6)^{\frac{3}{2}} = 64$$

A) $\{11, -2\}$ B) $\{-16, 11\}$

C) $\{11, -12\}$ D) $\{11\}$

74) Simplify the below

$$m^3 \cdot (2m^3 p^4 q^0)^4$$

A) $\dfrac{2mp^3}{q^2}$ B) $\dfrac{p^2}{m^4 q^7}$

C) $16m^{15}p^{16}$ D) $\dfrac{1}{4p^7}$

75) Simplify the below :

$$\sqrt{192m^4 n^3}$$

A) $7m\sqrt{5mn}$ B) $7n\sqrt{5m}$

C) $8m^2 n\sqrt{3n}$ D) $5m^2 n\sqrt{6n}$

76) The grade 10 class of MK Academy is cakes at a bake sale for a fundraiser. On Saturday they sold 3 French cheese cakes and 8 chocolate cheese cakes and raised $162. On Sunday they sold 10 French cheese cakes and 11 chocolate cheese cakes and raised $305. Find the cost of a French cheese cake and a chocolate cheese cake?

77) Find the solution for the below system of equations using method of elimination.

$$-72\,x + 99\,y = 31$$
$$32\,x - 44\,y = -12$$

A) $(\,10\,,-3\,)$

B) Infinite number of solutions

C) No solution

D) $(\,10\,,3\,)$

78) Simplify the below

$$u^{-3}v^{-4} \cdot uv^{-3}$$

A) $\dfrac{1}{u^2 v^7}$ B) $\dfrac{4}{v^2}$

C) $9v^6 u^3$ D) $12v^4 u^6$

79) Solve the below equation

$$n^{\frac{6}{5}} = 64$$

A) $\{32,-32\}$ B) $\{-5,5\}$

C) $\{-5,-32\}$ D) $\{5,-32\}$

80) Find the value of x that satisfies the below equation

$$-2 + \sqrt{2x+27} = 1$$

A) $\{-9,9\}$ B) $\{5\}$

C) $\{-9\}$ D) $\{9\}$

81) Find the value of n that satisfies the below equation

$$\sqrt{2n} = \sqrt{3n-2}$$

A) $\{2\}$ B) $\{2,-5\}$

C) $\{-2,-5\}$ D) $\{-5\}$

82) Simplify the below :

$$\dfrac{\sqrt{7} - 3\sqrt{3}}{8\sqrt{20}}$$

A) $\dfrac{\sqrt{35} - 3\sqrt{15}}{80}$

B) $\dfrac{7 + 6\sqrt{7}}{14}$

C) $\dfrac{-4\sqrt{35} - 12\sqrt{15}}{5}$

D) $\dfrac{3\sqrt{29} - 2\sqrt{174}}{29}$

83) Find the solution for the below system of equations using method of elimination.

$$-8x + 5y = -22$$
$$10x + 3y = -28$$

A) $(-1, -6)$ B) $(1, -6)$

C) $(-6, -1)$ D) $(-6, 1)$

84) Find the value of x that satisfies the below equation

$$\sqrt{13 - x} = \sqrt{2x - 2}$$

A) $\{-5\}$ B) $\{-5, -8\}$

C) $\{5\}$ D) $\{-1\}$

85) Find the value of n that satisfies the below equation

$$-49 = -7\sqrt{n + 8}$$

A) $\{-1\}$ B) $\{1\}$

C) $\{-41\}$ D) $\{41\}$

86) Solve the below equation

$$r^{\frac{1}{3}} = 3$$

A) $\{27\}$ B) $\{7, 9\}$

C) $\{7\}$ D) $\{-7, 7\}$

87) Simplify the below :

$$-3\sqrt{2} \cdot \sqrt{14}$$

A) 28   B) $2\sqrt{7}$

C) $-6\sqrt{7}$ D) 4

88) Simplify the below :

$$\sqrt{6b} \cdot -5\sqrt{15b^2}$$

A) $-15b\sqrt{10b}$ B) 90

C) $\sqrt{21}$ D) $3\sqrt{10}$

89) Find the value of b that satisfies the below equation

$$\sqrt{3b - 34} = \sqrt{b - 10}$$

A) $\{-12\}$ B) $\{9, 12\}$

C) $\{-8\}$ D) $\{12\}$

90) Find the value of m that satisfies the below equation

$$\sqrt{1 - 9m} - 1 = 7$$

A) $\{10\}$ B) $\{-7\}$

C) $\{-3\}$ D) $\{-9\}$

91)   Simplify the below :

$$\frac{\sqrt{7x} + 4}{\sqrt{3x}}$$

A) $\dfrac{4\sqrt{5} - 25}{10}$

B) $\dfrac{x\sqrt{21} + 4\sqrt{3x}}{3x}$

C) $\dfrac{-2\sqrt{13} + 2\sqrt{91}}{39}$

D) $\dfrac{4\sqrt{23} + \sqrt{115}}{23}$

92)   Simplify the below :

$$3\sqrt{10k} \cdot \sqrt{15k^2}$$

A) 150   B) $15k\sqrt{6k}$

C) 5   D) $5\sqrt{6}$

93) Solve the below equation

$$616 = x^{\frac{4}{3}} - 9$$

A) $\{1, -4\}$ B) $\{-7, -125\}$

C) $\{125, -125\}$ D) $\{1, -125\}$

94) Find the value of x that satisfies the below equation

$$7 = \sqrt{x + 3}$$

A) $\{-2, -4\}$ B) $\{46, -4\}$

C) $\{-4\}$ D) $\{46\}$

95)   Simplify the below

$$8r^0 \cdot 6p^7r^4 \cdot 7qr^7$$

96)   Simplify the below

$$\frac{x^7y^{-8}z^7}{4x^0}$$

A) $\dfrac{9}{2x^7y^7z^3}$ B) $\dfrac{10y^2z^{10}}{7x^8}$

C) $\dfrac{10z^5}{7x^2y^8}$ D) $\dfrac{x^7z^7}{4y^8}$

97)   Simplify the below :

$$\sqrt{27v^4}$$

A) $10\sqrt{2v}$ B) $14v\sqrt{2}$

C) $16v^2$ D) $3v^2\sqrt{3}$

www.math-knots.com | www.a4ace.com

98) Solve the below equation

$$12 = 5(a+9)^{\frac{1}{3}} + 2$$

A) $\{-1, -2\}$ B) $\{-2\}$

C) $\{-1, 12\}$ D) $\{-1\}$

99) Simplify the below :

$$\frac{\sqrt{5u^2v^2}}{5\sqrt{6u^4v^4}}$$

A) $|u| \cdot |v| \sqrt{30}$ B) $-\frac{\sqrt{7}}{28}$

C) $\frac{8\sqrt{10}}{5}$ D) $\frac{\sqrt{30}}{30uv}$

100) Simplify the below :

$$\frac{-8 - \sqrt{6}}{2\sqrt{12}}$$

A) $\frac{-8\sqrt{3} - 3\sqrt{2}}{12}$

B) $\frac{8\sqrt{17} - \sqrt{85}}{51}$

C) $\frac{-16\sqrt{3} + 6\sqrt{2}}{29}$

D) $\frac{-5\sqrt{6} + 12}{12}$

101) Simplify the below

$$2x^{-4} \cdot 4x^{-1}y^4$$

A) $\frac{12x^5}{y^2}$ B) $\frac{2}{xy^7}$

C) $\frac{8y^4}{x^5}$ D) $4x^8y^5$

102) Simplify the below

$$\frac{4r\,p^0q^9}{p^{10}q^{-7}r^6}$$

A) $\frac{q}{pr}$ B) $\frac{8q^4}{3r^{17}}$

C) $\frac{4q^{16}}{p^{10}r^5}$ D) $\frac{5r^7qp}{8}$

103) Solve the below equation

$$v^{\frac{1}{4}} = 3$$

A) $\{11, 0\}$ B) $\{0\}$

C) $\{81\}$ D) $\{-5, 0\}$

104) Simplify the below

$$4x^4y^2z^{10} \cdot x^{10}y^6z^7$$

www.math-knots.com | www.a4ace.com

105) Simplify the below :

$$\frac{4\sqrt{7n} - \sqrt{2n^4}}{\sqrt{18n^4}}$$

A) $\dfrac{3\sqrt{42} - \sqrt{14}}{98}$

B) $\dfrac{2\sqrt{14n} - n^2}{3n^2}$

C) $\dfrac{-6\sqrt{5} + 5}{60}$

D) $\dfrac{\sqrt{3} + 3\sqrt{2}}{12}$

106) Simplify the below :

$$\sqrt{18}$$

A) $3\sqrt{5}$ B) $3\sqrt{2}$

C) $3\sqrt{7}$ D) $2\sqrt{7}$

107) Simplify the below

$$5x^9 y^2 z^{-1} \cdot 5x^{-9} z^4$$

108) Solve the below equation

$$1 - 4x^{\frac{3}{2}} = -31$$

A) $\{3\}$ B) $\{3, -3\}$

C) $\{4\}$ D) $\{4, -3\}$

109) Simplify the below :

$$\sqrt{216}$$

A) 10   B) $6\sqrt{6}$

C) $3\sqrt{3}$ D) $6\sqrt{2}$

1) Identify the polynomial by degree and number of terms. Choose the right option.

$$-4$$

A) constant monomial

B) constant trinomial

C) linear polynomial with 0 terms

D) linear monomial

2) Identify the polynomial by degree and number of terms. Choose the right option.

$$6a^2 - 5a^6 - 8a^5$$

A) cubic polynomial with six terms

B) quintic trinomial

C) sixth degree trinomial

D) quintic monomial

3) Identify the polynomial by degree and number of terms. Choose the right option.

$$-6 + 4x$$

A) linear monomial

B) quadratic monomial

C) linear binomial

D) quartic binomial

4) Identify the polynomial by degree and number of terms. Choose the right option.

$$9a^4$$

A) cubic monomial

B) quartic trinomial

C) quartic monomial

D) linear monomial

5) Identify the polynomial by degree and number of terms. Choose the right option.

$$b - 7$$

A) quartic binomial

B) quadratic monomial

C) cubic binomial

D) linear binomial

6) Identify the polynomial by degree and number of terms. Choose the right option.

$$-3r - 2$$

A) linear monomial

B) quadratic binomial

C) linear binomial

D) quartic binomial

7) Identify the polynomial by degree and number of terms. Choose the right option.

$$-6x^2 - 8x^3$$

A) linear monomial

B) quadratic trinomial

C) quadratic binomial

D) cubic binomial

8) Identify the polynomial by degree and number of terms. Choose the right option.

$$10n^4 - 7n^3$$

A) quintic binomial

B) quadratic polynomial with four terms

C) quartic monomial

D) quartic binomial

9) Identify the polynomial by degree and number of terms. Choose the right option.

$$3x - 2x^5$$

A) quadratic binomial

B) quintic binomial

C) **quadratic polynomial with five terms**

D) cubic binomial

10) Identify the polynomial by degree and number of terms. Choose the right option.

$$5n^3 - 10n - 7 - 6n^2$$

A) quartic trinomial

B) constant trinomial

C) cubic trinomial

D) cubic polynomial with four terms

11) Identify the polynomial by degree and number of terms. Choose the right option.

$$-3b$$

A) quadratic monomial

B) linear binomial

C) linear monomial

D) quartic binomial

12) Identify the polynomial by degree and number of terms. Choose the right option.

$$-2r^4$$

A) constant binomial

B) quadratic monomial

C) quartic monomial

D) linear polynomial with four terms

13) Identify the polynomial by degree and number of terms. Choose the right option.

$$-4x$$

A) linear monomial

B) cubic binomial

C) linear binomial

D) linear trinomial

14) Identify the polynomial by degree and number of terms. Choose the right option.

$$4b$$

A) quartic monomial

B) quadratic monomial

C) linear monomial

D) quadratic binomial

15) Identify the polynomial by degree and number of terms. Choose the right option.

$$4$$

A) quartic trinomial

B) linear polynomial with 0 terms

C) constant monomial

D) quintic monomial

16) Identify the polynomial by degree and number of terms. Choose the right option.

$$-6x^2$$

A) cubic monomial

B) quartic monomial

C) quadratic monomial

D) quadratic binomial

www.math-knots.com | www.a4ace.com

17) Which expression is equivalent to the below.

$$\left(a^5 + 2a^3 + 8a^4\right) + \left(12a^5 - a^3\right)$$

A) $13a^5 + 8a^4 - 4a^3$

B) $13a^5 + 8a^4 - 5a^3$

C) $13a^5 + 8a^4 - 16a^3$

D) $13a^5 + 8a^4 + a^3$

18) Which expression is equivalent to the below.

$$\left(5x^4 - 1 - x^5\right) + \left(2 + x^4\right)$$

A) $-x^5 + 2x^4 + 1$

B) $-2x^5 + 2x^4 + 5$

C) $-x^5 + 2x^4 + 5$

D) $-x^5 + 6x^4 + 1$

19) Which expression is equivalent to the below.

$$\left(9v^5 - 5v^2 + 10v^3\right) + \left(7v^2 - 10v^3\right)$$

A) $9v^5 + 2v^2$

B) $9v^5 + 8v^2 - 11v^3$

C) $9v^5 + 8v^2 - 8v^3$

D) $9v^5 + 2v^2 - 11v^3$

20) Which expression is equivalent to the below.

$$\left(7 - 7m^2 - m^3\right) + \left(5m^2 + 6m^3\right)$$

A) $5m^3 - 11m^2 + 8$

B) $5m^3 - 2m^2 + 7$

C) $5m^3 - 2m^2 + 8$

D) $13m^3 - 11m^2 + 8$

21) Which expression is equivalent to the below.

$$\left(11b^3 + 4 - 5b\right) + \left(9 + 3b\right)$$

A) $b^3 - 2b + 8$

B) $11b^3 - 2b + 13$

C) $11b^3 - 2b + 8$

D) $b^3 - 2b + 12$

22) Which expression is equivalent to the below.

$$\left(9x^4 + 6x^3 - 11x^2\right) + \left(x^3 - 5x^4\right)$$

A) $4x^4 + 17x^3 - 11x^2$

B) $4x^4 + 7x^3 - 11x^2$

C) $7x^4 + 7x^3 - 11x^2$

D) $7x^4 + 17x^3 - 11x^2$

23) Which expression is equivalent to the below.

$$\left(7n^5 + 9n^4 - 3n\right) + \left(4n^4 + 10n^5\right)$$

A) $15n^5 + 13n^4 - 8n$

B) $17n^5 + 13n^4 - 3n$

C) $15n^5 + 13n^4 - 11n$

D) $15n^5 + 13n^4 - 3n$

24) Which expression is equivalent to the below.

$$\left(2k^3 - 11k^4 - 12k^2\right) + \left(4k^3 + 3k^2\right)$$

A) $-11k^4 + 6k^3 - 16k^2$

B) $-11k^4 + 6k^3 - 9k^2$

C) $-20k^4 + 6k^3 - 28k^2$

D) $-11k^4 + 6k^3 - 28k^2$

25) Which expression is equivalent to the below.

$$(7 + 8a^4 + 3a^2) + (10a^2 + 9a^4)$$

A) $17a^4 + 25a^2 + 9$

B) $17a^4 + 13a^2 + 9$

C) $17a^4 + 25a^2 + 18$

D) $17a^4 + 13a^2 + 7$

26) Which expression is equivalent to the below.

$$(3k^5 - 10k^3 - 6) + (12k^5 + 3)$$

A) $15k^5 - 3k^3 - 4$

B) $15k^5 - 3k^3 - 14$

C) $15k^5 - 10k^3 - 3$

D) $15k^5 - 3k^3 - 3$

27) Which expression is equivalent to the below.

$$(10x^2 - 3 + 4x^3) + (12x^3 + 6)$$

A) $3x^3 + 16x^2 + 3$

B) $16x^3 + 10x^2 + 3$

C) $13x^3 + 10x^2 + 3$

D) $3x^3 + 10x^2 + 3$

28) Which expression is equivalent to the below.

$$(6n + n^3 - 3n^5) + (8n - 11n^3)$$

A) $-3n^5 - 10n^3 + 14n$

B) $2n^5 - 10n^3 + 24n$

C) $2n^5 - 10n^3 + 14n$

D) $2n^5 - 9n^3 + 24n$

29) Which expression is equivalent to the below.

$$(8x^5 - 11x^2 - x^4) + (2x^4 - 3x^2)$$

A) $8x^5 - 7x^4 - 14x^2$

B) $8x^5 + 3x^4 - 14x^2$

C) $8x^5 + x^4 - 14x^2$

D) $8x^5 - 6x^4 - 14x^2$

30) Which expression is equivalent to the below.

$$(6b^3 + 3b^2 - 5b) + (10b^3 + 12b^2)$$

A) $16b^3 + 25b^2 - 5b$

B) $16b^3 + 15b^2 - 5b$

C) $16b^3 + 17b^2 - 5b$

D) $16b^3 + 6b^2 - 5b$

31) Which expression is equivalent to the below.

$$(10a^2 - a^4 - 4) + (4a^4 + 3a^2)$$

A) $-4a^4 + 13a^2 - 4$

B) $-4a^4 + 13a^2 + 6$

C) $3a^4 + 13a^2 - 4$

D) $-4a^4 + 13a^2 + 17$

32) Which expression is equivalent to the below.

$$(4 - 11k^4 - 4k^5) + (11 + 12k^5)$$

A) $8k^5 - 20k^4 + 9$

B) $8k^5 - 20k^4 + 15$

C) $8k^5 - 11k^4 + 15$

D) $18k^5 - 20k^4 + 9$

33) Which expression is equivalent to the below.

$$(5b + 5b^3 - 8b^2) + (14b^3 - 13b - 9b^2)$$

A) $19b^3 - 17b^2 - 8b$

B) $33b^3 - 17b^2 - 19b$

C) $19b^3 - 17b^2 - 22b$

D) $19b^3 - 17b^2 - 19b$

34) Which expression is equivalent to the below.

$$(4r^5 - 3r^3 + 14r^2) + (10r^3 - 11r^5 + 12r^2)$$

A) $-17r^5 + 5r^3 + 26r^2$

B) $-7r^5 + 5r^3 + 26r^2$

C) $-7r^5 + 7r^3 + 26r^2$

D) $-17r^5 + 5r^3 + 27r^2$

35) Which expression is equivalent to the below.

$$(9n + 6n^3 + 3n^4) + (8n^5 + 7n + 7n^4)$$

A) $6n^5 + 15n^4 + 6n^3 + 25n$

B) $8n^5 + 10n^4 + 6n^3 + 16n$

C) $8n^5 + 15n^4 + 6n^3 + 25n$

D) $8n^5 + 10n^4 + 6n^3 + 25n$

36) Which expression is equivalent to the below.

$$(7a^5 - 9 - 12a) + (10a^5 - 8a^4 - 6)$$

A) $17a^5 - 19a^4 - 12a - 15$

B) $17a^5 - 19a^4 - 6a - 15$

C) $17a^5 - 19a^4 - a - 15$

D) $17a^5 - 8a^4 - 12a - 15$

37) Which expression is equivalent to the below.

$$(6 - 8x^5 - 3x^3) + (5 + 4x^3 - x^5)$$

A) $-9x^5 + x^3 + 11$

B) $-10x^5 + 4x^3 + 11$

C) $-10x^5 - 7x^3 + 11$

D) $-9x^5 - 7x^3 + 11$

38) Which expression is equivalent to the below.

$$(12p + 3p^4 - 9p^5) + (2p - 9p^4 + 8p^5)$$

A) $-p^5 - 30p^4 + 14p$

B) $-p^5 - 17p^4 + 14p$

C) $-p^5 - 6p^4 + 14p$

D) $-p^5 - 20p^4 + 14p$

39) Which expression is equivalent to the below.

$$(10v^2 - 8v^3 - 2v^5) + (10v^3 - 6v^2 - 9v^5)$$

A) $-11v^5 - 5v^3 + 4v^2$

B) $-11v^5 + 2v^3 + 4v^2$

C) $-12v^5 - 5v^3 + 4v^2$

D) $-11v^5 - v^3 + 4v^2$

40) Which expression is equivalent to the below.

$$(11v^2 - v^3 - 11v^4) + (v^3 + 7v^5 + 5v^4)$$

A) $-6v^4 + 11v^2$

B) $-2v^4 + 11v^2 + 6v^3$

C) $-6v^4 + 11v^2 + 6v^3$

D) $7v^5 - 6v^4 + 11v^2$

41) Which expression is equivalent to the below.

$$(8x^5 + 14 + 10x^4) + (x^4 - 2 + 8x^5)$$

A) $16x^5 + 11x^4 + 12$

B) $11x^5 + 11x^4 + 12$

C) $11x^5 + 2x^4 + 12$

D) $24x^5 + 11x^4 + 12$

42) Which expression is equivalent to the below.

$$(10x - 3 + 12x^4) + (11 + 4x^4 - 11x^3)$$

A) $16x^4 - 11x^3 + 2x + 8$

B) $16x^4 - 7x^3 + 13x + 8$

C) $16x^4 - 11x^3 + 10x + 8$

D) $16x^4 - 11x^3 + 13x + 8$

43) Which expression is equivalent to the below.

$$(14b^3 - 2b^2 - 2) + (7b^2 - 6 - 11b^3)$$

A) $-b^3 + 5b^2 - 17$

B) $4b^3 + 5b^2 - 17$

C) $3b^3 + 5b^2 - 8$

D) $-b^3 + 5b^2 - 8$

44) Which expression is equivalent to the below.

$$(14v^5 + 3 + 3v) + (11v^3 - 4v - 7v^5)$$

A) $5v^5 + 11v^3 - v + 8$

B) $5v^5 + 11v^3 - v + 3$

C) $v^5 + 11v^3 - v + 3$

D) $7v^5 + 11v^3 - v + 3$

45) Which expression is equivalent to the below.

$$(8n + n^5 + 13n^3) + (7n^5 - 13n + 12n^2)$$

A) $8n^5 - n^3 + 12n^2 - 5n$

B) $6n^5 - n^3 + 12n^2 - 5n$

C) $8n^5 + 13n^3 + 12n^2 - 5n$

D) $6n^5 + 11n^3 + 12n^2 - 5n$

46) Which expression is equivalent to the below.

$$(14 - 10x - 14x^2) + (5 - 4x + 4x^2)$$

A) $4x^2 - 25x + 19$

B) $2x^2 - 25x + 19$

C) $-10x^2 - 14x + 19$

D) $4x^2 - 14x + 19$

47) Which expression is equivalent to the below.

$$(9v^4 + 12v - 12) + (9v^5 - 3 - 9v)$$

A) $9v^5 + 4v^4 + 3v - 15$

B) $9v^5 + 4v^4 - 5v - 15$

C) $9v^5 + 9v^4 + 3v - 15$

D) $9v^5 + 10v^4 + 3v - 15$

48) Which expression is equivalent to the below.

$$(13x^2 + 11x^5 + 2x) + (7x - 11x^4 - 6x^5)$$

A) $5x^5 - 11x^4 + 13x^2 + 9x$

B) $-x^5 - 11x^4 + 13x^2 - 2x$

C) $5x^5 - 11x^4 + 13x^2 - 2x$

D) $-x^5 - 11x^4 + 13x^2 + x$

49) Which expression is equivalent to the below.

$$\left(8n^4 - 2n^2 + 11n^3\right) + \left(11n^4 - 8n^3 + 6n^2\right)$$

A) $15n^4 + 22n^3 + 4n^2$

B) $15n^4 + 11n^3 + 4n^2$

C) $19n^4 + 3n^3 + 4n^2$

D) $19n^4 + 11n^3 + 4n^2$

50) Which expression is equivalent to the below.

$$\left(12p^2 - 3p - 12p^3\right) + \left(5p - 13p^3 - 3p^2\right)$$

A) $-7p^3 + 9p^2 - 8p$

B) $-16p^3 + 9p^2 + 2p$

C) $-25p^3 + 9p^2 + 2p$

D) $-16p^3 + 9p^2 - 8p$

51) Which expression is equivalent to the below.

$$\left(14 - 13x - 4x^3\right) + \left(3x + 4x^3 - 8\right)$$

A) $-10x + 19 + 14x^5$

B) $-10x + 19 + 27x^5$

C) $-10x + 6$

D) $-10x + 19$

52) Which expression is equivalent to the below.

$$\left(11 + 11p^5 + 14p^3\right) + \left(14p^5 - 6p^2 - 8\right)$$

A) $25p^5 + 26p^3 - 6p^2 + 3$

B) $39p^5 + 27p^3 - 6p^2 + 3$

C) $25p^5 + 27p^3 - 6p^2 + 3$

D) $25p^5 + 14p^3 - 6p^2 + 3$

53) Which expression is equivalent to the below.

$$\left(16 + 12m^2 - 19m\right) - \left(19 + 14m^2\right)$$

A) $-12m^2 - 19m - 3$

B) $-2m^2 - 19m - 3$

C) $-12m^2 - 19m - 22$

D) $8m^2 - 19m - 3$

54) Which expression is equivalent to the below.

$$\left(11m^5 + 6m^2 + 5m\right) - \left(16m^2 + 7m^5\right)$$

A) $-11m^5 + 3m^2 + 5m$

B) $4m^5 - 10m^2 + 5m$

C) $-11m^5 + 3m^2 + 23m$

D) $4m^5 + 3m^2 + 5m$

55) Which expression is equivalent to the below.

$$\left(9 - 6n^5 + 7n^2\right) - \left(20 + 20n^2\right)$$

A) $-6n^5 - 13n^2 - 17$

B) $-6n^5 - 13n^2 - 11$

C) $-6n^5 - 20n^2 - 17$

D) $-6n^5 - 20n^2 - 13$

56) Which expression is equivalent to the below.

$$\left(4x^4 + 16x - 2\right) - \left(20x + 20\right)$$

A) $11x^4 + 9x - 22$

B) $3x^4 + 9x - 22$

C) $4x^4 - 4x - 22$

D) $11x^4 - 4x - 22$

Algebra 1

57) Which expression is equivalent to the below.

$$\left(12x + 1 - 11x^2\right) - \left(8x^2 - 18x\right)$$

A) $-8x^2 + 30x + 1$

B) $-23x^2 + 30x - 16$

C) $-19x^2 + 30x + 1$

D) $-8x^2 + 30x - 16$

58) Which expression is equivalent to the below.

$$\left(14x^2 + 9x + 12x^4\right) - \left(8x - 10x^2\right)$$

A) $12x^4 + 26x^2 + x$

B) $12x^4 + 10x^2 + x$

C) $12x^4 + 12x^2 + x$

D) $12x^4 + 24x^2 + x$

59) Which expression is equivalent to the below.

$$\left(6x + x^3 + 8x^2\right) - \left(2x^3 + 19x\right)$$

A) $-x^3 + 8x^2 - 13x$

B) $-x^3 + 8x^2 - 25x$

C) $-x^3 + 8x^2 - 41x$

D) $-x^3 + 8x^2 - 24x$

60) Which expression is equivalent to the below.

$$\left(20a^2 + 20 + 9a^4\right) - \left(19 + 8a^2\right)$$

A) $9a^4 + 12a^2 + 1$

B) $10a^4 + 6a^2 + 1$

C) $9a^4 + 6a^2 + 1$

D) $10a^4 + 6a^2 - 2$

61) Which expression is equivalent to the below.

$$\left(5p + 10p^4 - 16p^3\right) - \left(20p^4 - 20p\right)$$

A) $-10p^4 - 16p^3 + 25p$

B) $-21p^4 - 16p^3 + 25p$

C) $-23p^4 - 16p^3 + 25p$

D) $-21p^4 - 16p^3 + 23p$

62) Which expression is equivalent to the below.

$$\left(20x^2 + 20 + 12x\right) - \left(14 - 18x^2\right)$$

A) $38x^2 + 12x + 6$

B) $38x^2 - 7x + 6$

C) $38x^2 - 11x + 19$

D) $38x^2 - 11x + 6$

63) Which expression is equivalent to the below.

$$\left(14n^3 - n^2 - 14n^5\right) - \left(7n^5 + 2n^2\right)$$

A) $-21n^5 + 14n^3 - 3n^2$

B) $-21n^5 + 32n^3 + 16n^2$

C) $-21n^5 + 14n^3 + 16n^2$

D) $-21n^5 + 20n^3 + 16n^2$

64) Which expression is equivalent to the below.

$$\left(4m^4 + m + 3m^3\right) - \left(2m - 7m^3\right)$$

A) $-16m^4 + 10m^3 + 15m$

B) $4m^4 + 10m^3 - m$

C) $-12m^4 + 10m^3 + 15m$

D) $-12m^4 + 10m^3 - m$

Algebra 1

65) Which expression is equivalent to the below.

$$\left(5k - 18 - 12k^4\right) - \left(12 - 18k^4\right)$$

A) $-3k^4 + 5k - 30$

B) $6k^4 + 5k - 30$

C) $-9k^4 + 24k - 30$

D) $-3k^4 + 24k - 30$

66) Which expression is equivalent to the below.

$$\left(9m^5 - 7m^3 - 19m\right) - \left(19m^3 - 18m^5\right)$$

A) $27m^5 - 26m^3 - 19m$

B) $11m^5 - 10m^3 - 14m$

C) $11m^5 - 26m^3 - 19m$

D) $11m^5 - 10m^3 - 19m$

67) Which expression is equivalent to the below.

$$\left(12 + 14a^2 + 19a^5\right) - \left(18 - 17a^2\right)$$

A) $19a^5 + 31a^2 - 6$

B) $2a^5 + 31a^2 - 6$

C) $-15a^5 + 31a^2 - 6$

D) $-13a^5 + 31a^2 - 6$

68) Which expression is equivalent to the below.

$$\left(13 + 5k^3 - 7k\right) - \left(19k + 17\right)$$

A) $-13k^3 - 26k - 4$

B) $-13k^3 - 26k + 15$

C) $-13k^3 - 6k + 15$

D) $5k^3 - 26k - 4$

69) Simplify the below expressions.

$$\left(6r + 8r^4 - 5r^2\right) - \left(10r - r^2 + 11r^4\right)$$

70) Simplify the below expressions.

$$\left(11 + 9r^2 + 10r^3\right) - \left(8r^2 + 8r^3 - 11\right)$$

71) Simplify the below expressions.

$$\left(5n^4 + 11 - 12n^3\right) - \left(9 + 5n^3 + 2n^4\right)$$

72) Simplify the below expressions.

$$\left(11b^2 - 4 + 7b^4\right) - \left(9b^2 + 12b^4 - 12\right)$$

73) Simplify the below expressions.

$$\left(12 - 9m^2 + 2m^4\right) - \left(4m^2 - 11 + 11m^4\right)$$

74) Simplify the below expressions.

$$\left(7x^4 - 4x^3 - 3\right) - \left(6x^4 - 6 - 6x^3\right)$$

Algebra 1

Vol 2
Week 29
Polynomials

75) Simplify the below expressions.

$$\left(5 - 8x^4 - 8x^2\right) - \left(2 + 6x^2 - 10x^4\right)$$

76) Simplify the below expressions.

$$\left(7x^3 - 11 - 4x\right) - \left(11 - 5x^3 - 4x\right)$$

77) Simplify the below expressions.

$$\left(3 - 6a^3 + 12a\right) - \left(6a^3 - 7 - 11a\right)$$

78) Simplify the below expressions.

$$\left(1 + 3m - 8m^3\right) - \left(7 - 6m - 2m^3\right)$$

79) Simplify the below expressions.

$$\left(8x^4 + 6 - 9x^2\right) - \left(12x^4 - 3 + 7x^2\right)$$

80) Simplify the below expressions.

$$\left(4n^4 - 10n^3 + 10\right) - \left(8n^3 - 5 - 7n\right)$$

81) Simplify the below expressions.

$$\left(5k^4 + 7k + 12k^2\right) - \left(4k^2 + 7k^4 - 5\right)$$

82) Simplify the below expressions.

$$\left(10x^2 + 6x^3 + 5x^4\right) - \left(8x^2 + 4x^3 + 6x^4\right)$$

83) Simplify the below expressions.

$$\left(14 - x^3 + 7x^2 + 20x\right) + \left(4x^4 + 12x^3 + 17 + 20x\right)$$

84) Simplify the below expressions.

$$\left(6x^4 + 2x^2 - 6 - 18x\right) - \left(3x^3 - x^2 + 18 - 11x\right)$$

85) Simplify the below expressions.

$$\left(6m^2 - 6 + 3m + 20m^4\right)$$
$$+$$
$$\left(15 - 12m - 9m^4 + 18m^2\right)$$

86) Simplify the below expressions.

$$\left(13x^3 - 11x^2 + 10x^4 + 11x\right)$$
$$+$$
$$\left(20x - 2x^2 + 14x^3 + 10x^4\right)$$

303

87) Simplify the below expressions.

$$\left(2x^4 - 7x^3 + 6 - 5x\right)$$
$$+$$
$$\left(3x^4 + 17x^2 - 20x - 19\right)$$

88) Simplify the below expressions.

$$\left(20n^2 + 13n + 13 - 9n^4\right)$$
$$-$$
$$\left(10 + 5n^2 - 13n^4 - 8n\right)$$

89) Simplify the below expressions.

$$\left(13 + 8r - 7r^2 + 9r^4\right)$$
$$-$$
$$\left(15r^2 - 17r^3 + 15r - 4r^4\right)$$

90) Simplify the below expressions.

$$\left(9n + 2n^3 + 3n^2 + 5n^4\right)$$
$$+$$
$$\left(16n - 16 - 7n^3 - 16n^4\right)$$

91) Simplify the below expressions.

$$\left(2 + 15b - 7b^4 + 6b^2\right)$$
$$+$$
$$\left(13 + 5b + 5b^4 - 16b^2\right)$$

92) Simplify the below expressions.

$$\left(20x^2 - 13x + 11x^4 - 8\right)$$
$$-$$
$$\left(17x^4 + 9 + 5x^2 - 17x\right)$$

93) Find the product of the below.

$$8\left(-3x + 11\right)$$

A) $-24x + 88$ B) $-49x^2 - 21x$

C) $27x + 21$ D) $112x + 42$

94) Find the product of the below.

$$-16\left(-3r + 14\right)$$

A) $-12r + 90$ B) $48r - 224$

C) $-187r^2 + 170r$ D) $-112r - 48$

95) Find the product of the below.

$$14\left(-20r + 10\right)$$

A) $-280r + 140$ B) $-24r^3 + 6r^2$

C) $-168r^2 - 84r$ D) $-96r + 114$

96) Find the product of the below.

$$-20n\left(-15n + 14\right)$$

A) $340n^4 + 85n^3$ B) $-160n - 220$

C) $300n^2 - 280n$ D) $300n^2 - 280n$

97) Find the product of the below.

$$4\left(-13p + 9\right)$$

A) $45p - 210$ B) $-52p + 36$

C) $-104p^2 - 56p$ D) $-216p - 144$

98) Find the product of the below.

$$11k^2\left(-3k - 4\right)$$

A) $110k^2 - 88k$ B) $-195k - 195$

C) $-33k^3 - 44k^2$ D) $323k + 153$

99) Find the product of the below.

$$-15(-16x+3)$$

A) $-49x^2-21x$ B) $-15x+24$

C) $240x-45$ D) $-60x^2-40x$

100) Find the product of the below.

$$-18(-12n+4)$$

A) $-270n^2-45n$ B) $-170n-34$

C) $-20n+2$ D) $216n-72$

101) Find the product of the below.

$$-18k(-15k-18)$$

A) $270k+72$ B) $270k^2+324k$

C) $-240k+300$ D) $130k+65$

102) Find the product of the below.

$$-(14a-15)$$

A) $30a^2+24a$ B) $-288a^3+18a^2$

C) $-340a-136$ D) $-14a+15$

103) Find the product of the below.

$$-9(5k+8)$$

A) $-45k-72$ B) $88k^3-104k^2$

C) $-120k+144k$ D) $-320k^2+144k$

104) Find the product of the below.

$$19(18b-13)$$

A) $-36b^4-12b^3$ B) $54b+117$

C) $342b-247$ D) $220b-400$

105) Find the product of the below.

$$-18(n+16)$$

A) $24n+14$ B) $-18n-288$

C) $6n^3-21n^2$ D) $48n+54$

106) Find the product of the below.

$$19v^2(4v+3)$$

A) $51v^5+187v^4$ B) $76v^3+57v^2$

C) $300v+400$ D) $-35v^2-30v$

107) Find the product of the below.

$$16(-3n+11)$$

A) $-15n^3-20n^2$ B) $9n-33$

C) $90n^2+90n$ D) $-48n+176$

108) Find the product of the below.

$$-16x(20x+14)$$

A) $-320x^2-224x$ B) $-60x+33$

C) $-78x^3-108x^2$ D) $304x^5+342x^4$

Algebra 1

109) Find the product of the below.

$$(8a + 13)(-14a + 9)$$

A) $-112a^2 - 254a - 117$

B) $-3a^2 - 22a - 40$

C) $-112a^2 + 117$

D) $-112a^2 - 110a + 117$

110) Find the product of the below.

$$(-3x + 1)(13x - 8)$$

A) $-39x^2 - 11x + 8$

B) $-14x^2 - 12$

C) $-39x^2 + 37x - 8$

D) $-14x^2 + 86x - 12$

111) Find the product of the below.

$$(-13m - 2)(8m + 1)$$

A) $-104m^2 - 3m + 2$

B) $-104m^2 + 3m + 2$

C) $-104m^2 - 29m - 2$

D) $-104m^2 - 2$

112) Find the product of the below.

$$(-3x + 4)(-2x + 9)$$

A) $6x^2 - 35x + 36$

B) $-16x^2 - 28x + 18$

C) $6x^2 + 36$

D) $-16x^2 + 18$

113) Find the product of the below.

$$(6x - 10)(7x - 2)$$

A) $55x^2 - 40x - 140$

B) $42x^2 - 82x + 20$

C) $54x^2 - 51x - 45$

D) $42x^2 + 20$

114) Find the product of the below.

$$(-10x + 12)(10x - 1)$$

A) $-100x^2 - 12$

B) $-100x^2 + 110x + 12$

C) $-100x^2 + 130x - 12$

D) $-100x^2 - 110x + 12$

115) Find the product of the below.

$$(3x + 14)(11x - 4)$$

A) $54x^2 + 111x + 56$

B) $54x^2 + 15x - 56$

C) $33x^2 + 142x - 56$

D) $54x^2 - 56$

116) Find the product of the below.

$$(5v - 10)(6v - 12)$$

A) $30v^2 - 120v + 120$

B) $30v^2 + 120$

C) $-10v^2 + 123v - 36$

D) $30v^2 - 120$

117) Find the product of the below.

$$(- 8\,n - 8)\,(- 2\,n - 4)$$

A) $104n^2 - 69n - 54$

B) $16n^2 + 48n + 32$

C) $16n^2 - 16n - 32$

D) $16n^2 + 16n - 32$

118) Find the product of the below.

$$(- 10\,p + 3)\,(14\,p + 10)$$

A) $- 6\,p^2 + 34p - 20$

B) $- 140\,p^2 - 142\,p - 30$

C) $- 140\,p^2 - 58\,p + 30$

D) $- 140\,p^2 + 142\,p - 30$

119) Find the product of the below.

$$(11\,b + 11)\,(10\,b + 11)$$

A) $24b^2 - 112$

B) $110b^2 + 231\,b + 121$

C) $24b^2 - 136\,b + 112$

D) $24b^2 - 88\,b - 112$

120) Find the product of the below.

$$(4\,v - 9)\,(- 10\,v - 10)$$

A) $-40v^2 + 130v - 90$

B) $-39v^2 - 58v + 77$

C) $-40v^2 + 50v + 90$

D) $-39v^2 + 77$

121) Find the product of the below.

$$(- 3\,m + 11)\,(- 12\,m + 7)$$

A) $36m^2 - 153m + 77$

B) $36m^2 + 111\,m - 77$

C) $-60m^2 + 22\,m + 14$

D) $36m^2 + 77$

122) Find the product of the below.

$$(13\,b + 7)\,(- 4\,b + 12)$$

A) $- 52b^2 + 128b + 84$

B) $- 20b^2 - 28$

C) $- 20b^2 - 136b + 28$

D) $- 20b^2 - 144b - 28$

123) Find the product of the below.

$$(- 12\,v - 5)\,(v - 4)$$

A) $- 12v^2 - 53v - 20$

B) $- 12v^2 + 20$

C) $- 12v^2 + 53v - 20$

D) $- 12v^2 + 43v + 20$

124) Find the product of the below.

$$(7\,k - 6)\,(- 5\,k - 6)$$

A) $-112k^2 + 126$

B) $-112k^2 + 268k - 126$

C) $-35k^2 - 12k + 36$

D) $-112k^2 + 124k + 126$

 **Algebra 1**

125)   Find the product of the below.

$$(7x - 4)(6x^2 - 6x - 6)$$

126)   Find the product of the below.

$$(2x + 1)(4x^2 + 4x + 5)$$

127)   Find the product of the below.

$$(8m + 1)(4m^2 + 3m + 3)$$

128)   Find the product of the below.

$$(3n + 2)(5n^2 + n + 3)$$

129)   Find the product of the below.

$$(7b - 4)(4b^2 - 4b + 7)$$

130)   Find the product of the below.

$$(3x + 2)(6x^2 + 5x - 3)$$

131)   Find the product of the below.

$$(6a - 6)(2a^2 - a - 3)$$

132)   Find the product of the below.

$$(7n + 8)(7n^2 + 6n + 4)$$

133)   Find the product of the below.

$$(2n + 1)(4n^2 - 5n + 8)$$

134)   Find the product of the below.

$$(6a + 5)(5a^2 - 8a + 1)$$

135)   Find the product of the below.

$$(3k - 4)(5k^2 + 6k - 2)$$

136)   Find the product of the below.

$$(5n + 6)(5n^2 - 2n - 8)$$

137)   Find the product of the below.

$$(2a + 6)(6a^2 + 7a - 4)$$

138)   Find the product of the below.

$$(x + 3)(7x^2 - 8x - 6)$$

139)   Find the product of the below.

$$(7m - 1)(7m^2 - 2m - 3)$$

140)   Find the product of the below.

$$(7n^2 - 3n + 8)(9n^2 + 9n - 3)$$

           www.math-knots.com | www.a4ace.com

141) Find the product of the below.

$$(10\,n^2 + 2\,n + 9)(7\,n^2 - 8\,n + 8)$$

142) Find the product of the below.

$$(6\,v^2 + 9\,v - 4)(7\,v^2 - 8\,v + 4)$$

143) Find the product of the below.

$$(4\,a^2 + a + 4)(9\,a^2 - 4\,a + 3)$$

144) Find the product of the below.

$$(2\,r^2 + 9\,r + 6)(8\,r^2 + 9\,r - 8)$$

1) Simplify the below.

$$\left(7n^5 + 36n^4 + 7n^3\right) \div 12n$$

A) $\dfrac{n}{6} + \dfrac{1}{3} + \dfrac{4}{n}$

B) $\dfrac{3n}{2} + 1 + \dfrac{1}{n}$

C) $\dfrac{1}{4} + \dfrac{3}{n} + \dfrac{1}{8n^2}$

D) $\dfrac{7n^4}{12} + 3n^3 + \dfrac{7n^2}{12}$

2) Simplify the below.

$$\left(6n^3 + 24n^2 + 4n\right) \div 12n^2$$

A) $\dfrac{2n}{7} + \dfrac{1}{7} + \dfrac{2}{n}$

B) $n^2 + \dfrac{n}{4} + \dfrac{1}{2}$

C) $\dfrac{n}{2} + 2 + \dfrac{1}{3n}$

D) $\dfrac{5}{8} + \dfrac{1}{2n} + \dfrac{1}{4n^2}$

3) Simplify the below.

$$\left(k^3 + 5k^2 + 70k\right) \div 14k^2$$

A) $\dfrac{k}{14} + \dfrac{5}{14} + \dfrac{5}{k}$

B) $\dfrac{1}{3} + \dfrac{6}{k} + \dfrac{1}{3k^2}$

C) $3k^2 + \dfrac{3k}{14} + \dfrac{1}{14}$

D) $\dfrac{k}{2} + \dfrac{5}{4} + \dfrac{1}{2k}$

4) Simplify the below.

$$\left(4r^3 + 7r^2 + 4r\right) \div 4r^3$$

A) $\dfrac{r^2}{4} + \dfrac{5r}{4} + 6$

B) $3r^2 + \dfrac{r}{2} + \dfrac{1}{2}$

C) $1 + \dfrac{7}{4r} + \dfrac{1}{r^2}$

D) $r^2 + \dfrac{5r}{4} + \dfrac{1}{2}$

www.math-knots.com | www.a4ace.com

5)   Simplify the below.

$$\left(4x^3 + 4x^2 + 40x\right) \div 8x^3$$

A) $\dfrac{7x^5}{4} + x^4 + 5x^3$

B) $\dfrac{x^2}{5} + x + \dfrac{1}{10}$

C) $\dfrac{x^4}{3} + \dfrac{x^3}{3} + 5x^2$

D) $\dfrac{1}{2} + \dfrac{1}{2x} + \dfrac{5}{x^2}$

6)   Simplify the below.

$$\left(4m^3 + 24m^2 + 3m\right) \div 6m$$

A) $\dfrac{m}{2} + 6 + \dfrac{1}{2m}$

B) $7m + \dfrac{1}{2} + \dfrac{1}{m}$

C) $\dfrac{2m^2}{3} + 4m + \dfrac{1}{2}$

D) $\dfrac{1}{6} + \dfrac{4}{m} + \dfrac{6}{m^2}$

7)   Simplify the below.

$$\left(3x^3 + 24x^2 + 4x\right) \div 4x$$

A) $\dfrac{3x^2}{4} + 6x + 1$

B) $\dfrac{x^2}{5} + x + \dfrac{1}{2}$

C) $3x^2 + x + \dfrac{3}{14}$

D) $\dfrac{x^2}{2} + x + 2$

8)   Simplify the below.

$$\left(5x^3 + 60x^2 + 60x\right) \div 12x$$

A) $4 + \dfrac{5}{9x} + \dfrac{7}{x^2}$

B) $\dfrac{x^2}{4} + \dfrac{3x}{8} + 5$

C) $6x^2 + 5x + 1$

D) $\dfrac{5x^2}{12} + 5x + 5$

9)   Simplify the below.

$$\left(7v^3 + 2v^2 + 3v\right) \div 14v^3$$

A) $\dfrac{v}{2} + 3 + \dfrac{2}{v}$

B) $\dfrac{7}{4} + \dfrac{1}{2v} + \dfrac{1}{2v^2}$

C) $\dfrac{7v}{8} + \dfrac{1}{4} + \dfrac{1}{8v}$

D) $\dfrac{1}{2} + \dfrac{1}{7v} + \dfrac{3}{14v^2}$

10)   Simplify the below.

$$\left(4n^4 + n^3 + 48n^2\right) \div 8n$$

A) $\dfrac{5n^2}{14} + 5n + 2$

B) $5n + 7 + \dfrac{1}{3n}$

C) $\dfrac{3n}{5} + 2 + \dfrac{2}{5n}$

D) $\dfrac{n^3}{2} + \dfrac{n^2}{8} + 6n$

www.math-knots.com | www.a4ace.com

11) Simplify the below.

$$\left(2b^3 + 7b^2 + 12b\right) \div 6b^3$$

A) $\dfrac{1}{5} + \dfrac{3}{b} + \dfrac{7}{10b^2}$

B) $6b^2 + b + \dfrac{1}{2}$

C) $\dfrac{1}{3} + \dfrac{7}{6b} + \dfrac{2}{b^2}$

D) $4b^3 + \dfrac{b^2}{3} + b$

12) Simplify the below.

$$\left(70x^3 + 56x^2 + 2x\right) \div 14x^3$$

A) $x^2 + 6x + 2$

B) $7x^6 + \dfrac{7x^5}{8} + 7x^4$

C) $\dfrac{5x^2}{14} + \dfrac{3x}{14} + \dfrac{1}{2}$

D) $5 + \dfrac{4}{x} + \dfrac{1}{7x^2}$

13) Simplify the below.

$$\left(48x^3 + 8x^2 + 24x\right) \div 8x^3$$

A) $3x + 4 + \dfrac{1}{2x}$

B) $\dfrac{x^4}{3} + \dfrac{x^3}{3} + \dfrac{x^2}{3}$

C) $6 + \dfrac{1}{x} + \dfrac{3}{x^2}$

D) $3 + \dfrac{4}{x} + \dfrac{6}{x^2}$

14) Simplify the below.

$$\left(6n^3 + 42n^2 + 3n\right) \div 6n^3$$

A) $n^3 + \dfrac{7n^2}{12} + 6n$

B) $2n^3 + \dfrac{n^2}{2} + 2n$

C) $7n^2 + \dfrac{n}{10} + 4$

D) $1 + \dfrac{7}{n} + \dfrac{1}{2n^2}$

15) Simplify the below.

$$\left(4k^3 + 3k^2 + 54k\right) \div 9k$$

A) $\dfrac{k^2}{2} + 7k + \dfrac{3}{10}$

B) $\dfrac{4k^2}{9} + \dfrac{k}{3} + 6$

C) $3k + 7 + \dfrac{1}{k}$

D) $6k + 1 + \dfrac{1}{4k}$

16) Simplify the below.

$$\left(72n^5 + 12n^4 + 3n^3\right) \div 12n^3$$

A) $\dfrac{3n^3}{4} + \dfrac{5n^2}{4} + \dfrac{n}{2}$

B) $1 + \dfrac{1}{4n} + \dfrac{1}{2n^2}$

C) $6n^2 + n + \dfrac{1}{4}$

D) $\dfrac{7}{8} + \dfrac{1}{n} + \dfrac{7}{n^2}$

17) Simplify the below.

$$(n^2 + 18n + 81) \div (n + 9)$$

A) $n + 4 - \dfrac{7}{n+9}$

B) $n + 9$

C) $n + 6 - \dfrac{9}{n+9}$

D) $n + 6 - \dfrac{2}{n+9}$

18) Simplify the below.

$$(8r^2 - 38r + 24) \div (r - 4)$$

A) $8r - 7 - \dfrac{6}{r-4}$

B) $8r - 6$

C) $8r - 5 - \dfrac{3}{r-4}$

D) $8r - 8 - \dfrac{4}{r-4}$

19) Simplify the below.

$$(p^2 + 10p + 24) \div (p + 4)$$

A) $p + 6$

B) $p + 11 + \dfrac{4}{p+4}$

C) $p + 9 + \dfrac{5}{p+4}$

D) $p + 9 + \dfrac{1}{p+4}$

20) Simplify the below.

$$(6a^2 - 50a + 56) \div (a - 7)$$

A) $6a - 6 + \dfrac{5}{a-7}$

B) $6a - 8$

C) $6a - 4$

D) $6a - 7 + \dfrac{2}{a-7}$

21) Simplify the below.

$$(n^2 - 3n + 2) \div (n - 1)$$

A) $n - 3 + \dfrac{4}{n-1}$

B) $n - 5 + \dfrac{3}{n-1}$

C) $n - 2$

D) $n - 4 + \dfrac{5}{n-1}$

22) Simplify the below.

$$(b^2 - b - 90) \div (b + 9)$$

A) $b - 16 + \dfrac{1}{b+9}$

B) $b - 10$

C) $b - 19 + \dfrac{5}{b+9}$

D) $b - 13 - \dfrac{3}{b+9}$

23) Simplify the below.

$$(6m^2 - 53m - 9) \div (m - 9)$$

A) $6m + 3 + \dfrac{4}{m - 9}$

B) $6m + 2 - \dfrac{5}{m - 9}$

C) $6m + 1$

D) $6m + 4$

24) Simplify the below.

$$(b^2 + 16b + 63) \div (b + 9)$$

A) $b + 11 - \dfrac{6}{b + 9}$

B) $b + 7$

C) $b + 9 - \dfrac{3}{b + 9}$

D) $b + 12 - \dfrac{1}{b + 9}$

25) Simplify the below.

$$(10n^2 + 36n + 18) \div (n + 3)$$

A) $10n + 4$

B) $10n + 6$

C) $10n + 6 - \dfrac{3}{n + 3}$

D) $10n + 5 - \dfrac{4}{n + 3}$

26) Simplify the below.

$$(8r^2 + 13r - 6) \div (r + 2)$$

A) $8r - 6$

B) $8r - 4 - \dfrac{5}{r + 2}$

C) $8r - 5 + \dfrac{5}{r + 2}$

D) $8r - 3$

27) Simplify the below.

$$(p^2 + 9p + 20) \div (p + 4)$$

A) $p + 4 + \dfrac{4}{p + 4}$

B) $p + 4 + \dfrac{6}{p + 4}$

C) $p + 1 + \dfrac{2}{p + 4}$

D) $p + 5$

28) Simplify the below.

$$(9x^2 - 46x - 48) \div (x - 6)$$

A) $9x + 8$

B) $9x + 10 - \dfrac{4}{x - 6}$

C) $9x + 6 - \dfrac{5}{x - 6}$

D) $9x + 9 - \dfrac{7}{x - 6}$

 **Algebra 1**

29) Simplify the below.

$$\left(3p^2 - 27p + 54\right) \div \left(p - 6\right)$$

A) $3p - 12 + \dfrac{4}{p-6}$

B) $3p - 14 + \dfrac{2}{p-6}$

C) $3p - 11 + \dfrac{1}{p-6}$

D) $3p - 9$

30) Simplify the below.

$$\left(x^2 - 8x + 15\right) \div \left(x - 5\right)$$

A) $x - 3$ B) $x - 3 - \dfrac{8}{x-5}$

C) $x - \dfrac{1}{x-5}$ D) $x - 1 - \dfrac{5}{x-5}$

31) Simplify the below.

$$\left(8x^2 - 64x + 56\right) \div \left(x - 7\right)$$

A) $8x - 10$

B) $8x - 8$

C) $8x - 13 - \dfrac{2}{x-7}$

D) $8x - 11 + \dfrac{2}{x-7}$

32) Simplify the below.

$$\left(n^2 - 9n + 18\right) \div \left(n - 6\right)$$

A) $n - 3$

B) $n - 1 - \dfrac{6}{n-6}$

C) $n - 1 - \dfrac{4}{n-6}$

D) $n + 1 - \dfrac{5}{n-6}$

33) Simplify the below.

$$\left(x^2 - 6x - 7\right) \div \left(x + 1\right)$$

A) $x - 7$

B) $x - 2 + \dfrac{6}{x+1}$

C) $x - 5 + \dfrac{2}{x+1}$

D) $x - 4 + \dfrac{9}{x+1}$

34) Simplify the below.

$$\left(v^2 - 2v - 15\right) \div \left(v - 5\right)$$

A) $v + 3$

B) $v + 6 - \dfrac{1}{v-5}$

C) $v + 3 + \dfrac{1}{v-5}$

D) $v + 4 - \dfrac{1}{v-5}$

35) Simplify the below.

$$(a^2 + 9a - 10) \div (a + 10)$$

A) $a - 2 + \dfrac{2}{a + 10}$

B) $a - 7 + \dfrac{4}{a + 10}$

C) $a - 4 + \dfrac{7}{a + 10}$

D) $a - 1$

36) Simplify the below.

$$(k^2 + 4k + 4) \div (k + 2)$$

A) $k + 4 + \dfrac{6}{k + 2}$

B) $k + 2$

C) $k + 7 + \dfrac{11}{k + 2}$

D) $k + 3 + \dfrac{4}{k + 2}$

37) Simplify the below.

$$(14x^3 - 222x^2 - 289x + 282) \div (x - 17)$$

A) $14x^2 + 16x - 11 - \dfrac{8}{x - 17}$

B) $14x^2 + 16x - 12 - \dfrac{7}{x - 17}$

C) $14x^2 + 16x - 14 - \dfrac{10}{x - 17}$

D) $14x^2 + 16x - 17 - \dfrac{7}{x - 17}$

38) Simplify the below.

$$(p^3 + 30p^2 + 186p - 273) \div (p + 20)$$

A) $p^2 + 10p - 17 - \dfrac{4}{p + 20}$

B) $p^2 + 10p - 19 - \dfrac{1}{p + 20}$

C) $p^2 + 10p - 17 + \dfrac{3}{p + 20}$

D) $p^2 + 10p - 14 + \dfrac{7}{p + 20}$

39) Simplify the below.

$$(x^3 - 16x^2 - 32x + 246) \div (x - 17)$$

A) $x^2 + x - 18 - \dfrac{18}{x - 17}$

B) $x^2 + x - 15 - \dfrac{9}{x - 17}$

C) $x^2 + x - 21 - \dfrac{13}{x - 17}$

D) $x^2 + x - 18 - \dfrac{14}{x - 17}$

40) Simplify the below.

$$(m^3 + 10m^2 - 93m + 42) \div (m + 16)$$

A) $m^2 - 6m + 4 - \dfrac{7}{m + 16}$

B) $m^2 - 6m + 2 - \dfrac{11}{m + 16}$

C) $m^2 - 6m + 1 - \dfrac{6}{m + 16}$

D) $m^2 - 6m + 3 - \dfrac{6}{m + 16}$

www.math-knots.com | www.a4ace.com

41) Simplify the below.

$$\left(n^3 - 2n^2 - 9n + 28\right) \div \left(n - 3\right)$$

A) $n^2 + n - 11 + \dfrac{2}{n - 3}$

B) $n^2 + n - 8 + \dfrac{1}{n - 3}$

C) $n^2 + n - 9 + \dfrac{5}{n - 3}$

D) $n^2 + n - 6 + \dfrac{10}{n - 3}$

42) Simplify the below.

$$\left(x^3 + 3x^2 - 114x + 48\right) \div \left(x - 9\right)$$

A) $x^2 + 12x - 6 - \dfrac{13}{x - 9}$

B) $x^2 + 12x - 7 - \dfrac{8}{x - 9}$

C) $x^2 + 12x - 6 - \dfrac{6}{x - 9}$

D) $x^2 + 12x - 4 - \dfrac{12}{x - 9}$

43) Simplify the below.

$$\left(3r^3 - 14r^2 + 15r - 15\right) \div \left(r - 4\right)$$

A) $3r^2 - 2r + 6 + \dfrac{7}{r - 4}$

B) $3r^2 - 2r + 9 + \dfrac{8}{r - 4}$

C) $3r^2 - 2r + 7 + \dfrac{13}{r - 4}$

D) $3r^2 - 2r + 4 + \dfrac{2}{r - 4}$

44) Simplify the below.

$$\left(n^3 - 18n^2 + 51n + 59\right) \div \left(n - 14\right)$$

A) $n^2 - 4n - 8 - \dfrac{10}{n - 14}$

B) $n^2 - 4n - 7 - \dfrac{8}{n - 14}$

C) $n^2 - 4n - 6 - \dfrac{14}{n - 14}$

D) $n^2 - 4n - 5 - \dfrac{11}{n - 14}$

45) Simplify the below.

$$\left(k^3 - 9k^2 - 94k - 270\right) \div \left(k - 16\right)$$

A) $k^2 + 7k + 18 + \dfrac{18}{k - 16}$

B) $k^2 + 7k + 14 + \dfrac{22}{k - 16}$

C) $k^2 + 7k + 15 + \dfrac{25}{k - 16}$

D) $k^2 + 7k + 17 + \dfrac{22}{k - 16}$

46) Simplify the below.

$$\left(3b^3 - 47b^2 + b - 267\right) \div \left(b - 16\right)$$

A) $3b^2 + b + 16 + \dfrac{3}{b - 16}$

B) $3b^2 + b + 18 + \dfrac{1}{b - 16}$

C) $3b^2 + b + 17 - \dfrac{1}{b - 16}$

D) $3b^2 + b + 17 + \dfrac{5}{b - 16}$

47) Simplify the below.

$$\left(n^3 + 18n^2 + 46n - 238\right) \div \left(n + 13\right)$$

A) $n^2 + 5n - 15 + \dfrac{4}{n + 13}$

B) $n^2 + 5n - 16 + \dfrac{7}{n + 13}$

C) $n^2 + 5n - 18 + \dfrac{6}{n + 13}$

D) $n^2 + 5n - 19 + \dfrac{9}{n + 13}$

48) Simplify the below.

$$\left(n^3 - 28n^2 + 173n + 285\right) \div \left(n - 16\right)$$

A) $n^2 - 12n - 19 - \dfrac{19}{n - 16}$

B) $n^2 - 12n - 18 - \dfrac{20}{n - 16}$

C) $n^2 - 12n - 23 - \dfrac{13}{n - 16}$

D) $n^2 - 12n - 20 - \dfrac{18}{n - 16}$

49) Simplify the below.

$$\left(x^3 + 21x^2 + 78x - 28\right) \div \left(x + 16\right)$$

A) $x^2 + 5x - 2 + \dfrac{8}{x + 16}$

B) $x^2 + 5x - 3 + \dfrac{9}{x + 16}$

C) $x^2 + 5x - 2 + \dfrac{4}{x + 16}$

D) $x^2 + 5x + \dfrac{7}{x + 16}$

50) Simplify the below.

$$\left(r^3 - 16r^2 - 55r + 338\right) \div \left(r - 18\right)$$

A) $r^2 + 2r - 19 - \dfrac{4}{r - 18}$

B) $r^2 + 2r - 20 + \dfrac{2}{r - 18}$

C) $r^2 + 2r - 18 + \dfrac{7}{r - 18}$

D) $r^2 + 2r - 22 - \dfrac{2}{r - 18}$

51) Simplify the below.

$$\left(3x^3 - 42x^2 - 197x - 356\right) \div \left(x - 18\right)$$

A) $3x^2 + 12x + 15 - \dfrac{12}{x - 18}$

B) $3x^2 + 12x + 13 - \dfrac{11}{x - 18}$

C) $3x^2 + 12x + 16 - \dfrac{13}{x - 18}$

D) $3x^2 + 12x + 19 - \dfrac{14}{x - 18}$

52) Simplify the below.

$$\left(n^3 - 9n^2 + 6n + 55\right) \div \left(n - 5\right)$$

A) $n^2 - 4n - 20 - \dfrac{13}{n - 5}$

B) $n^2 - 4n - 17 - \dfrac{12}{n - 5}$

C) $n^2 - 4n - 18 - \dfrac{16}{n - 5}$

D) $n^2 - 4n - 14 - \dfrac{15}{n - 5}$

53) Simplify the below.

$$\left(n^3 + 7n^2 + 28n + 83\right) \div \left(n + 5\right)$$

A) $n^2 + 2n + 19 - \dfrac{5}{n + 5}$

B) $n^2 + 2n + 20 - \dfrac{2}{n + 5}$

C) $n^2 + 2n + 16 - \dfrac{3}{n + 5}$

D) $n^2 + 2n + 18 - \dfrac{7}{n + 5}$

54) Simplify the below.

$$\left(n^3 - 3n^2 + 7n - 10\right) \div \left(n - 2\right)$$

A) $n^2 - n + 5$

B) $n^2 - n + 4 + \dfrac{6}{n - 2}$

C) $n^2 - n + 2 + \dfrac{3}{n - 2}$

D) $n^2 - n + 7 + \dfrac{10}{n - 2}$

55) Simplify the below.

$$\left(x^3 - 9x^2 - 20x - 172\right) \div \left(x - 12\right)$$

A) $x^2 + 3x + 18 + \dfrac{21}{x - 12}$

B) $x^2 + 3x + 17 + \dfrac{16}{x - 12}$

C) $x^2 + 3x + 16 + \dfrac{20}{x - 12}$

D) $x^2 + 3x + 16 + \dfrac{16}{x - 12}$

56) Simplify the below.

$$\left(n^3 + 19n^2 + 57n - 26\right) \div \left(n + 4\right)$$

A) $n^2 + 15n - 9 - \dfrac{9}{n + 4}$

B) $n^2 + 15n - 6 - \dfrac{11}{n + 4}$

C) $n^2 + 15n - 3 - \dfrac{14}{n + 4}$

D) $n^2 + 15n - 12 - \dfrac{4}{n + 4}$

57) Simplify the below.

$$\left(x^3 + 2x^2 - 371x + 188\right) \div \left(x - 18\right)$$

A) $x^2 + 20x - 11 - \dfrac{10}{x - 18}$

B) $x^2 + 20x - 10 - \dfrac{5}{x - 18}$

C) $x^2 + 20x - 8 - \dfrac{4}{x - 18}$

D) $x^2 + 20x - 7 - \dfrac{2}{x - 18}$

58) Simplify the below.

$$\left(13x^3 - 45x^2 - 105x + 30\right) \div \left(x - 5\right)$$

A) $13x^2 + 20x - 4 + \dfrac{6}{x - 5}$

B) $13x^2 + 20x - 6 + \dfrac{13}{x - 5}$

C) $13x^2 + 20x - 3 + \dfrac{8}{x - 5}$

D) $13x^2 + 20x - 5 + \dfrac{5}{x - 5}$

59) Simplify the below.

$$\left(p^3 - 11p^2 - 16p - 10\right) \div \left(p + 1\right)$$

A) $p^2 - 12p - 5 - \dfrac{5}{p + 1}$

B) $p^2 - 12p - 2 - \dfrac{8}{p + 1}$

C) $p^2 - 12p - 2 - \dfrac{10}{p + 1}$

D) $p^2 - 12p - 4 - \dfrac{6}{p + 1}$

60) Simplify the below.

$$\left(n^3 - 8n^2 - 95n - 136\right) \div \left(n - 15\right)$$

A) $n^2 + 7n + 10 + \dfrac{14}{n - 15}$

B) $n^2 + 7n + 9 + \dfrac{21}{n - 15}$

C) $n^2 + 7n + 9 + \dfrac{12}{n - 15}$

D) $n^2 + 7n + 7 + \dfrac{17}{n - 15}$

61) Find the product of the below

$$(p + 12)^2$$

A) $p^2 - 144$

B) $p^2 + 24p + 144$

C) $p^2 - 20p + 100$

D) $p^2 + 144$

62) Find the product of the below

$$(r - 19)^2$$

A) $r^2 + 361$

B) $r^2 - 38r + 361$

C) $r^2 + 12r + 36$

D) $r^2 - 361$

63) Find the product of the below

$$(x - 17)^2$$

A) $x^2 - 30x + 225$

B) $x^2 - 34x + 289$

C) $x^2 - 289$

D) $x^2 + 289$

64) Find the product of the below

$$(b + 1)^2$$

A) $b^2 + 1$ B) $b^2 - 18b + 81$

C) $b^2 - 1$ D) $b^2 + 2b + 1$

65) Find the product of the below

$$(n - 18)^2$$

A) $n^2 + 324$

B) $n^2 - 324$

C) $n + 324$

D) $n^2 - 36n + 324$

66) Find the product of the below

$$(p + 3)^2$$

A) $p^2 + 9$

B) $p^2 - 9$

C) $p^2 + 12p + 36$

D) $p^2 + 6p + 9$

67) Find the product of the below

$$(n + 9)^2$$

A) $n^2 + 40n + 400$ B) $n^2 + 81$

C) $n^2 + 18n + 81$ D) $n^2 - 81$

68) Find the product of the below

$$(b - 2)^2$$

A) $b + 4$ B) $b^2 - 4b + 4$

C) $b^2 - 4$ D) $b^2 + 4$

69) Find the product of the below

$$(a + 6)^2$$

A) $a^2 + 36$

B) $a^2 - 36$

C) $a^2 + 12a + 36$

D) $a^2 + 28a + 196$

70) Find the product of the below

$$(n - 1)^2$$

A) $n^2 - 20n + 100$

B) $n^2 - 2n + 1$

C) $n^2 + 1$

D) $n^2 - 1$

71) Find the product of the below

$$(p + 13)^2$$

A) $p^2 - 169$

B) $p^2 + 169$

C) $p^2 + 169$

D) $p^2 + 26p + 169$

72) Find the product of the below

$$(x + 5)^2$$

A) $x^2 + 25$ B) $x^2 + 10x + 25$

C) $x^2 - 25$ D) $x^2 - 24x + 144$

73) Find the product of the below

$$(x-9)^2$$

A) $x^2 - 81$ B) $x^2 + 81$

C) $x + 81$ D) $x^2 - 18x + 81$

74) Find the product of the below

$$(p-4)^2$$

A) $p^2 + 16$ B) $p^2 - 16$

C) $p^2 - 8p + 16$ D) $p + 16$

75) Find the product of the below

$$(a-14)^2$$

A) $a^2 + 196$

B) $a^2 - 196$

C) $a + 196$

D) $a^2 - 28a + 196$

76) Find the product of the below

$$(x+11)^2$$

A) $x^2 - 121$ B) $x^2 + 121$

C) $x^2 + 22x + 121$ D) $x + 121$

77) Find the product of the below

$$(k-3)^2$$

A) $k^2 - 9$

B) $k^2 + 12k + 36$

C) $k^2 - 6k + 9$

D) $k^2 + 9$

78) Find the product of the below

$$(n+10)^2$$

A) $n^2 - 100$ B) $n + 100$

C) $n^2 + 20n + 100$ D) $n^2 + 100$

79) Find the product of the below

$$(x+7)^2$$

A) $x^2 + 14x + 49$ B) $x^2 - 49$

C) $x^2 - 40x + 400$ D) $x^2 + 49$

80) Find the product of the below

$$(n+15)^2$$

A) $n^2 + 225$

B) $n + 225$

C) $n^2 - 225$

D) $n^2 + 30n + 225$

81) Find the product of the below

$$(n - 6)^2$$

A) $n^2 + 36$ B) $n^2 - 36 - 12n$

C) $n + 36$ D) $n^2 - 12n + 36$

82) Find the product of the below

$$(n - 12)^2$$

A) $n^2 + 144$ B) $n^2 - 144$

C) $n^2 - 24n + 144$ D) $n + 144$

83) Find the product of the below

$$(v + 5)(v - 5)$$

A) $v^2 - 10v + 25$ B) $v^2 - 121$

C) $v^2 + 10v + 25$ D) $v^2 - 25$

84) Find the product of the below

$$(a - 5)(a + 5)$$

A) $a^2 - 10a + 25$ B) $a^2 - 25$

C) $a^2 - 1$ D) $1 - 64a^2$

85) Find the product of the below

$$(1 - 3p)(1 + 3p)$$

A) $p^2 + 22p + 121$

B) $p^2 - 121$

C) $1 - 9p^2$

D) $1 - 6p + 9p^2$

86) Find the product of the below

$$(1 - 10n)(1 + 10n)$$

A) $1 + 20n + 100n^2$

B) $1 - 100n^2$

C) $1 - 20n + 100n^2$

D) $1 - 64n^2$

87) Find the product of the below

$$(v + 3)(v - 3)$$

A) $v^2 + 6v + 9$ B) $v^2 - 9$

C) $v^2 - 6v + 9$ D) $1 - 4v^2$

88) Find the product of the below

$$(n - 4)(n + 4)$$

A) $n^2 - 9$ B) $n^2 - 8n + 16$

C) $n^2 - 16$ D) $1 - 4n^2$

Algebra 1

Week 30
Quadratic
Equations

89) Find the product of the below

$$(r - 3)(r + 3)$$

A) $r^2 - 6r + 9$ B) $r^2 - 9$

C) $r^2 - 4$ D) $r^2 + 6r + 9$

90) Find the product of the below
$$(1 + x)(1 - x)$$

A) $1 - 36x^2$

B) $1 + 2x + x^2$

C) $1 - x^2$

D) $1 - 12x + 36x^2$

91) Find the product of the below
$$(n - 6)(n + 6)$$

A) $n^2 - 36$

B) $n^2 - 81$

C) $n^2 - 18n + 81$

D) $n^2 - 12n + 36$

92) Find the product of the below

$$(x + 7)(x - 7)$$

A) $1 - 9x^2$ B) $1 - 49x^2$

C) $x^2 - 49$ D) $x^2 + 14x + 49$

93) Find the product of the below

$$(x - 10)(x + 10)$$

A) $x^2 - 14x + 49$

B) $x^2 - 20x + 100$

C) $x^2 - 49$

D) $x^2 - 100$

94) Find the product of the below

$$(x - 12)(x + 12)$$

A) $x^2 - 22x + 121$

B) $x^2 - 121$

C) $x^2 - 144$

D) $x^2 - 24x + 144$

95) Find the product of the below
$$(1 - 2v)(1 + 2v)$$

A) $1 - 4v + 4v^2$ B) $1 - 4v^2$

C) $1 + 4v + 4v^2$ D) $1 - 49v^2$

96) Find the product of the below
$$(1 - 4n)(1 + 4n)$$

A) $1 + 8n + 16n^2$

B) $1 - 8n + 16n^2$

C) $n^2 - 1$

D) $1 - 16n^2$

©All rights reserved-Math-Knots LLC., VA-USA   324   www.math-knots.com | www.a4ace.com

97) Find the product of the below

$$(1 - 6n)(1 + 6n)$$

A) $1 - 12n + 36n^2$

B) $1 + 12n + 36n^2$

C) $1 - 36n^2$

D) $1 - 64n^2$

98) Find the product of the below

$$(n - 9)(n + 9)$$

A) $n^2 - 18n + 81$ B) $n^2 - 64$

C) $n^2 - 16n + 64$ D) $n^2 - 81$

99) Find the product of the below

$$(n - 8)(n + 8)$$

A) $n^2 - 64$

B) $n^2 - 81$

C) $n^2 + 16n + 64$

D) $n^2 - 16n + 64$

100) Find the product of the below

$$(1 + 12x)(1 - 12x)$$

A) $1 - 24x + 144x^2$

B) $1 - 100x^2$

C) $1 - 144x^2$

D) $1 + 24x + 144x^2$

101) Find the product of the below

$$(x + 8)(x - 8)$$

A) $x^2 + 8x + 16$

B) $x^2 - 16$

C) $x^2 - 64$

D) $x^2 + 16x + 64$

102) Find the product of the below

$$(n + 9)(n - 9)$$

A) $n^2 + 24n + 144$

B) $n^2 - 81$

C) $n^2 + 18n + 81$

D) $n^2 - 144$

103) Find the product of the below

$$(p + 1)(p - 1)$$

A) $p^2 - 1$

B) $p^2 - 36$

C) $p^2 + 2p + 1$

D) $p^2 - 2p + 1$

104) Find the product of the below

$$(1 + 11v)(1 - 11v)$$

A) $1 - 121v^2$

B) $1 - 22v + 121v^2$

C) $1 - 4v^2$

D) $1 + 22v + 121v^2$

105) Find the product of the below

$$(-9a+8)^2$$

A) $81a^2 - 64$

B) $4a^2 - 24a + 36$

C) $81a^2 + 64$

D) $81a^2 - 144a + 64$

106) Find the product of the below

$$(-4-3x)^2$$

A) $16 + 9x^2$

B) $64 - 128x + 64x^2$

C) $16 + 24x + 9x^2$

D) $16 - 9x^2$

107) Find the product of the below

$$(12-5x)^2$$

A) $144 + 25x^2$

B) $25 + 30x + 9x^2$

C) $144 - 120x + 25x^2$

D) $144 - 25x^2$

108) Find the product of the below

$$(10n-3)^2$$

A) $10n + 9$

B) $100n^2 - 9$

C) $100n^2 + 9$

D) $100n^2 - 60n + 9$

109) Find the product of the below

$$(-3-5v)^2$$

A) $9 + 30v + 25v^2$

B) $81v^2 - 90v + 25$

C) $9 - 25v^2$

D) $9 + 25v^2$

110) Find the product of the below

$$(n+6)^2$$

A) $n^2 + 12n + 36$ B) $n + 36$

C) $n^2 - 36$ D) $n^2 + 36$

111) Find the product of the below

$$(3v+4)^2$$

A) $9v^2 + 24v + 16$ B) $9v^2 + 16$

C) $3v + 16$ D) $9v^2 - 16$

112) Find the product of the below

$$(-9p+10)^2$$

A) $81p^2 + 100$

B) $81p^2 - 100$

C) $16p^2 + 40p + 25$

D) $81p^2 - 180p + 100$

113) Find the product of the below

$$(-5+7x)^2$$

A) $25+49x^2$

B) $25-70x+49x^2$

C) $25-49x^2$

D) $9+60x+100x^2$

114) Find the product of the below

$$(-1+10n)^2$$

A) $1-20n+100n^2$

B) $121+66n+9n^2$

C) $1+100n^2$

D) $1-100n^2$

115) Find the product of the below

$$(7n-8)^2$$

A) $49n^2-112n+64$

B) $7n+64$

C) $49n^2-64$

D) $49n^2+64$

116) Find the product of the below

$$(-v+7)^2$$

A) $v^2-14v+49$ B) v^2+49

C) $-v+49$ D) v^2-49

117) Find the product of the below

$$(4x+5)^2$$

A) $16x^2-25$

B) $4x+25$

C) $16x^2+25$

D) $16x^2+40x+25$

118) Find the product of the below

$$(10-12x)^2$$

A) $100-144x^2$

B) $100-240x+144x^2$

C) $10+144x^2$

D) $100+144x^2$

119) Find the product of the below

$$(-11a+9)^2$$

A) $121a^2-81$

B) $121a^2-198a+81$

C) $121a^2+81$

D) a^2-2a+1

120) Find the product of the below

$$(-x-8)^2$$

A) x^2-64

B) $64x^2+112x+49$

C) x^2+64

D) $x^2+16x+64$

121) Find the product of the below

$$(6x - 12)^2$$

A) $36x^2 - 144$

B) $4x^2 + 40x + 100$

C) $36x^2 + 144$

D) $36x^2 - 144x + 144$

122) Find the product of the below

$$(11x - 10)^2$$

A) $121x^2 + 100$

B) $11x + 100$

C) $121x^2 - 220x + 100$

D) $121x^2 - 100$

123) Find the product of the below

$$(2a - 2)^2$$

A) $4a^2 + 4$ B) $4a^2 - 4$

C) $4a^2 - 8a + 4$ D) $2a + 4$

124) Find the product of the below

$$(-3k - 11)^2$$

A) $-3k + 121$

B) $9k^2 + 121$

C) $9k^2 + 66k + 121$

D) $9k^2 - 121$

125) Find the product of the below

$$(v + 9)^2$$

A) $v^2 - 81$

B) $v^2 + 81$

C) $9 - 72v + 144v^2$

D) $v^2 + 18v + 81$

126) Find the product of the below

$$(n + 1)^2$$

A) $n^2 + 2n + 1$ B) $n^2 + 1$

C) $n^2 - 1$ D) $n + 1$

127) Find the product of the below

$$(-14n + 5)(-14n - 5)$$

A) $49n - 49$

B) $196n^2 - 25$

C) $196n^2 + 140n + 25$

D) $196n^2 - 140n + 25$

128) Find the product of the below

$$(-10 + 5b)(-10 - 5b)$$

A) $100 - 100b + 25b^2$

B) $100 + 100b + 25b^2$

C) $144 - 100b^2$

D) $100 - 25b^2$

129) Find the product of the below

$$(-p - 2)(-p + 2)$$

A) $144p^2 - 144$

B) $p^2 - 4p + 4$

C) $p^2 + 4p + 4$

D) $p^2 - 4$

130) Find the product of the below

$$(13n + 2)(13n - 2)$$

A) $144n^2 - 1$

B) $169n^2 - 4$

C) $169n^2 + 52n + 4$

D) $169n^2 - 52n + 4$

131) Find the product of the below

$$(-5b + 11)(-5b - 11)$$

A) $25b^2 - 110b + 121$

B) $25b^2 + 110b + 121$

C) $25b^2 - 121$

D) $25b^2 - 25$

132) Find the product of the below

$$(9m + 4)(9m - 4)$$

A) $81m^2 - 16$

B) $4 - 36m^2$

C) $81m^2 + 72m + 16$

D) $49 - 144m^2$

133) Find the product of the below

$$(-11x - 6)(-11x + 6)$$

A) $121x^2 + 132x + 36$

B) $9x^2 - 30x + 25$

C) $9x^2 - 25$

D) $121x^2 - 36$

134) Find the product of the below

$$(2 + 7m)(2 - 7m)$$

A) $169 - 4m^2$

B) $4 - 28m + 49m^2$

C) $4 + 28m + 49m^2$

D) $4 - 49m^2$

135) Find the product of the below

$$(-5x - 14)(-5x + 14)$$

A) $x^2 - 196$

B) $25x^2 - 196$

C) $25x^2 + 140x + 196$

D) $x^2 - 28x + 196$

136) Find the product of the below

$$(10 - 10x)(10 + 10x)$$

A) $100 - 200x + 100x^2$

B) $100 - 100x^2$

C) $169 - 25x^2$

D) $100 + 200x + 100x^2$

137) Find the product of the below

$(11 a + 4) (11 a - 4)$

A) $36 - 84a + 49a^2$

B) $36 - 49a^2$

C) $121a^2 + 88a + 16$

D) $121a^2 - 16$

138) Find the product of the below

$(- 9 v - 8) (- 9 v + 8)$

A) $81v^2 - 16$

B) $81v^2 - 64$

C) $81v^2 + 144v + 64$

D) $81v^2 + 72v + 16$

139) Find the product of the below

$(6 x + 13) (6 x - 13)$

A) $144x^2 - 100$

B) $36x^2 - 169$

C) $36x^2 - 156x + 169$

D) $36x^2 + 156x + 169$

140) Find the product of the below

$(6 p + 14) (6 p - 14)$

A) $49p^2 - 196p + 196$

B) $36p^2 + 168p + 196$

C) $49p^2 - 196$

D) $36p^2 - 196$

141) Find the product of the below

$(- 14 x + 4) (- 14 x - 4)$

A) $196x^2 - 112x + 16$

B) $196x^2 - 16$

C) $121x^2 - 169$

D) $121x^2 - 286x + 169$

142) Find the product of the below

$(- 2 - 14 x) (- 2 + 14 x)$

A) $4 - 169x^2$

B) $25 - 4x^2$

C) $4 - 196x^2$

D) $4 + 56x + 196x^2$

143) Find the product of the below

$(- 10 k - 8) (- 10 k + 8)$

A) $100k^2 - 160k + 64$

B) $49k^2 - 81$

C) $100k^2 - 64$

D) $100k^2 + 160k + 64$

144) Find the product of the below

$(- 9 r - 12) (- 9 r + 12)$

A) $64r^2 - 81$

B) $81r^2 - 216r + 144$

C) $81r^2 + 216r + 144$

D) $81r^2 - 144$

145) Find the product of the below

$$(7 + 3v)(7 - 3v)$$

A) $49 - 9v^2$

B) $25 - 100v^2$

C) $25 + 100v + 100v^2$

D) $49 + 42v + 9v^2$

146) Find the product of the below

$$(-7n - 1)(-7n + 1)$$

A) $49n^2 - 1$

B) $100n^2 - 16$

C) $49n^2 + 14n + 1$

D) $49n^2 - 14n + 1$

147) Find the product of the below

$$(11 - 5n)(11 + 5n)$$

A) $121 - 25n^2$

B) $100 - 121n^2$

C) $16n^2 - 25$

D) $121 - 110n + 25n^2$

148) Find the product of the below

$$(-11r - 10)(-11r + 10)$$

A) $121r^2 - 100$

B) $121r^2 + 220r + 100$

C) $100r^2 - 16$

D) $100r^2 + 80r + 16$

149) Find the product of the below

$$(8k - 5)(8k + 5)$$

A) $36k^2 + 12k + 1$

B) $36k^2 - 1$

C) $64k^2 - 80k + 25$

D) $64k^2 - 25$

 **Algebra 1**

1) Simplify the below expression

$$\frac{12}{5n} + \frac{13m}{12m^3}$$

A) $\dfrac{13 + m + 12m^3}{5n + 13m}$

B) $\dfrac{144m^2 + 65n}{60nm^2}$

C) $\dfrac{169 + 13m + 60nm^2 - 5n}{65n}$

D) $\dfrac{156 + 13m + 60m^2 n}{65n}$

2) Simplify the below expression

$$\frac{6x}{4} + \frac{9x}{3x^3}$$

A) $\dfrac{3x^3 + 6}{2x^2}$

B) $\dfrac{18x + 3 + 4x^2}{12}$

C) $\dfrac{6x^3 + x^2 + 12}{4x^2}$

D) $\dfrac{21x + 3 + 4x^2}{12}$

3) Simplify the below expression

$$\frac{4}{7x} + \frac{6x}{8}$$

A) $\dfrac{16 + 21x^2}{28x}$

B) $\dfrac{20 + 4x + 21x^2}{28x}$

C) $\dfrac{40 + 3x - x^2}{42x^2}$

D) $\dfrac{20 + 21x^2}{28x}$

4) Simplify the below expression

$$\frac{7}{11x^2 y} - \frac{9x}{9x^4}$$

A) $\dfrac{20 + 2x}{13yx}$

B) $\dfrac{10 + x}{26yx(4 - x)}$

C) $\dfrac{-104yx + 26yx^2 + 10 + x}{26yx}$

D) $\dfrac{7x - 11y}{11x^3 y}$

5) Simplify the below expression

$$\frac{12}{2a^4} + \frac{8}{12b}$$

A) $\dfrac{18b + 2a^4}{3a^4 b}$

B) $\dfrac{20 - a}{2(a^4 + 6b)}$

C) $\dfrac{21 - a}{2(a^4 + 6b)}$

D) $\dfrac{56 - 7a}{12a^4 b}$

7) Simplify the below expression

$$\frac{4}{9x} + \frac{14x}{13xy}$$

A) $\dfrac{4y + 9x}{9xy}$

B) $\dfrac{52y - 4}{117x}$

C) $\dfrac{52y + 126x}{117xy}$

D) $\dfrac{52 + 117xy - 18x}{117x}$

6) Simplify the below expression

$$\frac{8y}{11y} + \frac{x + 12y}{11}$$

A) $\dfrac{35x + 35y}{12}$

B) $\dfrac{8yx + 96y^2 + x + 133y}{11y(x + 12y)}$

C) $\dfrac{8 + x + 12y}{11}$

D) $\dfrac{15y^2 + 7x + 7y}{6y}$

8) Simplify the below expression

$$\frac{8u + 4v}{12uv^2} + \frac{9v}{14u}$$

A) $\dfrac{28u + 14v + 27v^3}{42uv^2}$

B) $\dfrac{63u + 28v + 54v^3}{84uv^2}$

C) $\dfrac{20v + 35u^2}{14u^2}$

D) $\dfrac{20v + 2u + 35u^2}{14u^2}$

9) Simplify the below expression

$$\frac{5x}{5y} - \frac{6}{8x}$$

A) $\dfrac{6x - 6}{5y - 8x}$

B) $\dfrac{4x^2 - 3y}{4yx}$

C) $\dfrac{7x}{9}$

D) $\dfrac{9x + 7}{9}$

11) Simplify the below expression

$$\frac{11x}{6} + \frac{9y}{4x^3}$$

A) $\dfrac{22x^4 + 27y}{12x^3}$

B) $\dfrac{33xy + 8x^3}{18y}$

C) $\dfrac{11x + 9y}{2(3 + 2x^3)}$

D) $\dfrac{18xy + 4x^3}{9y}$

10) Simplify the below expression

$$\frac{6v}{v - 10} + \frac{10v}{v - 3}$$

A) $\dfrac{17v^2 - 114v}{(v - 10)(v - 3)}$

B) $\dfrac{16v^2 - 118v}{(v - 10)(v - 3)}$

C) $\dfrac{16v}{2v - 13}$

D) $\dfrac{54v^2}{(v - 10)(v - 3)}$

12) Simplify the below expression

$$\frac{6}{7k - 1} + \frac{6k}{k + 8}$$

A) $\dfrac{6 + 6k}{8k + 7}$

B) $\dfrac{36k}{(7k - 1)(k + 8)}$

C) $\dfrac{48 + 42k^2}{(7k - 1)(k + 8)}$

D) $\dfrac{46k - 4}{3k(7k - 1)}$

www.math-knots.com | www.a4ace.com

 **Algebra 1**

13) Simplify the below expression

$$\frac{11}{3(x+3)} - \frac{12}{x-11}$$

A) $\dfrac{-25x-229}{3(x-11)(x+3)}$

B) $\dfrac{77+19x}{12(x+3)}$

C) $\dfrac{77+12x-x^2}{12(x+3)}$

D) $\dfrac{77+16x-x^2}{12(x+3)}$

14) Simplify the below expression

$$\frac{9}{m+10} + \frac{m+1}{12(2m+3)}$$

A) $\dfrac{251m+370+m^2}{12(m+10)(2m+3)}$

B) $\dfrac{12+m}{25m+46}$

C) $\dfrac{299m+442+m^2}{12(m+10)(2m+3)}$

D) $\dfrac{227m+334+m^2}{12(m+10)(2m+3)}$

15) Simplify the below expression

$$\frac{6n}{n+9} + \frac{9}{n-12}$$

A) $\dfrac{54n+9}{(n-12)(n+9)}$

B) $\dfrac{60n-90+n^2}{9(n+9)}$

C) $\dfrac{6n^2-63n+81}{(n-12)(n+9)}$

D) $\dfrac{60n-99+n^2}{9(n+9)}$

16) Simplify the below expression

$$\frac{v-1}{(5v-9)(v-6)} + \frac{10v}{4}$$

A) $\dfrac{8+6v}{2v+13}$

B) $\dfrac{11v-1}{(v-2)(5v-29)}$

C) $\dfrac{28}{3v(v+6)}$

D) $\dfrac{272v-2+25v^3-195v^2}{2(v-6)(5v-9)}$

17) Simplify the below expression

$$\frac{9}{5(11v-6)(v+11)} - 9$$

A) $\dfrac{81-9v}{5(11v-6)(v+11)}$

B) $\dfrac{16v^2 - 1101v + 596 + 11v^3}{(11v-6)(v+11)}$

C) $\dfrac{-495v^2 - 5175v + 2979}{5(11v-6)(v+11)}$

D) $\dfrac{1+2v}{55v^2 + 575v - 331}$

18) Simplify the below expression

$$\frac{12x}{x+7} - \frac{10x}{x+6}$$

A) $\dfrac{2x^2 + 2x}{(x+7)(x+6)}$

B) $\dfrac{11x-4}{-9x+7}$

C) $\dfrac{2x^2 + 3x + 6}{(x+7)(x+6)}$

D) $\dfrac{119x^2 - 3x - 42}{10x(x+7)}$

19) Simplify the below expression

$$\frac{8v}{3(v+3)} + \frac{2v}{v-4}$$

A) $\dfrac{14v^2 - 14v}{3(v-4)(v+3)}$

B) $\dfrac{15v^2 - 16v - 8}{3(v-4)(v+3)}$

C) $\dfrac{10v+1}{4v+5}$

D) $\dfrac{15v^2 - 17v - 4}{3(v-4)(v+3)}$

20) Simplify the below expression

$$8 + \frac{b-5}{(5b+4)(b+11)}$$

A) $\dfrac{68b - 1 + 5b^2}{b-5}$

B) $\dfrac{40b^2 + 473b + 347}{(b+11)(5b+4)}$

C) $\dfrac{62b - 1 + 6b^2}{b-5}$

D) $\dfrac{67b - 1 + 5b^2}{b-5}$

21) Simplify the below expression

$$\frac{8k}{10k(k+3)} - \frac{4}{5k}$$

A) $\dfrac{k-24}{10k(k+3)}$

B) $\dfrac{18k+2-25k^3-75k^2}{20k(k+3)}$

C) $-\dfrac{12}{5k(k+3)}$

D) $\dfrac{18}{25k(k+3)}$

22) Simplify the below expression

$$10+\frac{12r}{11r(r-6)}$$

A) $\dfrac{110r-648}{11(r-6)}$

B) $\dfrac{10}{r-6}$

C) $\dfrac{10+11r}{1+11r^2-66r}$

D) $\dfrac{11r-65}{r-6}$

23) Simplify the below expression

$$\frac{10x}{3(x+2)} + \frac{6x}{12}$$

A) $\dfrac{28x+3x^2}{6(x+2)}$ B) $\dfrac{54x+5x^2}{12(x+2)}$

C) $\dfrac{26x+3x^2}{6(x+2)}$ D) $\dfrac{16x}{3(x+6)}$

24) Simplify the below expression

$$\frac{9x}{3x} - \frac{x-11}{30(x-2)}$$

A) $\dfrac{-81x^2+82x-11}{3x(x-11)}$

B) $\dfrac{89x-169}{30(x-2)}$

C) $\dfrac{-99x-11}{90x(x-2)}$

D) $\dfrac{8x+11}{3(-9x+20)}$

25) Simplify the below expression

$$\frac{9m-3}{m+9} - \frac{8m}{10}$$

A) $\dfrac{9m-15-4m^2}{5(m+9)}$

B) $\dfrac{64m^2-22m-90}{7m(m+9)}$

C) $\dfrac{11m-13}{3(-2m+3)}$

D) $\dfrac{27m-30-7m^2}{10(m+9)}$

26) Simplify the below expression

$$\frac{3}{5} - \frac{5}{4n(2n-1)}$$

A) $\dfrac{32n^2-16n-25}{20n(2n-1)}$

B) $\dfrac{5}{4n(2n-1)}$

C) $\dfrac{24n^2-12n-25}{20n(2n-1)}$

D) $\dfrac{36n^2-20n+8n^3-25}{20n(2n-1)}$

27) Simplify the below expression

$$\frac{12}{3} - \frac{r-4}{4(r+8)}$$

A) $\dfrac{15r+132}{4(r+8)}$

B) $-\dfrac{4}{r+8}$

C) $\dfrac{13r+107}{3(r+8)}$

D) $\dfrac{4r+4+4r^2}{(r-2)(3r+1)}$

28) Simplify the below expression

$$\frac{12}{4n(2n+7)} + \frac{8n}{2}$$

A) $\dfrac{32n^3+112n^2+13}{4n(2n+7)}$

B) $\dfrac{14}{2n+7}$

C) $\dfrac{16n^3+56n^2+7}{2n(2n+7)}$

D) $\dfrac{8n^3+28n^2+3}{n(2n+7)}$

29) Simplify the below expression

$$\frac{11}{11x-6} - \frac{12x}{x+9}$$

A) $\dfrac{x+9}{x(11x-6)}$

B) $\dfrac{84x+108-132x^2}{(11x-6)(x+9)}$

C) $\dfrac{144x}{(11x-6)(x+9)}$

D) $\dfrac{83x+99-132x^2}{(11x-6)(x+9)}$

30) Simplify the below expression

$$\frac{9x^2}{3x+11} \cdot \frac{3x^2+35x+88}{9x^2}$$

A) $\dfrac{x-8}{x+10}$ B) $x+1$

C) $\dfrac{11x}{4}$ D) $x+8$

31) Simplify the below expression

$$(7a-7) \cdot \frac{10a-110}{14-14a}$$

A) $-5(a-11)$ B) $\dfrac{2(a+11)}{a-7}$

C) $\dfrac{a+11}{4}$ D) 1

32) Simplify the below expression

$$\frac{k+12}{2k(11k-9)} + \frac{11}{3}$$

A) $\dfrac{-8k-54}{(k+6)(k+4)}$

B) $\dfrac{-195k+36+242k^2}{6k(11k-9)}$

C) $\dfrac{11k+143}{6k(11k-9)}$

D) $\dfrac{-9k-58}{(k+6)(k+4)}$

33) Simplify the below expression

$$\frac{24m^3-24m^2}{5} \cdot \frac{1}{2m-2}$$

A) $\dfrac{m-3}{7}$ B) $\dfrac{12m^2}{5}$

C) $\dfrac{12m^2}{m-11}$ D) $\dfrac{12m}{(m-9)(m-5)}$

34) Simplify the below expression

$$\frac{2x^2-3x-27}{9} \cdot \frac{9}{2x^2-27x+81}$$

A) $\dfrac{5}{3x}$ B) $\dfrac{x+3}{x-9}$

C) $\dfrac{12x^2}{x-12}$ D) $\dfrac{x+7}{x-5}$

35) Simplify the below expression

$$\frac{21x^2 - 65x + 50}{7x - 10} \cdot \frac{x + 5}{3x - 5}$$

A) $x + 5$

B) $\dfrac{x + 5}{x + 9}$

C) $x + 11$

D) $\dfrac{x + 1}{7(x + 5)}$

36) Simplify the below expression

$$\frac{7}{56p^2 - 96p} \left(56p^2 - 96p\right)$$

A) 7

B) $(p - 3)(p + 2)$

C) $\dfrac{10}{p + 5}$

D) $\dfrac{10p}{p - 12}$

37) Simplify the below expression

$$\frac{11v - 5}{v - 12} \cdot \frac{5v^2 - 37v + 14}{-55v^2 + 47v - 10}$$

A) $v - 8$

B) $\dfrac{5v^2}{v - 2}$

C) $\dfrac{v + 6}{3(v - 12)}$

D) $\dfrac{-v + 7}{v - 12}$

38) Simplify the below expression

$$\frac{7p}{2p} \cdot \frac{5p^2 - 44p + 32}{35p^2 - 28p}$$

A) $\dfrac{p - 8}{2p}$

B) $\dfrac{9}{8}$

C) 6

D) $\dfrac{(p - 9)(p + 7)}{7}$

39) Simplify the below expression

$$\frac{55n^2 - 106n - 33}{11n + 3} \cdot \frac{4n^2}{5n^2 - 26n + 33}$$

A) $\dfrac{4n^2}{n - 3}$

B) 2

C) $\dfrac{n + 11}{3n}$

D) $\dfrac{11}{2(n - 8)}$

40) Simplify the below expression

$$\frac{21n^2 - 113n + 132}{24n^3 - 88n^2} \cdot \frac{n + 10}{7n - 12}$$

A) $\dfrac{8}{11}$

B) $\dfrac{3(n - 5)}{n - 10}$

C) $\dfrac{n + 4}{11n^2}$

D) $\dfrac{n + 10}{8n^2}$

41) Simplify the below expression

$$\frac{x+8}{2x-9} \cdot \frac{8x^2-36x}{4x}$$

A) $\dfrac{5x}{6}$

B) $x-9$

C) $\dfrac{x+7}{x+2}$

D) $x+8$

42) Simplify the below expression

$$\frac{56n+64}{n-4} \cdot \frac{1}{7n+8}$$

A) $\dfrac{n+3}{n-2}$

B) $\dfrac{8}{n-4}$

C) $n+5$

D) $\dfrac{n-1}{n+7}$

43) Simplify the below expression

$$\frac{1}{r-3} \cdot \frac{10r^2+40r}{2r+8}$$

A) $\dfrac{r+7}{6r^2}$

B) $\dfrac{r+3}{4r^2}$

C) $\dfrac{r+5}{r-6}$

D) $\dfrac{5r}{r-3}$

44) Simplify the below expression

$$\frac{1}{3n-11} \cdot \frac{6n-22}{7}$$

A) $\dfrac{4}{n^2}$

B) $\dfrac{6n^2(n-9)}{n-6}$

C) 24

D) $\dfrac{2}{7}$

45) Simplify the below expression

$$\frac{5b+1}{5b^2-59b-12}(b-9)$$

A) $30b^3$

B) $\dfrac{b+6}{b+11}$

C) $\dfrac{b-9}{b-12}$

D) $\dfrac{b-7}{6}$

46) Simplify the below expression

$$\frac{24p^2-108p}{2p^2-p-36}(p+1)$$

A) $\dfrac{12p(p+1)}{p+4}$

B) $5p^2$

C) $\dfrac{4p}{p+9}$

D) $\dfrac{12}{(p-7)(p-4)}$

47) Simplify the below expression

$$\frac{36 - 28k}{k + 11} \cdot \frac{5}{35k - 45}$$

A) $\dfrac{6}{5}$ 　　B) $-\dfrac{4}{k + 11}$

C) $\dfrac{k + 5}{10k^2}$ 　　D) $\dfrac{4k^2}{k + 7}$

48) Simplify the below expression

$$\frac{7r^3 + 28r^2}{r - 4} \cdot \frac{4}{7r^2}$$

A) $\dfrac{4(r + 4)}{r - 4}$ 　　B) $r - 8$

C) $(r + 9)(r + 7)$ 　　D) $\dfrac{10}{r + 6}$

49) Simplify the below expression

$$\frac{6n^3 + 48n^2}{n^2 - 64} \cdot \frac{n - 8}{3n + 3}$$

A) $\dfrac{2n^2}{n + 1}$ 　　B) $n - 9$

C) $\dfrac{8(n - 8)}{7n}$ 　　D) $5(n + 2)$

50) Simplify the below expression

$$\frac{2b + 4}{2b^2 + 6b + 4} \cdot 3b$$

A) $b - 8$ 　　B) $b - 6$

C) $\dfrac{2b}{b + 1}$ 　　D) $\dfrac{3b}{b + 1}$

51) Simplify the below expression

$$(10p - 5) \cdot \frac{8}{10p - 5}$$

A) $\dfrac{1}{p^2}$ 　　B) 8

C) $11p$ 　　D) $\dfrac{10}{3}$

52) Simplify the below expression

$$\frac{k^2 - 5k + 6}{k - 3} \cdot \frac{3k^3 - 24k^2}{k^2 - 10k + 16}$$

A) $\dfrac{k + 1}{(k - 3)(k + 9)}$ 　　B) $\dfrac{2k(k - 2)}{5}$

C) $\dfrac{2k^2}{(k + 9)(k + 4)}$ 　　D) $3k^2$

53) Simplify the below expression

$$\frac{24x+32}{6x^3+4x^2} \cdot \frac{24x+16}{24x+32}$$

A) 1

B) $\dfrac{4}{x^2}$

C) $\dfrac{18}{x+10}$

D) $\dfrac{90\,x^3}{x+5}$

56) Simplify the below expression

$$\frac{8x^2+56x}{8x} \cdot \frac{4x^3-12x^2}{4x^3+28x^2}$$

A) $\dfrac{3}{7(x+1)}$

B) $x-3$

C) $x+4$

D) $\dfrac{6}{x+1}$

54) Simplify the below expression

$$\frac{6b^3+48b^2}{30b^3-6b^2} \cdot \frac{35b^3-7b^2}{b^2+4b-32}$$

A) $\dfrac{7b^2}{b-4}$

B) $\dfrac{b+4}{28b^2}$

C) $6(b-9)$

D) 1

57) Simplify the below expression

$$\frac{4x-14}{x+4} \cdot \frac{5x+20}{14x-49}$$

A) $\dfrac{x+8}{(x+2)(x-2)}$

B) $\dfrac{10(x-6)}{x-3}$

C) $\dfrac{10}{7}$

D) $x-2$

55) Simplify the below expression

$$\frac{2n^2+14n}{2n^2+18n} \cdot \frac{2n+20}{n^2+17n+70}$$

A) $40n$

B) $\dfrac{2}{n+9}$

C) $\dfrac{n+8}{n-8}$

D) $\dfrac{n+6}{n+2}$

58) Simplify the below expression

$$\frac{6x+36}{x^2+15x+54} \cdot \frac{x^2+17x+72}{6x+48}$$

A) $\dfrac{5}{2x(x-10)}$

B) 1

C) $\dfrac{5}{3}$

D) $\dfrac{x+10}{6x(x-7)}$

59) Simplify the below expression

$$\frac{3x^2 + 12x}{3x + 27} \cdot \frac{x^2 - 81}{x^2 - 5x - 36}$$

A) $\dfrac{2x^2}{(x-6)(x+5)}$ B) x

C) $\dfrac{3}{x-10}$ D) $\dfrac{x-5}{2x(x-2)}$

62) Simplify the below expression

$$\frac{30x + 42}{6x - 48} \cdot \frac{6x}{45x^3 + 63x^2}$$

A) $\dfrac{9x}{x-8}$ B) $8x$

C) $\dfrac{5x}{x-5}$ D) $\dfrac{2}{3x(x-8)}$

60) Architect Mary is designing an office tower where the ground floor (floor 0) is the largest and the floor space is reduced by 5% per floor. If the ground floor has an area of 10,400.0 ft², what is the area of the 9th floor?

A) $10400 \cdot 0.97^9 \approx 7,906.4$ ft²

B) $10400 \cdot 0.95^9 \approx 6,554.6$ ft²

C) $41600 \cdot 0.99^9 \approx 38,002.3$ ft²

D) $20800 \cdot 0.85^9 \approx 4,817.6$ ft²

63) For tax purposes, a racer car rental company assumes each car in their fleet depreciates by 8% per year. If the initial value of a car is $29,600.00, what will the value be when the car is 9 years old?

A) $29600 \cdot 0.92^9 \approx \$13,975.98$

B) $88800 \cdot 0.92^9 \approx \$41,927.93$

C) $88800 \cdot 0.96^9 \approx \$61,497.02$

D) $29600 \cdot 0.97^9 \approx \$22,502.84$

61) New York state pledges to reduce its annual CO_2 emissions by 2% per year. If the emissions in 2022 are 5,020 Mt (metric megatons), what are the maximum allowable emissions in the year 2027?

A) $5020 \cdot 0.92^5 \approx 3,309$ Mt

B) $5020 \cdot 0.96^5 \approx 4,093$ Mt

C) $5020 \cdot 0.99^5 \approx 4,774$ Mt

D) $5020 \cdot 0.98^5 \approx 4,538$ Mt

64) California state pledges to reduce its annual CO_2 emissions by 4% per year. If the emissions in 2022 are 5,780 Mt (metric megatons), what are the maximum allowable emissions in the year 2031?

A) $5780 \cdot 0.99^9 \approx 5,280$ Mt

B) $5780 \cdot 0.98^9 \approx 4,819$ Mt

C) $5780 \cdot 0.84^9 \approx 1,203$ Mt

D) $5780 \cdot 0.96^9 \approx 4,003$ Mt

65) Architect Ben is designing a tapered office tower where the ground floor (floor 0) is the largest and the floor space is reduced by 6% per floor. If the ground floor has an area of 1,080.0 m², what is the area of the 6th floor?

A) $1080 \cdot 0.98^6 \approx 956.7$ m²

B) $4320 \cdot 0.94^6 \approx 2{,}980.2$ m²

C) $4320 \cdot 0.76^6 \approx 832.5$ m²

D) $1080 \cdot 0.94^6 \approx 745.1$ m²

66) Virginia state pledges to reduce its annual CO_2 emissions by 2% per year. If the emissions in 2022 are 6,900 Mt (metric megatons), what are the maximum allowable emissions in the year 2031?

A) $27600 \cdot 0.98^9 \approx 23{,}011$ Mt

B) $6900 \cdot 0.98^9 \approx 5{,}753$ Mt

C) $20700 \cdot 0.99^9 \approx 18{,}910$ Mt

D) $6900 \cdot 0.99^9 \approx 6{,}303$ Mt

67) Snap Chat is increasing its user base by approximately 4% per month. If the site currently has 29,440 users, what will the approximate user base be 10 months from now?

A) $9813 \cdot 1.04^{10} \approx 14{,}526$ users

B) $29440 \cdot 1.04^{10} \approx 43{,}578$ users

C) $29440 \cdot 1.01^{10} \approx 32{,}520$ users

D) $29440 \cdot 1.02^{10} \approx 35{,}887$ users

68) Fiona pledges to donate 13% of a fund each year. If the fund initially has $763,000.00, how much will the fund have after 10 years?

A) $763000 \cdot 0.74^{10} \approx \$37{,}570.05$

B) $1526000 \cdot 0.97^{10} \approx \$1{,}125{,}309.22$

C) $763000 \cdot 0.87^{10} \approx \$189{,}547.07$

D) $763000 \cdot 0.61^{10} \approx \$5{,}442.81$

69) Trenton, NJ pledges to reduce its annual CO_2 emissions by 3% per year. If the emissions in 2022 are 7,620 Mt (metric megatons), what are the maximum allowable emissions in the year 2027?

A) $7620 \cdot 0.88^5 \approx 4{,}021$ Mt

B) $7620 \cdot 0.91^5 \approx 4{,}755$ Mt

C) $30480 \cdot 0.88^5 \approx 16{,}085$ Mt

D) $7620 \cdot 0.97^5 \approx 6{,}544$ Mt

70) GK electronics promises to release a new smartphone model every month. Each model's battery life will be 6% longer than the previous model's. If the current model's battery life is 634.0 minutes, what will the latest model's battery life be 6 months from now?

A) $634 \cdot 1.03^6 \approx 757.0$ minutes

B) $634 \cdot 1.06^6 \approx 899.3$ minutes

C) $634 \cdot 1.18^6 \approx 1{,}711.5$ minutes

D) $634 \cdot 1.02^6 \approx 714.0$ minutes

71) A bouncy ball is dropped from a height of 108.00 inches onto a hard flat floor. After each bounce, the ball returns to a height that is 20% less than the previous maximum height. What is the maximum height reached after the 10th bounce?

A) $108 \cdot 0.6^{10} \approx 0.65$ inches

B) $108 \cdot 0.8^{10} \approx 11.60$ inches

C) $216 \cdot 0.8^{10} \approx 23.19$ inches

D) $108 \cdot 0.9^{10} \approx 37.66$ inches

72) Tom's savings account balance is compounded annually.
If the interest rate is 2% per year and the current balance is $1,267.00, what will the balance be 8 years from now?

A) $1267 \cdot 1.01^{8} \approx \$1,371.98$

B) $1267 \cdot 1.08^{8} \approx \$2,345.13$

C) $316.75 \cdot 1.08^{8} \approx \$586.28$

D) $1267 \cdot 1.02^{8} \approx \$1,484.49$

73) Instagram site is increasing its user base by approximately 6% per month. If the site currently has 30,790 users, what will the approximate user base be 6 months from now?

A) $30790 \cdot 1.03^{6} \approx 36,765$ users

B) $10263 \cdot 1.06^{6} \approx 14,558$ users

C) $30790 \cdot 1.24^{6} \approx 111,928$ users

D) $30790 \cdot 1.06^{6} \approx 43,676$ users

74) A bouncy ball is dropped from a height of 165.00 cm onto a hard flat floor. After each bounce, the ball returns to a height that is 20% less than the previous maximum height. What is the maximum height reached after the 9th bounce?

A) $165 \cdot 0.9^{9} \approx 63.92$ cm

B) $165 \cdot 0.8^{9} \approx 22.15$ cm

C) $165 \cdot 0.6^{9} \approx 1.66$ cm

D) $165 \cdot 0.95^{9} \approx 103.99$ cm

75) Happy homes high rise apartment building get more expensive higher up, since the views get better. The ground floor (floor 1) rent is $1,900.00. The rent increases 3% per floor. What is the rent on the 10th floor?

A) $1900 \cdot 1.09^{10} \approx \$4,497.99$

B) $1900 \cdot 1.06^{10} \approx \$3,402.61$

C) $1900 \cdot 1.01^{10} \approx \$2,098.78$

D) $1900 \cdot 1.03^{10} \approx \$2,553.44$

76) Lola receives a 2% raise once per year. If the employee's initial salary is $64,000.00, what will the employee's salary be after 8 years?

A) $64000 \cdot 1.04^{8} \approx \$87,588.42$

B) $64000 \cdot 1.02^{8} \approx \$74,986.20$

C) $21333.33 \cdot 1.02^{8} \approx \$24,995.40$

D) $32000 \cdot 1.02^{8} \approx \$37,493.10$

77) Rentals in any high rise apartment building get more expensive higher up, since the views get better. The ground floor (floor 1) rent is $1,690.00. The rent increases 2% per floor. What is the rent on the 6th floor?

A) $1690 \cdot 1.02^6 \approx \$1,903.21$

B) $1690 \cdot 1.04^6 \approx \$2,138.39$

C) $1690 \cdot 1.01^6 \approx \$1,793.97$

D) $422.5 \cdot 1.02^6 \approx \$475.80$

78) MK Electronics promises to release a new smartphone model every month. Each model's battery life will be 4% longer than the previous model's. If the current model's battery life is 744.0 minutes, what will the latest model's battery life be 9 months from now?

A) $744 \cdot 1.12^9 \approx 2,063.2$ minutes

B) $744 \cdot 1.16^9 \approx 2,829.4$ minutes

C) $744 \cdot 1.08^9 \approx 1,487.3$ minutes

D) $744 \cdot 1.04^9 \approx 1,058.9$ minutes

79) Solve the below equation

$$\frac{1}{n+14} + \frac{1}{n^2+14n} = \frac{2}{n^2+14n}$$

A) $\{1\}$ B) $\{-13\}$

C) $\left\{-\dfrac{11}{8}\right\}$ D) $\{13\}$

80) Jack receives a 2% raise once per year. If the employee's initial salary is $59,100.00, what will the employee's salary be after 9 years?

A) $59100 \cdot 1.01^9 \approx \$64,636.80$

B) $59100 \cdot 1.08^9 \approx \$118,141.17$

C) $59100 \cdot 1.02^9 \approx \$70,629.97$

D) $14775 \cdot 1.08^9 \approx \$29,535.29$

81) A savings account balance is compounded annually. If the interest rate is 4% per year and the current balance is $1,768.00, what will the balance be 8 years from now?

A) $1768 \cdot 1.04^8 \approx \$2,419.63$

B) $1768 \cdot 1.16^8 \approx \$5,796.24$

C) $1768 \cdot 1.08^8 \approx \$3,272.44$

D) $1768 \cdot 1.02^8 \approx \$2,071.49$

82) Solve the below equation

$$\frac{3}{x^2-17x+72} = \frac{9}{x^2-17x+72} - \frac{1}{x-8}$$

A) $\{12\}$ B) $\{-9\}$

C) $\{10\}$ D) $\{15\}$

83) Solve the below equation

$$\frac{1}{a-12} = \frac{3}{a-12} - 1$$

A) $\{14\}$ B) $\{-3\}$

C) $\{-13\}$ D) $\{-11\}$

84) Solve the below equation

$$\frac{1}{5p} + 1 = \frac{2}{p}$$

A) $\{1, 5\}$ B) $\{7\}$

C) $\{-6\}$ D) $\left\{\frac{9}{5}\right\}$

85) Solve the below equation

$$\frac{1}{x^2 + 8x} = \frac{4}{x^2 + 8x} + \frac{1}{x}$$

A) $\left\{\frac{7}{6}\right\}$ B) $\{6\}$

C) $\left\{\frac{7}{6}, -\frac{13}{6}\right\}$ D) $\{-11\}$

86) Solve the below equation

$$\frac{5}{x^2 - 13x} = \frac{1}{x-13} + \frac{1}{x^2 - 13x}$$

A) $\{4\}$ B) $\{1\}$

C) $\left\{\frac{78}{11}, 12\right\}$ D) $\left\{\frac{78}{11}\right\}$

87) Solve the below equation

$$\frac{7}{v-5} - \frac{8}{v^2 + 4v - 45} = \frac{1}{v-5}$$

A) $\{-6\}$ B) $\left\{-\frac{23}{3}\right\}$

C) $\left\{-\frac{7}{4}\right\}$ D) $\left\{\frac{23}{3}\right\}$

88) Solve the below equation

$$\frac{1}{n^2 - 14n + 40} = \frac{1}{n-10} - \frac{11}{n^2 - 14n + 40}$$

A) $\{16\}$ B) $\{6\}$

C) $\{-6\}$ D) $\{-9\}$

89) Solve the below equation

$$\frac{1}{x} = \frac{9}{7x} + 1$$

A) $\{5\}$ B) $\{11\}$

C) $\left\{\frac{2}{7}\right\}$ D) $\left\{-\frac{2}{7}\right\}$

92) Solve the below equation

$$\frac{4}{2m+5} = 1 + \frac{6}{2m+5}$$

A) $\{-8\}$ B) $\{8\}$

C) $\{-6\}$ D) $\left\{-\frac{7}{2}\right\}$

90) Solve the below equation

$$\frac{3}{2k} = \frac{1}{4k} - 1$$

A) $\left\{-\frac{5}{4}\right\}$ B) $\left\{\frac{5}{4}\right\}$

C) $\{10\}$ D) $\{1\}$

93) Solve the below equation

$$\frac{1}{4} = \frac{11}{4} + \frac{n-2}{4n-48}$$

A) $\left\{\frac{122}{11}\right\}$ B) $\{-12\}$

C) $\{12\}$ D) $\{3\}$

91) Solve the below equation

$$7 = \frac{2x-18}{x+6} - \frac{10}{x+6}$$

A) $\left\{-\frac{3}{4}\right\}$ B) $\{3\}$

C) $\left\{\frac{21}{8}\right\}$ D) $\{-14\}$

94) Solve the below equation

$$\frac{b+3}{b^2+4b} = \frac{1}{b} - \frac{8}{b+4}$$

A) $\{-7\}$ B) $\left\{\frac{1}{8}\right\}$

C) $\left\{9, \frac{1}{8}\right\}$ D) $\{9\}$

Algebra 1

95) Solve the below equation

$$\frac{1}{6x} - \frac{1}{6} = \frac{3x+5}{x^2}$$

A) $\{-4,-6\}$ B) $\{-4,-2\}$

C) $\{4,-6\}$ D) $\{-15,-2\}$

96) Solve the below equation

$$\frac{p+1}{2p} = \frac{3p+3}{p^2} + \frac{p-3}{2p}$$

A) $\{-6,-4\}$ B) $\{-2,-4\}$

C) $\{-3\}$ D) $\{-3,-4\}$

97) Solve the below equation

$$\frac{m+2}{m} + \frac{1}{m} = \frac{m^2-5m+4}{m^2}$$

A) $\{-5,1\}$ B) $\{3\}$

C) $\left\{\frac{1}{2}\right\}$ D) $\{3,1\}$

98) Solve the below equation

$$\frac{n-6}{5} + \frac{1}{5n} = \frac{n^2+2n-15}{5n}$$

A) $\left\{-3,-\frac{7}{2}\right\}$ B) $\left\{-\frac{8}{5},4\right\}$

C) $\{2\}$ D) $\{-3,4\}$

99) Solve the below equation

$$6 + \frac{n-4}{n^2} = \frac{1}{n^2}$$

A) $\left\{\frac{5}{6},-1\right\}$ B) $\left\{\frac{5}{6},\frac{7}{3}\right\}$

C) $\left\{\frac{5}{6},\frac{1}{4}\right\}$ D) $\left\{\frac{5}{6}\right\}$

100) Solve the below equation

$$\frac{x^2-9x+20}{x^3} + \frac{6}{x^2} = \frac{x^2+3x-18}{x^3}$$

A) $\left\{-\frac{19}{3},\frac{19}{3}\right\}$ B) $\left\{-\frac{19}{3},5\right\}$

C) $\left\{\frac{19}{3}\right\}$ D) $\left\{2,\frac{19}{3}\right\}$

101) Solve the below equation

$$\frac{1}{2x} + \frac{1}{6} = \frac{x^2 - 3x - 18}{x^2}$$

A) $\left\{ 2, -\frac{7}{4} \right\}$ B) $\left\{ -3, -\frac{7}{4} \right\}$

C) $\left\{ -\frac{7}{4} \right\}$ D) $\left\{ -3, \frac{36}{5} \right\}$

102) Solve the below equation

$$\frac{3n^2 + 16n + 5}{4n} + \frac{3}{n} = \frac{1}{4n}$$

A) $\left\{ -4, -\frac{4}{3} \right\}$ B) $\left\{ -4, -\frac{7}{2} \right\}$

C) $\left\{ -\frac{7}{2} \right\}$ D) $\{ -4, 2 \}$

103) Solve the below equation

$$1 - \frac{1}{v - 2} = \frac{2}{v - 2}$$

A) $\{ 3 \}$ B) $\{ -5 \}$

C) $\{ 10 \}$ D) $\{ 5 \}$

104) Solve the below equation

$$\frac{a^2 - 7a + 12}{a^2} + \frac{3}{2a^2} = \frac{1}{2a}$$

A) $\{ 2, 1 \}$ B) $\{ 2, -1 \}$

C) $\left\{ 3, \frac{9}{2} \right\}$ D) $\left\{ 2, \frac{9}{2} \right\}$

105) Solve the below equation

$$\frac{v^2 - 4v + 3}{v^2} - \frac{1}{v} = \frac{3v + 12}{v^2}$$

A) $\left\{ \frac{3}{5}, -1 \right\}$ B) $\{ 9, -1 \}$

C) $\{ 3, -1 \}$ D) $\left\{ \frac{3}{5}, -2 \right\}$

106) Solve the below equation

$$\frac{1}{a + 1} = \frac{1}{6a + 6} + \frac{13a + 182}{6a + 6}$$

A) $\{ -10 \}$ B) $\{ -13 \}$

C) $\{ 10 \}$ D) $\left\{ -\frac{177}{13} \right\}$

107) Solve the below equation

$$\frac{12x+14}{5x-10} - \frac{12x-3}{5x-10} = 1$$

A) $\left\{ \frac{27}{5} \right\}$ B) $\left\{ -\frac{27}{5} \right\}$

C) $\{-2\}$ D) $\{0\}$

108) Solve the below equation

$$\frac{1}{10x+8} - 1 = \frac{2}{5x+4}$$

A) $\{-5\}$ B) $\{10\}$

C) $\{-3\}$ D) $\left\{ -\frac{11}{10} \right\}$

109) Solve the below equation

$$\frac{8}{r+4} + \frac{2}{r^2-5r-36} = \frac{1}{r^2-5r-36}$$

A) $\{7\}$ B) $\{-6\}$

C) $\{6\}$ D) $\left\{ \frac{71}{8} \right\}$

110) Solve the below equation

$$\frac{5}{n+6} = \frac{n+4}{n^2+6n} + \frac{1}{n+6}$$

A) $\{-11\}$ B) $\left\{ \frac{4}{3} \right\}$

C) $\left\{ -\frac{4}{3} \right\}$ D) $\{13\}$

90) Simplify the below and state the excluded values.

$$\frac{30}{25x-35}$$

91) Simplify the below and state the excluded values.

$$\frac{p+10}{9p+90}$$

92) Simplify the below and state the excluded values.

$$\frac{2n^2+16n}{n+8}$$

93) Simplify the below and state the excluded values.

$$\frac{6n+12}{n+2}$$

94) Simplify the below and state the excluded values.

$$\frac{x+4}{7x^2+28x}$$

95) Simplify the below and state the excluded values.

$$\frac{10r}{6r+16}$$

96) Simplify the below and state the excluded values.

$$\frac{42k-12}{54k^2}$$

97) Simplify the below and state the excluded values.

$$\frac{30n^2}{10n+35}$$

98) Simplify the below and state the excluded values.

$$\frac{m-6}{m^2-5m-6}$$

99) Simplify the below and state the excluded values.

$$\frac{x^2-6x-27}{x+3}$$

100) Simplify the below and state the excluded values.

$$\frac{27x^2-81x}{63x}$$

101) Simplify the below and state the excluded values.

$$\frac{r+1}{r^2+10r+9}$$

102) Simplify the below and state the excluded values.

$$\frac{m+3}{m^2+13m+30}$$

103) Simplify the below and state the excluded values.

$$\frac{1-b}{2b-2}$$

104) Simplify the below and state the excluded values.

$$\frac{20}{15n + 30}$$

105) Simplify the below and state the excluded values.

$$\frac{56k}{21k + 35}$$

106) Simplify the below and state the excluded values.

$$\frac{a^2 + a - 2}{a + 2}$$

107) Simplify the below and state the excluded values.

$$\frac{3 - x}{2x^2 - 6x}$$

108) Simplify the below and state the excluded values.

$$\frac{n - 9}{n^2 - 81}$$

109) Simplify the below and state the excluded values.

$$\frac{x - 2}{7x^2 - 14x}$$

110) Simplify the below and state the excluded values.

$$\frac{15x^2 - 50x}{50x}$$

111) Simplify the below and state the excluded values.

$$\frac{4n^2 - 8n}{n - 2}$$

112) Simplify the below and state the excluded values.

$$\frac{6x + 8}{20}$$

113) Simplify the below and state the excluded values.

$$\frac{v^2 + 24v + 143}{v^2 + 22v + 117}$$

114) Simplify the below and state the excluded values.

$$\frac{99n - 90}{99n - 9}$$

115) Simplify the below and state the excluded values.

$$\frac{n^2 + 2n - 120}{n^2 + 13n + 12}$$

116) Simplify the below and state the excluded values.

$$\frac{n^2 - 10n + 24}{n^2 - 16}$$

117) Simplify the below and state the excluded values.

$$\frac{33n + 39}{9n - 27}$$

118) Simplify the below and state the excluded values.

$$\frac{11p + 22}{p^2 - 10p - 24}$$

119) Simplify the below and state the excluded values.

$$\frac{60v^2 + 144v}{36v^2 - 108v}$$

120) Simplify the below and state the excluded values.

$$\frac{r^2 - 21r + 110}{r^2 - 18r + 77}$$

121) Simplify the below and state the excluded values.

$$\frac{5 - 5x}{2x - 2}$$

122) Simplify the below and state the excluded values.

$$\frac{39x^2 - 104x}{91x^2 + 39x}$$

123) Simplify the below and state the excluded values.

$$\frac{x^2 - 16}{7x^3 + 28x^2}$$

124) Simplify the below and state the excluded values.

$$\frac{65\,a + 130}{91\,a - 182}$$

125) Simplify the below and state the excluded values.

$$\frac{x^2 - 5\,x - 126}{x^2 - 20\,x + 84}$$

126) Simplify the below and state the excluded values.

$$\frac{10\,p - 50}{p^2 + 6\,p - 55}$$

127) Simplify the below and state the excluded values.

$$\frac{65x^2 - 13x}{26x^2 + 156x}$$

128) Simplify the below and state the excluded values.

$$\frac{x^2 + 20x + 91}{x^2 + 17x + 70}$$

129) Simplify the below and state the excluded values.

$$\frac{n^2 + 6n - 112}{11n - 88}$$

130) Simplify the below and state the excluded values.

$$\frac{11\,b + 110}{b^2 - 100}$$

131) Simplify the below and state the excluded values.

$$\frac{x^2 - 12\,x - 28}{x^2 + 12\,x + 20}$$

132) Simplify the below and state the excluded values.

$$\frac{x^2 - 10x + 21}{14\,x - 98}$$

1) Solve the below quadratic equation using the method of factorization.

$$k^2 - 9k - 162 = 0$$

A) (9 , − 18)   B)   (-9 , 18)
C) { 10 , 0)   D) (16 , 18)

2) Solve the below quadratic equation using the method of factorization.

$$k^2 + 8k - 105 = 0$$

A) { − 2 , 0 }   B) { − 15 , 7 }
C) { − 15 , − 7 }   D) { − 12 , 19 }

3) Solve the below quadratic equation using the method of factorization.

$$n^2 - 12n + 32 = 0$$

A) { 8 , 1 }   B) { − 8 , − 4 }
C) { 2 , 18 }   D) { 8 , 4 }

4) Solve the below quadratic equation using the method of factorization.

$$v^2 - 3v - 238 = 0$$

A) { 15 , − 8 }   B) { − 14 , 17 }
C) { 11 , 9 }   D) { − 20 , − 1 }

5) Solve the below quadratic equation using the method of factorization.

$$b^2 + 15b - 34 = 0$$

A) { − 17 , 2 }   B)   { 13 , -2}
C) { 20 , 4 }   D)   { 12 , 10 }

6) Solve the below quadratic equation using the method of factorization.

$$r^2 + 26r + 133 = 0$$

A) { − 7 , − 19 }   B) { 4 , 7 }
C) { 9 , 0 }   D) { − 19 , − 4 }

7) Solve the below quadratic equation using the method of factorization.

$$k^2 - 5k - 24 = 0$$

A) { − 20 , 8 }   B) { − 16 , 3 }
C) { 3 , − 8 }   D) { − 3 , 7 }

8) Solve the below quadratic equation using the method of factorization.

$$r^2 - 13r + 12 = 0$$

A) { − 10 , 0 }   B) { − 7 , 0 }
C) { 12 , 1 }   D) { − 16 , 1 }

9) Solve the below quadratic equation using the method of factorization.

$$k^2 - 2k - 224 = 0$$

A) { − 14 , 16 } B) { 4 , 12 }

C) { 14 , − 4 } D) { 16 , 12 }

10) Solve the below quadratic equation using the method of factorization.

$$x^2 - 3x - 208 = 0$$

A) { − 10 , 20 } B) { − 16 , 13 }

C) { 16 , − 13 } D) { − 3 , 0 }

11) Solve the below quadratic equation using the method of factorization.

$$b^2 + 18b + 72 = 0$$

A) { − 6 , 5 } B) { − 6 , − 12 }

C) { − 10 , − 4 } D) { 13 , − 16 }

12) Solve the below quadratic equation using the method of factorization.

$$b^2 - 8b - 9 = 0$$

A) { − 9 , − 3 } B) { − 1 }

C) { 9 , − 1 } D) { 13 , − 1 }

13) Solve the below quadratic equation using the method of factorization.

$$v^2 - 29v + 190 = 0$$

A) { − 19 , − 10 } B) { − 19 , 17 }

C) { 19 , 10 } D) { 7 , − 10 }

14) Solve the below quadratic equation using the method of factorization.

$$n^2 + 16n - 57 = 0$$

A) { 3 , − 19 } B) { 4 , 16 }

C) { 19 , 0 } D) { − 3 , 19 }

15) Solve the below quadratic equation using the method of factorization.

$$x^2 - 3x - 4 = 0$$

A) { − 1 , 4 } B) { − 12 , − 19 }

C) { − 15 , − 4 } D) { − 7 , 0 }

16) Solve the below quadratic equation using the method of factorization.

$$n^2 + 30n + 209 = 0$$

A) { 2 , 0 } B) { − 16 , 17 }

C) { 20 , 19 } D) { − 11 , − 19 }

www.math-knots.com | www.a4ace.com

17) Solve the below quadratic equation using the method of factorization.

$$k^2 + 6k - 160 = 0$$

A) { 10 , − 3 }   B) { 10 , − 16 }

C) { 13 , 20 }   D) { 4 , 0 }

18) Solve the below quadratic equation using the method of factorization.

$$x^2 + 29x + 190 = 0$$

A) { 4 , − 18 }   B) { 19 , 10 }

C) { − 19 , − 10 }   D) { 7 , 11 }

19) Solve the below quadratic equation using the method of factorization.

$$x^2 - 18x + 81 = 0$$

A) { − 4 , 12 }   B) { 9 }

C) { − 9 }   D) { − 8 , − 17 }

20) Solve the below quadratic equation using the method of factorization.

$$x^2 + 14x - 15 = 0$$

A) { − 15 , 1 }   B) { 15 , 20 }

C) { 20 , − 6 }   D) { 15 , − 7 }

21) Solve the below quadratic equation using the method of factorization.

$$n^2 + 20n + 99 = 0$$

A) { 9 , 4 }   B) { 9 , − 10 }

C) { − 9 , − 11 }   D) { − 7 , 3 }

22) Solve the below quadratic equation using the method of factorization.

$$m^2 + m = 0$$

A) { 1 , 0 }   B) { − 10 , − 4 }

C) { − 15 , − 11 }   D) { − 1 , 0 }

23) Solve the below quadratic equation using the method of factorization.

$$x^2 - 2x = 0$$

A) { − 2 , 19 }   B) { − 3 , 19 }

C) { 2 , − 16 }   D) { 2 , 0 }

24) Solve the below quadratic equation using the method of factorization.

$$p^2 + 3p + 2 = 0$$

A) { 1 , 2 }   B) { − 1 , − 14 }

C) { − 1 , − 2 }   D) { 14 , − 12 }

www.math-knots.com | www.a4ace.com

25) Solve the below quadratic equation using the method of factorization.

$$m^2 - 21\,m + 54 = 0$$

A) { 1 , – 16 }   B) { – 18 , 16 }

C) { 18 , 3 }   D) { 18 , – 10 }

26) Solve the below quadratic equation using the method of factorization.

$$n^2 + 34\,n + 280 = 0$$

A) { – 10 , 2 }   B) { 20 , 14 }

C) { – 20 , – 14 }   D) { – 3 , – 14 }

27) Solve the below quadratic equation using the method of factorization.

$$9\,k^2 - 45\,k + 56 = 0$$

A) $\left\{ -\dfrac{4}{5}, \dfrac{7}{3} \right\}$ B) $\left\{ \dfrac{2}{5}, 6 \right\}$

C) $\left\{ -\dfrac{4}{5}, 8 \right\}$ D) $\left\{ \dfrac{8}{3}, \dfrac{7}{3} \right\}$

28) Solve the below quadratic equation using the method of factorization.

$$5\,x^2 + 48\,x + 64 = 0$$

A) $\left\{ \dfrac{6}{7}, 0 \right\}$ B) $\left\{ -\dfrac{8}{5}, -8 \right\}$

C) $\left\{ \dfrac{8}{5}, 2 \right\}$ D) $\left\{ -\dfrac{7}{2}, 8 \right\}$

29) Solve the below quadratic equation using the method of factorization.

$$7a^2 + 23\,a - 20 = 0$$

A) $\left\{ -\dfrac{6}{7}, 2 \right\}$ B) $\left\{ \dfrac{5}{7}, -4 \right\}$

C) $\left\{ -\dfrac{5}{7}, -8 \right\}$ D) $\left\{ -\dfrac{2}{3}, \dfrac{1}{7} \right\}$

30) Solve the below quadratic equation using the method of factorization.

$$5\,r^2 - 13\,r - 6 = 0$$

A) $\left\{ \dfrac{4}{5}, -6 \right\}$ B) $\left\{ -\dfrac{2}{5}, 3 \right\}$

C) $\left\{ -\dfrac{1}{5}, 0 \right\}$ D) $\left\{ -\dfrac{4}{5}, 3 \right\}$

31) Solve the below quadratic equation using the method of factorization.

$$35\,n^2 - 34\,n + 8 = 0$$

A) $\left\{ \dfrac{2}{5}, \dfrac{4}{7} \right\}$ B) $\left\{ -\dfrac{4}{5}, \dfrac{4}{7} \right\}$

C) $\left\{ -\dfrac{1}{6}, 2 \right\}$ D) $\left\{ -\dfrac{2}{5}, 4 \right\}$

32) Solve the below quadratic equation using the method of factorization.

$$6\,k^2 + 23\,k + 7 = 0$$

A) $\left\{ \dfrac{1}{7}, -2 \right\}$ B) $\left\{ \dfrac{1}{3}, 4 \right\}$

C) $\left\{ -\dfrac{2}{7}, \dfrac{7}{2} \right\}$ D) $\left\{ -\dfrac{1}{3}, -\dfrac{7}{2} \right\}$

33) Solve the below quadratic equation using the method of factorization.

$$8n^2 - 19n - 15 = 0$$

A) $\left\{\dfrac{7}{2}, 3\right\}$ B) $\left\{\dfrac{5}{3}, -1\right\}$

C) $\left\{-\dfrac{5}{8}, 3\right\}$ D) $\left\{\dfrac{5}{8}, \dfrac{7}{5}\right\}$

34) Solve the below quadratic equation using the method of factorization.

$$5k^2 + 18k - 8 = 0$$

A) $\left\{-\dfrac{2}{5}, 4\right\}$ B) $\left\{-\dfrac{4}{7}, -4\right\}$

C) $\left\{\dfrac{2}{5}, -4\right\}$ D) $\left\{-\dfrac{8}{3}, -8\right\}$

35) Solve the below quadratic equation using the method of factorization.

$$5k^2 - 26k - 24 = 0$$

A) $\left\{-\dfrac{1}{5}, -6\right\}$ B) $\left\{\dfrac{4}{5}, -6\right\}$

C) $\left\{-\dfrac{4}{5}, -3\right\}$ D) $\left\{-\dfrac{4}{5}, 6\right\}$

36) Solve the below quadratic equation using the method of factorization.

$$5n^2 + 21n - 20 = 0$$

A) $\left\{\dfrac{4}{5}, -5\right\}$ B) $\left\{\dfrac{7}{5}, 6\right\}$

C) $\left\{-\dfrac{4}{5}, 5\right\}$ D) $\left\{\dfrac{6}{5}, 3\right\}$

37) Solve the below quadratic equation using the method of factorization.

$$7n^2 + 50n + 48 = 0$$

A) $\left\{-\dfrac{8}{7}, -6\right\}$ B) $\left\{\dfrac{5}{2}, -6\right\}$

C) $\left\{-\dfrac{4}{3}, 1\right\}$ D) $\left\{\dfrac{8}{7}, \dfrac{1}{2}\right\}$

38) Solve the below quadratic equation using the method of factorization.

$$2n^2 + n - 28 = 0$$

A) $\left\{-\dfrac{7}{2}, 5\right\}$ B) $\left\{-\dfrac{7}{2}, 7\right\}$

C) $\left\{-\dfrac{3}{7}, -6\right\}$ D) $\left\{\dfrac{7}{2}, -4\right\}$

39) Solve the below quadratic equation using the method of factorization.

$$5n^2 - 42n + 49 = 0$$

A) $\left\{\dfrac{7}{2}, 3\right\}$ B) $\left\{-\dfrac{4}{3}, \dfrac{3}{2}\right\}$

C) $\left\{\dfrac{2}{5}, 7\right\}$ D) $\left\{\dfrac{7}{5}, 7\right\}$

40) Solve the below quadratic equation using the method of factorization.

$$5x^2 + 34x - 7 = 0$$

A) $\left\{-\dfrac{6}{7}, -1\right\}$ B) $\left\{\dfrac{1}{5}, -7\right\}$

C) $\left\{\dfrac{7}{3}, -4\right\}$ D) $\left\{\dfrac{7}{3}, -7\right\}$

41) Solve the below quadratic equation using the method of factorization.

$$7x^2 - 6x - 1 = 0$$

A) $\left\{-\dfrac{5}{7}, -3\right\}$ B) $\left\{-\dfrac{1}{7}, 1\right\}$

C) $\left\{-\dfrac{1}{7}, \dfrac{1}{2}\right\}$ D) $\left\{\dfrac{1}{7}, -1\right\}$

42) Solve the below quadratic equation using the method of factorization.

$$3x^2 - 16x - 35 = 0$$

A) $\left\{-\dfrac{5}{3}, 7\right\}$ B) $\left\{-\dfrac{7}{3}, -\dfrac{1}{7}\right\}$

C) $\left\{\dfrac{1}{5}, -\dfrac{4}{7}\right\}$ D) $\left\{-\dfrac{2}{3}, -4\right\}$

43) Solve the below quadratic equation using the method of factorization.

$$7n^2 + 3n - 4 = 0$$

A) $\left\{-\dfrac{4}{7}, 0\right\}$ B) $\left\{\dfrac{4}{7}, -1\right\}$

C) $\left\{\dfrac{1}{3}, 1\right\}$ D) $\left\{\dfrac{1}{2}, -\dfrac{4}{7}\right\}$

44) Solve the below quadratic equation using the method of factorization.

$$3x^2 + 5x - 8 = 0$$

A) $\left\{-\dfrac{8}{3}, 1\right\}$ B) $\left\{\dfrac{8}{5}, -2\right\}$

C) $\left\{-\dfrac{8}{3}, -5\right\}$ D) $\left\{\dfrac{7}{8}, -4\right\}$

45) Solve the below quadratic equation using the method of factorization.

$$21p^2 + 38p + 16 = 0$$

A) $\left\{\dfrac{7}{5}, -\dfrac{8}{7}\right\}$ B) $\left\{-\dfrac{2}{3}, -2\right\}$

C) $\left\{\dfrac{3}{5}, 2\right\}$ D) $\left\{-\dfrac{2}{3}, -\dfrac{8}{7}\right\}$

46) Solve the below quadratic equation using the method of factorization.

$$5x^2 + 7x + 2 = 0$$

A) $\left\{-\dfrac{2}{5}, -1\right\}$ B) $\left\{\dfrac{1}{7}, -1\right\}$

C) $\left\{-\dfrac{1}{2}, -\dfrac{6}{7}\right\}$ D) $\left\{\dfrac{2}{5}, 1\right\}$

47) Solve the below quadratic equation using the method of factorization.

$$7b^2 - 25b - 12 = 0$$

A) $\left\{\dfrac{1}{7}, -\dfrac{4}{7}\right\}$ B) $\left\{\dfrac{7}{2}, 0\right\}$

C) $\left\{-\dfrac{3}{7}, 4\right\}$ D) $\left\{\dfrac{3}{7}, -4\right\}$

48) Solve the below quadratic equation using the method of factorization.

$$6x^2 - 17x + 10 = 0$$

A) $\left\{-\dfrac{3}{2}, -8\right\}$ B) $\left\{\dfrac{8}{7}, -8\right\}$

C) $\left\{\dfrac{5}{2}, 4\right\}$ D) $\left\{\dfrac{5}{6}, 2\right\}$

49) Solve the below quadratic equation using the method of factorization.

$$8k^2 + 19k - 15 = 0$$

A) $\left\{ \dfrac{1}{6}, 3 \right\}$ B) $\left\{ -\dfrac{5}{8}, 0 \right\}$

C) $\left\{ \dfrac{1}{5}, \dfrac{8}{5} \right\}$ D) $\left\{ \dfrac{5}{8}, -3 \right\}$

50) Solve the below quadratic equation using the method of factorization.

$$7r^2 - 3r = 0$$

A) $\left\{ -\dfrac{7}{2}, 0 \right\}$ B) $\left\{ \dfrac{2}{5}, -8 \right\}$

C) $\left\{ \dfrac{4}{5}, 2 \right\}$ D) $\left\{ \dfrac{3}{7}, 0 \right\}$

51) Solve the below quadratic equation using the method of factorization.

$$3v^2 + v - 14 = 0$$

A) $\left\{ -\dfrac{7}{3}, 1 \right\}$ B) $\left\{ -\dfrac{8}{5}, -2 \right\}$

C) $\left\{ -\dfrac{7}{3}, 2 \right\}$ D) $\left\{ -\dfrac{1}{2}, -4 \right\}$

52) Solve the below quadratic equation using the method of factorization.

$$14p^2 + 27p + 9 = 0$$

A) $\left\{ -\dfrac{4}{3}, -5 \right\}$ B) $\left\{ \dfrac{3}{2}, 6 \right\}$

C) $\left\{ \dfrac{8}{7}, 0 \right\}$ D) $\left\{ -\dfrac{3}{2}, -\dfrac{3}{7} \right\}$

53) Solve the below quadratic equation using the method of factorization.

$$3x^2 + 16x + 21 = 0$$

A) $\left\{ -\dfrac{7}{3}, -3 \right\}$ B) $\left\{ -\dfrac{1}{3}, -3 \right\}$

C) $\left\{ \dfrac{7}{3}, 3 \right\}$ D) $\left\{ \dfrac{5}{3}, -8 \right\}$

54) Solve the below quadratic equation using the method of factorization.

$$7n^2 + 13n + 6 = 0$$

A) $\left\{ -\dfrac{6}{7}, -1 \right\}$ B) $\left\{ \dfrac{3}{2}, \dfrac{4}{5} \right\}$

C) $\left\{ \dfrac{4}{3}, -3 \right\}$ D) $\left\{ \dfrac{7}{2}, -7 \right\}$

55) Solve the below quadratic equation using the method of factorization.

$$5x^2 + 36x + 7 = 0$$

A) $\left\{ -\dfrac{1}{5}, -7 \right\}$ B) $\left\{ \dfrac{1}{5}, 7 \right\}$

C) $\left\{ -\dfrac{4}{5}, -\dfrac{8}{7} \right\}$ D) $\left\{ -\dfrac{1}{2}, -7 \right\}$

56) Solve the below quadratic equation using the method of factorization.

$$3n^2 + 23n + 14 = 0$$

A) $\left\{ -\dfrac{5}{2}, -3 \right\}$ B) $\left\{ \dfrac{2}{3}, 7 \right\}$

C) $\left\{ -\dfrac{2}{3}, -7 \right\}$ D) $\left\{ -\dfrac{7}{2}, -7 \right\}$

 **Algebra 1**

57) Solve the below quadratic equation using the method of factorization.

$$8x^2 - 15 = -19x$$

A) $\left\{-\dfrac{5}{8}, 3\right\}$ B) $\left\{\dfrac{8}{7}, 1\right\}$

C) $\left\{-\dfrac{7}{6}, 2\right\}$ D) $\left\{\dfrac{5}{8}, -3\right\}$

58) Solve the below quadratic equation using the method of factorization.

$$7b^2 = -55b + 8$$

A) $\left\{\dfrac{2}{5}, \dfrac{2}{7}\right\}$ B) $\left\{-\dfrac{4}{7}, 0\right\}$

C) $\left\{-\dfrac{1}{7}, 8\right\}$ D) $\left\{\dfrac{1}{7}, -8\right\}$

59) Solve the below quadratic equation using the method of factorization.

$$4x^2 - 35 = 4x$$

A) $\left\{-\dfrac{7}{5}, -1\right\}$ B) $\left\{-\dfrac{5}{7}, 3\right\}$

C) $\left\{\dfrac{7}{2}, -\dfrac{5}{2}\right\}$ D) $\left\{-\dfrac{2}{7}, -6\right\}$

60) Solve the below quadratic equation using the method of factorization.

$$5x^2 = 33x - 40$$

A) $\left\{-\dfrac{5}{2}, -\dfrac{2}{5}\right\}$ B) $\left\{\dfrac{8}{5}, 5\right\}$

C) $\left\{-\dfrac{6}{7}, 6\right\}$ D) $\left\{\dfrac{2}{7}, 1\right\}$

61) Solve the below quadratic equation using the method of factorization.

$$7n^2 + 40 = 61n$$

A) $\left\{-\dfrac{2}{5}, -1\right\}$ B) $\left\{\dfrac{5}{7}, 8\right\}$

C) $\left\{\dfrac{7}{5}, 8\right\}$ D) $\left\{-\dfrac{2}{5}, -8\right\}$

62) Solve the below quadratic equation using the method of factorization.

$$4n^2 - 7n = 15$$

A) $\left\{-\dfrac{5}{4}, 3\right\}$ B) $\left\{\dfrac{7}{5}, 0\right\}$

C) $\left\{-\dfrac{4}{5}, \dfrac{4}{5}\right\}$ D) $\left\{\dfrac{1}{7}, -\dfrac{4}{7}\right\}$

63) Solve the below quadratic equation using the method of factorization.

$$3v^2 - 13v = 10$$

A) $\left\{\dfrac{2}{5}, -1\right\}$ B) $\left\{-\dfrac{2}{3}, 5\right\}$

C) $\left\{\dfrac{2}{3}, -5\right\}$ D) $\left\{-\dfrac{2}{3}, 6\right\}$

64) Solve the below quadratic equation using the method of factorization.

$$15k^2 - 28 = -23k$$

A) $\left\{-\dfrac{7}{3}, 4\right\}$ B) $\left\{\dfrac{7}{3}, -\dfrac{4}{5}\right\}$

C) $\left\{-\dfrac{7}{3}, \dfrac{4}{5}\right\}$ D) $\left\{\dfrac{2}{3}, -\dfrac{4}{5}\right\}$

364

65) Solve the below quadratic equation using the method of factorization.

$$3\,b^2 + 8 = -\,11\,b$$

A) $\left\{-\dfrac{1}{3}, 6\right\}$ B) $\left\{-\dfrac{8}{3}, -1\right\}$

C) $\left\{-\dfrac{8}{3}, -7\right\}$ D) $\left\{\dfrac{3}{2}, -\dfrac{5}{7}\right\}$

66) Solve the below quadratic equation using the method of factorization.

$$7\,v^2 + 10 = 37\,v$$

A) $\left\{-\dfrac{2}{7}, 2\right\}$ B) $\left\{\dfrac{2}{7}, 5\right\}$

C) $\left\{-\dfrac{2}{5}, 3\right\}$ D) $\left\{-\dfrac{5}{7}, 5\right\}$

67) Solve the below quadratic equation using the method of factorization.

$$15\,x^2 + 16 = -\,32\,x$$

A) $\left\{-\dfrac{4}{3}, -\dfrac{4}{5}\right\}$ B) $\left\{-\dfrac{1}{7}, -6\right\}$

C) $\left\{\dfrac{4}{3}, 8\right\}$ D) $\left\{\dfrac{7}{3}, \dfrac{4}{5}\right\}$

68) Solve the below quadratic equation using the method of factorization.

$$6\,x^2 - 53\,x = -\,40$$

A) $\left\{\dfrac{4}{3}, 0\right\}$ B) $\left\{-\dfrac{4}{7}, -6\right\}$

C) $\left\{\dfrac{5}{6}, 8\right\}$ D) $\left\{-\dfrac{5}{2}, \dfrac{2}{3}\right\}$

69) Solve the below quadratic equation using the method of factorization.

$$2\,x^2 - 13\,x = -\,20$$

A) $\left\{\dfrac{5}{2}, 4\right\}$ B) $\left\{-\dfrac{4}{7}, -\dfrac{2}{5}\right\}$

C) $\left\{-\dfrac{5}{2}, -\dfrac{7}{5}\right\}$ D) $\left\{\dfrac{8}{5}, 4\right\}$

70) Solve the below quadratic equation using the method of factorization.

$$7\,x^2 = -\,54\,x + 16$$

A) $\left\{\dfrac{2}{7}, -8\right\}$ B) $\left\{-\dfrac{7}{5}, -7\right\}$

C) $\left\{-\dfrac{2}{7}, 8\right\}$ D) $\left\{-\dfrac{3}{7}, 4\right\}$

71) Solve the below quadratic equation using the method of factorization.

$$5\,m^2 + 36 = -\,36\,m$$

A) $\left\{\dfrac{2}{7}, -\dfrac{7}{5}\right\}$ B) $\left\{-\dfrac{3}{2}, \dfrac{6}{7}\right\}$

C) $\left\{\dfrac{6}{7}, 1\right\}$ D) $\left\{-\dfrac{6}{5}, -6\right\}$

72) Solve the below quadratic equation using the method of factorization.

$$3\,x^2 = 1 - 2\,x$$

A) $\left\{-\dfrac{7}{3}, 6\right\}$ B) $\left\{\dfrac{1}{3}, -1\right\}$

C) $\left\{-\dfrac{1}{3}, \dfrac{4}{7}\right\}$ D) $\left\{-\dfrac{1}{3}, 6\right\}$

 Algebra 1

73) Solve the below quadratic equation using the method of factorization.

$$15 x^2 - 10 = 19 x$$

A) $\left\{ \dfrac{1}{5}, 1 \right\}$ B) $\left\{ \dfrac{5}{3}, -\dfrac{2}{5} \right\}$

C) $\left\{ \dfrac{5}{8}, 3 \right\}$ D) $\left\{ \dfrac{4}{5}, -7 \right\}$

74) Solve the below quadratic equation using the method of factorization.

$$4 k^2 = - k$$

A) $\left\{ -\dfrac{2}{3}, 4 \right\}$ B) $\left\{ -\dfrac{1}{4}, 0 \right\}$

C) $\left\{ \dfrac{7}{2}, -3 \right\}$ D) $\left\{ -\dfrac{8}{5}, -7 \right\}$

75) Solve the below quadratic equation using the method of factorization.

$$3 b^2 = 14 b + 5$$

A) $\left\{ -\dfrac{1}{3}, 6 \right\}$ B) $\left\{ \dfrac{1}{5}, 0 \right\}$

C) $\left\{ -\dfrac{1}{3}, 5 \right\}$ D) $\left\{ -\dfrac{1}{5}, -2 \right\}$

76) Solve the below quadratic equation using the method of factorization.

$$7 n^2 + 3 n = 0$$

A) $\left\{ -\dfrac{3}{2}, 1 \right\}$ B) $\left\{ -\dfrac{4}{7}, -5 \right\}$

C) $\left\{ \dfrac{5}{4}, 2 \right\}$ D) $\left\{ -\dfrac{3}{7}, 0 \right\}$

77) Solve the below quadratic equation using the method of factorization.

$$5 v^2 + 41 v = - 8$$

A) $\left\{ -\dfrac{1}{5}, -8 \right\}$ B) $\left\{ \dfrac{1}{5}, 0 \right\}$

C) $\left\{ -\dfrac{6}{5}, \dfrac{7}{2} \right\}$ D) $\left\{ \dfrac{1}{5}, 8 \right\}$

78) Solve the below quadratic equation using the method of factorization.

$$7 n^2 = 41 n - 30$$

A) $\left\{ \dfrac{7}{5}, -5 \right\}$ B) $\left\{ \dfrac{6}{7}, 5 \right\}$

C) $\left\{ -\dfrac{5}{3}, 5 \right\}$ D) $\left\{ \dfrac{4}{5}, 0 \right\}$

79) Solve the below quadratic equation using the method of factorization.

$$7 p^2 = 14 + 47 p$$

A) $\left\{ -\dfrac{2}{7}, 7 \right\}$ B) $\left\{ \dfrac{6}{5}, -4 \right\}$

C) $\left\{ -\dfrac{3}{2}, 7 \right\}$ D) $\left\{ -\dfrac{8}{7}, \dfrac{3}{2} \right\}$

80) Solve the below quadratic equation using the method of factorization.

$$5 r^2 = 7 + 2 r$$

A) $\left\{ \dfrac{7}{5}, 8 \right\}$ B) $\left\{ \dfrac{7}{5}, -1 \right\}$

C) $\left\{ \dfrac{7}{8}, 2 \right\}$ D) $\left\{ \dfrac{7}{5}, -8 \right\}$

81) Solve the below quadratic equation using the method of factorization.

$$5\,a^2 + 6\,a = 8$$

A) $\left\{-\dfrac{1}{3},\ 2\right\}$ B) $\left\{\dfrac{4}{5},\ -2\right\}$

C) $\left\{\dfrac{1}{3},\ -6\right\}$ D) $\left\{-\dfrac{4}{5},\ 2\right\}$

82) Solve the below quadratic equation using the method of factorization.

$$5\,x^2 = -18 - 33\,x$$

A) $\left\{\dfrac{6}{5},\ 5\right\}$ B) $\left\{-\dfrac{8}{3},\ -6\right\}$

C) $\left\{-\dfrac{3}{5},\ -6\right\}$ D) $\left\{-\dfrac{7}{2},\ -6\right\}$

1) Find the discriminant value for the below quadratic equation. Also, state the number and type of solutions.

$$-6r^2 + 3r - 1 = 0$$

2) Find the discriminant value for the below quadratic equation. Also, state the number and type of solutions.

$$-8r^2 - 4r - 2 = 0$$

3) Find the discriminant value for the below quadratic equation. Also, state the number and type of solutions.

$$2k^2 - 8k + 8 = 0$$

4) Find the discriminant value for the below quadratic equation. Also, state the number and type of solutions.

$$k^2 - 6k + 9 = 0$$

5) Find the discriminant value for the below quadratic equation. Also, state the number and type of solutions.

$$-2p^2 - 4p = 0$$

6) Find the discriminant value for the below quadratic equation. Also, state the number and type of solutions.

$$-7x^2 - 5x + 2 = 0$$

7) Find the discriminant value for the below quadratic equation. Also, state the number and type of solutions.

$$-5x^2 - 10x - 8 = 0$$

8) Find the discriminant value for the below quadratic equation. Also, state the number and type of solutions.

$$-x^2 + 2x - 6 = 0$$

9) Find the discriminant value for the below quadratic equation. Also, state the number and type of solutions.

$$-2r^2 + 4r + 2 = 0$$

10) Find the discriminant value for the below quadratic equation. Also, state the number and type of solutions.

$$-10k^2 + 4k - 1 = 0$$

11) Find the discriminant value for the below quadratic equation. Also, state the number and type of solutions.

$$-3x^2 + 3x + 6 = 0$$

16) Find the discriminant value for the below quadratic equation. Also, state the number and type of solutions.

$$-6n^2 - 6 = 0$$

12) Find the discriminant value for the below quadratic equation. Also, state the number and type of solutions.

$$10k^2 + 7k - 6 = 0$$

17) Find the discriminant value for the below quadratic equation. Also, state the number and type of solutions.

$$-3x^2 + x - 8 = 0$$

13) Find the discriminant value for the below quadratic equation. Also, state the number and type of solutions.

$$8m^2 - 7m + 4 = 0$$

18) Find the discriminant value for the below quadratic equation. Also, state the number and type of solutions.

$$2v^2 + 3v = 0$$

14) Find the discriminant value for the below quadratic equation. Also, state the number and type of solutions.

$$-4x^2 + 2x + 2 = 0$$

19) Find the discriminant value for the below quadratic equation. Also, state the number and type of solutions.

$$-6n^2 - 5n - 6 = 0$$

15) Find the discriminant value for the below quadratic equation. Also, state the number and type of solutions.

$$4r^2 - 4r + 1 = 0$$

20) Find the discriminant value for the below quadratic equation. Also, state the number and type of solutions.

$$-5x^2 + 10x - 5 = 0$$

21) Find the discriminant value for the below quadratic equation. Also, state the number and type of solutions.

$$3v^2 - 8v + 9 = 0$$

22) Find the discriminant value for the below quadratic equation. Also, state the number and type of solutions.

$$6m^2 - m + 1 = 0$$

23) Find the discriminant value for the below quadratic equation. Also, state the number and type of solutions.

$$-6n^2 + 5n - 7 = 0$$

24) Find the discriminant value for the below quadratic equation. Also, state the number and type of solutions.

$$4x^2 + 8x + 8 = 0$$

25) Find the discriminant value for the below quadratic equation. Also, state the number and type of solutions.

$$4v^2 - 2v + 3 = 0$$

26) Find the discriminant value for the below quadratic equation. Also, state the number and type of solutions.

$$-14k^2 - 16 = 9k$$

27) Find the discriminant value for the below quadratic equation. Also, state the number and type of solutions.

$$2n^2 + 9n = -15$$

28) Find the discriminant value for the below quadratic equation. Also, state the number and type of solutions.

$$-3r^2 - r = 8$$

29) Find the discriminant value for the below quadratic equation. Also, state the number and type of solutions.

$$3k^2 = -12k - 12$$

30) Find the discriminant value for the below quadratic equation. Also, state the number and type of solutions.

$$-10x^2 + 12x = 5$$

31) Find the discriminant value for the below quadratic equation. Also, state the number and type of solutions.

$$3n^2 + 15 = -4n$$

32) Find the discriminant value for the below quadratic equation. Also, state the number and type of solutions.

$$-6n^2 = 10 - 11n$$

33) Find the discriminant value for the below quadratic equation. Also, state the number and type of solutions.

$$11x^2 - 14x = -14$$

34) Find the discriminant value for the below quadratic equation. Also, state the number and type of solutions.

$$-16x^2 + 5x = 6$$

35) Find the discriminant value for the below quadratic equation. Also, state the number and type of solutions.

$$-3v^2 - v = 0$$

36) Find the discriminant value for the below quadratic equation. Also, state the number and type of solutions.

$$10n^2 + 3 = -13n$$

37) Find the discriminant value for the below quadratic equation. Also, state the number and type of solutions.

$$-4n^2 + 16n = 16$$

38) Find the discriminant value for the below quadratic equation. Also, state the number and type of solutions.

$$-5n^2 - 7 = 12n$$

39) Find the discriminant value for the below quadratic equation. Also, state the number and type of solutions.

$$2r^2 - 2r = -10$$

40) Find the discriminant value for the below quadratic equation. Also, state the number and type of solutions.

$$-16v^2 = -7 + 9v$$

 **Algebra 1**

41) Find the discriminant value for the below quadratic equation. Also, state the number and type of solutions.

$$-9n^2 = 7 - 2n$$

42) Find the discriminant value for the below quadratic equation. Also, state the number and type of solutions.

$$-3m^2 = -m - 14$$

43) Find the discriminant value for the below quadratic equation. Also, state the number and type of solutions.

$$-8a^2 - 16 = -7a$$

44) Find the discriminant value for the below quadratic equation. Also, state the number and type of solutions.

$$-p^2 - 9 = -6p$$

45) Find the discriminant value for the below quadratic equation. Also, state the number and type of solutions.

$$-12p^2 - 3 = 12p$$

46) Find the discriminant value for the below quadratic equation. Also, state the number and type of solutions.

$$16n^2 = -1 + 8n$$

47) Find the value of c for the below to complete the square.

$$x^2 + 26x + c$$

48) Find the value of c for the below to complete the square.

$$x^2 - 34x + c$$

49) Find the value of c for the below to complete the square.

$$z^2 - 32z + c$$

50) Find the value of c for the below to complete the square.

$$x^2 + 10x + c$$

51) Find the value of c for the below to complete the square.

$$x^2 + 32x + c$$

52) Find the value of c for the below to complete the square.

$$x^2 - 28x + c$$

53) Find the value of c for the below to complete the square.

$$p^2 + 2p + c$$

54) Find the value of c for the below to complete the square.

$$z^2 + 28z + c$$

55) Find the value of c for the below to complete the square.

$$y^2 + 30y + c$$

56) Find the value of c for the below to complete the square.

$$x^2 + 16x + c$$

57) Find the value of c for the below to complete the square.

$$x^2 + 14x + c$$

58) Find the value of c for the below to complete the square.

$$x^2 + 40x + c$$

59) Find the value of c for the below to complete the square.

$$y^2 + 24y + c$$

60) Find the value of c for the below to complete the square.

$$x^2 - 20x + c$$

www.math-knots.com | www.a4ace.com

61) Find the value of c for the below to complete the square.

$$a^2 - 36\,a + c$$

66) Find the value of c for the below to complete the square.

$$z^2 - 42\,z + c$$

62) Find the value of c for the below to complete the square.

$$n^2 + 18\,n + c$$

67) Find the value of c for the below to complete the square.

$$r^2 + 36\,r + c$$

63) Find the value of c for the below to complete the square.

$$r^2 - 4\,r + c$$

68) Find the value of c for the below to complete the square.

$$x^2 + 11\,x + c$$

64) Find the value of c for the below to complete the square.

$$x^2 + 4\,x + c$$

69) Find the value of c for the below to complete the square.

$$r^2 - \frac{4}{3}\,r + c$$

65) Find the value of c for the below to complete the square.

$$x^2 - 24\,x + c$$

70) Find the value of c for the below to complete the square.

$$m^2 + 17\,m + c$$

Algebra 1

71) Find the value of c for the below to complete the square.

$$y^2 + 3y + c$$

72) Find the value of c for the below to complete the square.

$$m^2 + 5m + c$$

73) Find the value of c for the below to complete the square.

$$x^2 - x + c$$

74) Find the value of c for the below to complete the square.

$$x^2 + \frac{13}{8}x + c$$

75) Find the value of c for the below to complete the square.

$$z^2 + \frac{8}{5}z + c$$

76) Find the value of c for the below to complete the square.

$$z^2 + \frac{4}{3}z + c$$

77) Find the value of c for the below to complete the square.

$$x^2 - 7x + c$$

78) Find the value of c for the below to complete the square.

$$m^2 - 15m + c$$

79) Find the value of c for the below to complete the square.

$$z^2 + 9z + c$$

80) Find the value of c for the below to complete the square.

$$m^2 - \frac{43}{13}m + c$$

 **Algebra 1**

81) Find the value of c for the below to complete the square.

$$x^2 - 3x + c$$

82) Find the value of c for the below to complete the square.

$$p^2 + 15p + c$$

83) Find the value of c for the below to complete the square.

$$y^2 - 9y + c$$

84) Find the value of c for the below to complete the square.

$$n^2 + 19n + c$$

85) Find the value of c for the below to complete the square.

$$z^2 - \frac{3}{4}z + c$$

86) Find the value of c for the below to complete the square.

$$x^2 - 13x + c$$

87) Find the value of c for the below to complete the square.

$$x^2 + 7x + c$$

88) Find the value of c for the below to complete the square.

$$x^2 - 19x + c$$

89) Solve the below quadratic equation.

$$n^2 + 2n - 18 = -3$$

90) Solve the below quadratic equation.

$$m^2 - 14m - 24 = 8$$

 **Algebra 1**

91) Solve the below quadratic equation.

$$n^2 - 10n - 100 = -8$$

92) Solve the below quadratic equation

$$m^2 - 18m + 38 = -7$$

93) Solve the below quadratic equation.

$$n^2 + 10n - 84 = -9$$

94) Solve the below quadratic equation.

$$x^2 - 8x - 10 = 3$$

95) Solve the below quadratic equation.

$$n^2 + 10n - 104 = 8$$

96) Solve the below quadratic equation.

$$x^2 - 14x - 22 = -7$$

97) Solve the below quadratic equation.

$$a^2 - 6a - 65 = 7$$

98) Solve the below quadratic equation.

$$n^2 + 14n - 105 = -8$$

99) Solve the below quadratic equation.

$$x^2 - 10x - 47 = 9$$

100) Solve the below quadratic equation.

$$n^2 - 4n - 88 = 8$$

101) Solve the below quadratic equation.

$$b^2 + 16b - 76 = 4$$

102) Solve the below quadratic equation.

$$x^2 - 6x - 11 = -10$$

103) Solve the below quadratic equation.

$$b^2 - 18b + 80 = 3$$

104) Solve the below quadratic equation.

$$a^2 - 8a - 32 = 9$$

105) Solve the below quadratic equation.

$$a^2 - 8a - 43 = 5$$

106) Solve the below quadratic equation.

$$x^2 - 4x - 6 = 6$$

107) Solve the below quadratic equation.

$$n^2 + 16n - 53 = 4$$

 **Algebra 1**

Vol 2
Week 34
Direct &
Inverse
Variations

1) Does the below equation represent a direct or inverse variation ?

$$y = 14\,x$$

A) Direct   B) Inverse

2) Does the below equation represent a direct or inverse variation ?

$$y = 5\,x$$

A) Direct   B) Inverse

3) Does the below equation represent a direct or inverse variation ?

$$y = \frac{11}{x}$$

A) Inverse   B) Direct

4) Does the below equation represent a direct or inverse variation ?

$$y = \frac{2}{5x}$$

A) Direct   B) Inverse

5) Does the below equation represent a direct or inverse variation ?

$$y = -\frac{11}{x}$$

A) Inverse   B) Direct

6) Does the below equation represent a direct or inverse variation ?

$$y = \frac{3}{x}$$

A) Inverse   B) Direct

7) Does the below equation represent a direct or inverse variation ?

$$y = 2\,x$$

A) Inverse   B) Direct

8) Does the below equation represent a direct or inverse variation ?

$$y = \frac{6}{x}$$

A) Inverse   B) Direct

Algebra 1

Vol 2
Week 34
Direct &
Inverse
Variations

9) Does the below equation represent a direct or inverse variation ?

$$y = \frac{5}{x}$$

A) Inverse   B) Direct

10) Does the below equation represent a direct or inverse variation ?

$$y = \frac{12}{x}$$

A) Direct   B) Inverse

11) Does the below equation represent a direct or inverse variation ?

$$y = 12x$$

A) Inverse   B) Direct

12) Does the below equation represent a direct or inverse variation ?

$$y = \frac{14}{x}$$

A) Direct   B) Inverse

13) Does the below equation represent a direct or inverse variation ?

$$y = \frac{10}{x}$$

A) Inverse   B) Direct

14) Does the below equation represent a direct or inverse variation ?

$$y = -10x$$

A) Direct   B) Inverse

15) Does the below equation represent a direct or inverse variation ?

$$y = \frac{9}{x}$$

A) Inverse   B) Direct

16) Does the below equation represent a direct or inverse variation ?

$$y = 6x$$

A) Inverse   B) Direct

Algebra 1

**Vol 2
Week 34**
Direct &
Inverse
Variations

17) Does the below equation represent a direct or inverse variation ?

$$y = 4x$$

A) Direct   B) Inverse

18) Does the below equation represent a direct or inverse variation ?

$$y = \frac{4}{3}x$$

A) Direct   B) Inverse

19) Does the below equation represent a direct or inverse variation ?

$$y = -\frac{1}{2}x$$

A) Inverse   B) Direct

20) Does the below equation represent a direct or inverse variation ?

$$y = 8x$$

A) Direct   B) Inverse

21) Does the below equation represent a direct or inverse variation ?

$$y = \frac{7}{x}$$

A) Direct   B) Inverse

22) Find the constant of variation ?

$$y = \frac{15}{x}$$

A) 14   B) −12
C) 4   D) 15

23) Find the constant of variation ?

$$y = \frac{14}{x}$$

A) 14   B) −5
C) 7   D) 4

24) Find the constant of variation ?

$$y = -\frac{6}{x}$$

A) −6   B) 11
C) 3   D) 5

 **Algebra 1**

**Vol 2
Week 34**
Direct &
Inverse
Variations

25) Find the constant of variation ?

$$y = \frac{2}{x}$$

A) 8   B) 7

C) 2   D) 13

26) Find the constant of variation ?

$$y = \frac{8}{x}$$

A) 8   B) 13

C) 9   D) 3

27) Find the constant of variation ?

$$y = \frac{15}{2x}$$

A) $\frac{3}{2}$ B) $-\frac{5}{9}$

C) $\frac{15}{2}$ D) $\frac{5}{4}$

28) Find the constant of variation ?

$$y = 8x$$

A) 11   B) 8

C) 14   D) 13

29) Find the constant of variation ?

$$y = 10x$$

A) 11   B) −10

C) 5   D) 10

30) Find the constant of variation ?

$$y = \frac{9}{x}$$

A) −5   B) 9

C) 13   D) 7

31) Find the constant of variation ?

$$y = \frac{5}{7x}$$

A) $\frac{5}{7}$ B) $\frac{4}{3}$

C) $\frac{4}{5}$ D) 3

32) Find the constant of variation ?

$$y = 11x$$

A) 13   B) 9

C) 11   D) 15

       www.math-knots.com | www.a4ace.com

 **Algebra 1**

Vol 2
Week 34
Direct &
Inverse
Variations

33) Find the constant of variation ?

$$y = \frac{11}{x}$$

A) 11   B) −11

C) 15   D) −14

34) Find the constant of variation ?

$$y = 2x$$

A) 12   B) 3

C) 2   D) −13

35) Find the constant of variation ?

$$y = -\frac{5}{7}x$$

A) $\frac{3}{10}$ B) $\frac{8}{3}$

C) $-\frac{5}{7}$ D) $\frac{8}{9}$

36) Find the constant of variation ?

$$y = 6x$$

A) 4   B) 14

C) 15   D) 6

37) Find the constant of variation ?

$$y = 13x$$

A) −13   B) 2

C) 13   D) 8

38) Find the constant of variation ?

$$y = -\frac{5}{x}$$

A) −5   B) 6

C) 8   D) 5

39) Find the constant of variation ?

$$y = -\frac{8}{x}$$

A) 11   B) −8

C) 8   D) 5

40) Find the constant of variation ?

$$y = \frac{4}{x}$$

A) 13   B) 10

C) 8   D) 4

Algebra 1

41) Find the constant of variation ?

$$y = -4x$$

A) 15   B) 12

C) 5   D) −4

42) Find the constant of variation ?

$$y = 15x$$

A) 14   B) 12

C) 6   D) 15

43) Find the constant of variation ?

$$y = 8x$$

A) 11   B) 8

C) 28   D) 12

44) Find the constant of variation ?

$$y = \frac{5}{x}$$

A) −5   B) 5

C) 16   D) 10

45) Find the constant of variation ?

$$y = -\frac{18}{x}$$

A) 13   B) 17

C) −18   D) 6

46) Find the constant of variation ?

$$y = \frac{18}{x}$$

A) 18   B) 29

C) 20   D) 30

47) Find the constant of variation ?

$$y = \frac{19}{x^2}$$

A) 3   B) 19

C) 21   D) 27

48) Find the constant of variation ?

$$y = \frac{8}{x^2}$$

A) 25   B) 8

C) −28   D) 21

 **Algebra 1**

Vol 2
Week 34
Direct &
Inverse
Variations

49) Find the constant of variation ?

$$y = \frac{2}{x}$$

A) 22   B) 2

C) 27   D) 13

50) Find the constant of variation ?

$$y = 2x^2$$

A) 2   B) 23

C) -2   D) 19

51) Find the constant of variation ?

$$y = -\frac{5}{x}$$

A) 8   B) -5

C) 13   D) 5

52) Find the constant of variation ?

$$y = \frac{25}{x}$$

A) 17   B) 25

C) 7   D) 27

53) Find the constant of variation ?

$$y = -4x$$

A) 7   B) -4

C) 2   D) 11

54) Find the constant of variation ?

$$y = 23x$$

A) 6   B) -23

C) 17   D) 23

55) Find the constant of variation ?

$$y = -\frac{4}{x}$$

A) 15   B) 4

C) -4   D) 12

56) Find the constant of variation ?

$$y = \frac{27}{x}$$

A) -24   B) 3

C) 12   D) 27

 Algebra 1

Vol 2
Week 34
Direct &
Inverse
Variations

57) Find the constant of variation ?

$$y = 14x$$

A) 14   B) 7
C) 8   D) 11

58) Find the constant of variation ?

$$y = \frac{13}{x}$$

A) 18   B) 14
C) 27   D) 13

59) Find the constant of variation ?

$$y = \frac{15}{x^2}$$

A) 15   B) 2
C) 27   D) 11

60) Find the constant of variation ?

$$y = -29x$$

A) 13   B) 4
C) 6   D) -29

61) Find the constant of variation ?

$$y = 21x$$

A) 29   B) 24
C) -28   D) 21

62) Find the constant of variation ?

$$-12x + y = 0$$

A) 22   B) 12
C) 16   D) -12

63) Find the constant of variation ?

$$xy = 13$$

A) 13   B) 23
C) -19   D) 30

64) Find the constant of variation ?

$$y = \frac{20}{x}$$

A) 7   B) 8
C) 9   D) 20

Algebra 1

Vol 2
Week 34
Direct &
Inverse
Variations

65) Find the constant of variation ?

$$y = \frac{7}{x^2}$$

A) 12   B) 7

C) 20   D) 4

66) Find the constant of variation ?

$$\frac{y}{x} = 11$$

A) 24   B) 8

C) 11   D) 10

67) Find the constant of variation ?

$$-19x + y = 0$$

A) 19   B) −19

C) 6   D) 9

68) If y varies inversely as x, and y = 4 when x = 9, find y when x = $\frac{3}{2}$.

A) 29   B) 27

C) 24   D) 25

69) If y varies inversely as x, and y = 4 when x = 13, find y when x = 1.

A) 48   B) 50

C) 57   D) 52

70) If y varies inversely as x, and y = 15 when x = 10, find y when x = 14.

A) $\frac{75}{8}$ B) 15

C) $\frac{75}{7}$ D) $\frac{71}{3}$

71) If y varies inversely as x, and y = 7 when x = 5, find y when x = 15.

A) $\frac{7}{8}$ B) 7

C) $\frac{7}{3}$ D) $\frac{8}{3}$

72) If y varies directly as x, and y = 4 when x = 2, find y when x = 13.

A) 21   B) 26

C) 22   D) 27

www.math-knots.com | www.a4ace.com

 **Algebra 1**

Vol 2
Week 34
Direct &
Inverse
Variations

73) If y varies inversely as x, and y = 14 when x = 9, find y when x = 4.

   A) $\dfrac{63}{2}$ B) 21

   C) 33   D) $\dfrac{59}{2}$

74) If y varies inversely as x, and y = 14 when x = 5, find y when x = 2.

   A) 28   B) $\dfrac{28}{5}$

   C) $\dfrac{28}{9}$ D) $\dfrac{29}{5}$

75) If y varies inversely as x, and y = 6 when x = 10, find y when x = 5.

   A) 12   B) 15

   C) 9   D) 7

76) If y varies inversely as x, and y = 6 when x = 5, find y when x = 11 .

   A) $\dfrac{33}{2}$ B) $\dfrac{66}{5}$

   C) 35   D) 12

77) If y varies directly as x, and y = 7 when x = 2, find y when x = 6.

   A) 21   B) 24

   C) 19   D) 20

78) If y varies inversely as x, and y = 12 when x = 11, find y when x = 15.

   A) $\dfrac{42}{5}$ B) $\dfrac{44}{5}$

   C) $\dfrac{11}{2}$ D) $\dfrac{46}{5}$

79) If y varies inversely as x, and y = 14 when x = 3, find y when x = $\dfrac{1}{2}$.

   A) 84   B) 91

   C) 76   D) 85

80) If y varies directly as x, and y = 5 when x = 9, find y when x = 3.

   A) $\dfrac{7}{2}$ B) $\dfrac{1}{3}$

   C) $\dfrac{5}{3}$ D) $\dfrac{4}{3}$

Algebra 1

Vol 2
Week 34
Direct &
Inverse
Variations

81) If y varies inversely as x, and y = 9 when x = 10, find y when x = 5.

   A) 5   B) 18

   C) $\dfrac{4}{5}$ D) $\dfrac{9}{7}$

82) If y varies directly as x, and y = 13 when x = 4, find y when x = 5.

   A) $\dfrac{33}{2}$ B) $\dfrac{65}{4}$

   C) $\dfrac{59}{6}$ D) 13

83) If y varies inversely as x, and y = 3 when x = 9, find y when x = $\dfrac{9}{10}$.

   A) 25   B) 34

   C) 26   D) 30

84) If y varies directly as x, and y = 2 when x = 10, find y when x = $\dfrac{5}{9}$.

   A) $\dfrac{6}{5}$ B) $\dfrac{3}{7}$

   C) $\dfrac{1}{9}$ D) $\dfrac{1}{4}$

85) If y varies inversely as x, and y = 10 when x = 7, find y when x = 4.

   A) 7   B) $\dfrac{35}{2}$

   C) $\dfrac{39}{4}$ D) 10

86) If y varies inversely as x, and y = 11 when x = 9, find y when x = 5.

   A) $\dfrac{106}{5}$ B) 11

   C) $\dfrac{99}{5}$ D) 20

87) If y varies inversely as x, and y = 7 when x = 6, find y when x = $\dfrac{9}{4}$.

   A) $\dfrac{58}{3}$ B) $\dfrac{56}{3}$

   C) 8   D) 56

88) If y varies inversely as x, and y = 5 when x = 2, find y when x = 14.

   A) $\dfrac{4}{5}$ B) $\dfrac{1}{4}$

   C) $\dfrac{5}{7}$ D) $\dfrac{5}{11}$

www.math-knots.com | www.a4ace.com

 **Algebra 1**

89) If y varies inversely as x, and y = 6 when x = 15, find y when x = 4.

  A) $\dfrac{41}{2}$ B) 25

  C) $\dfrac{45}{2}$ D) $\dfrac{22}{3}$

90) If y varies directly as x, and y = 11 when x = 9, find y when x = 3.

  A) 33   B) 36

  C) 35   D) 38

91) If y varies inversely as x, and y = 5 when x = 3, find y when x = $\dfrac{4}{3}$.

  A) $\dfrac{45}{4}$ B) 10

  C) $\dfrac{15}{2}$ D) 45

92) If y varies inversely as x, and y = 7 when x = 14, find y when x = 10.

  A) $\dfrac{49}{8}$ B) 9

  C) $\dfrac{48}{5}$ D) $\dfrac{49}{5}$

93) If y varies inversely as x, and y = 9 when x = 3, find y when x = 10.

  A) 2.5   B) 3.3
  C) 32   D) 2.7

94) If y varies inversely as x, and y = 11 when x = 12, find y when x = 8.

  A) $\dfrac{11}{2}$ B) $\dfrac{33}{2}$

  C) 11   D) 33

95) If y varies directly as x, and y = 11 when x = 7, find y when x = 4.

  A) $\dfrac{23}{6}$ B) $\dfrac{44}{7}$

  C) $\dfrac{44}{5}$ D) $\dfrac{13}{3}$

96) If y varies directly as x, and y = 11 when x = 3, find y when x = 4.

  A) $\dfrac{44}{3}$ B) $\dfrac{40}{3}$

  C) 14   D) 22

      www.math-knots.com | www.a4ace.com

Algebra 1

Vol 2
Week 34
Direct &
Inverse
Variations

97) If y varies inversely as x, and y = 12 when x = 7, find y when x = 4.

A) 21   B) 20

C) 23   D) 16

98) If y varies inversely as x, and y = 12 when x = 2, find y when x = 5.

A) 7   B) $\dfrac{24}{5}$

C) 5   D) $\dfrac{8}{3}$

99) If y varies inversely as x, and y = 9 when x = 2, find y when x = 6.

A) 3   B) 12

C) 27   D) 6

100) If y varies directly as x, and y = 13 when x = 10, find y when x = 15.

A) 15   B) $\dfrac{13}{2}$

C) $\dfrac{37}{3}$ D) $\dfrac{26}{3}$

101) If y varies inversely as x, and y = 11 when x = 5, find y when x = $\dfrac{15}{4}$.

A) $\dfrac{44}{3}$ B) 16

C) 22   D) $\dfrac{11}{2}$

102) If y varies directly as x, and y = 12 when x = 2, find y when x = 4.

A) 20   B) 24

C) 29   D) 25

103) If y varies directly as x, and y = 7 when x = 2, find y when x = 12.

A) 46   B) 38

C) 42   D) 41

104) If y varies inversely as x, and y = 3 when x = 5, find y when x = 2.

A) $\dfrac{13}{2}$ B) $\dfrac{15}{2}$

C) 6   D) $\dfrac{14}{3}$

 **Algebra 1**

105) If y varies directly as x, and y = 14 when x = 12, find y when x = 9.

A) $\dfrac{23}{2}$ B) $\dfrac{21}{2}$

C) 13   D) 3

106) If y varies inversely as x, and y = 11 when x = 3, find y when x = 9.

A) $\dfrac{11}{3}$ B) $\dfrac{8}{3}$

C) $\dfrac{11}{6}$ D) $\dfrac{7}{3}$

107) If y varies inversely as x, and y = 11 when x = 9, find y when x = 2.

A) 98   B) $\dfrac{105}{2}$

C) $\dfrac{99}{2}$ D) $\dfrac{99}{4}$

1) Identify the polynomial by degree and number of terms. Choose the right option.

$$-6$$

A) constant monomial

B) constant binomial

C) constant trinomial

D) linear polynomial with 0 terms

2) Find the product of the below

$$(-x+9)(-x-9)$$

A) $x^2 - 18x + 81$

B) $x^2 - 81$

C) $4x^2 - 9$

D) $x^2 + 18x + 81$

3) Which expression is equivalent to the below.

$$\left(11 + 3n^2 + 15n^4\right) - \left(7n^2 + 5n^4\right)$$

A) $10n^4 + 6n^2 + 11$

B) $16n^4 + 6n^2 + 11$

C) $16n^4 + 6n^2 + 30$

D) $10n^4 - 4n^2 + 11$

4) Simplify the below expressions.

$$\left(5 + 2m^2 - 7m\right) - \left(11m - 8m^2 + 4m^3\right)$$

5) Solve the below quadratic equation using the method of factorization.

$$v^2 + 21v + 98 = 0$$

A) $\{-14, -7\}$ B) $\{-10, -9\}$

C) $\{-6, -7\}$ D) $\{-6, 0\}$

6) If y varies inversely as x, and y = 15 when x = 3, find y when x = 8.

A) $\dfrac{41}{12}$ B) 9

C) $\dfrac{45}{8}$ D) $\dfrac{15}{4}$

7) Find the value of c for the below to complete the square.

$$x^2 - 5x + c$$

8) Does the below equation represent a direct or inverse variation ?

$$y = 11x$$

A) Direct   B) Inverse

www.math-knots.com | www.a4ace.com

9) Which expression is equivalent to the below.

$$\left(9v^5 - 8v^4 + v^2\right) + \left(4v^5 + 7v^4\right)$$

A) $2v^5 - v^4 + 13v^2$

B) $2v^5 - v^4 + 4v^2$

C) $13v^5 - v^4 + v^2$

D) $2v^5 - v^4 + v^2$

10) Simplify the below.

$$\left(p^3 + 33p^2 + 254\ p - 309\right) \div \left(p + 17\right)$$

A) $p^2 + 16\ p - 17 - \dfrac{8}{p + 17}$

B) $p^2 + 16\ p - 18 - \dfrac{3}{p + 17}$

C) $p^2 + 16\ p - 15 - \dfrac{5}{p + 17}$

D) $p^2 + 16\ p - 16 - \dfrac{6}{p + 17}$

11) If y varies directly as x, and y = 2 when x = 12, find y when x = 15.

A) $\dfrac{8}{5}$ B) $\dfrac{7}{5}$

C) $\dfrac{13}{5}$ D) $\dfrac{3}{5}$

12) Solve the below quadratic equation.

$$r^2 + 16r + 48 = -8$$

13) Solve the below quadratic equation using the method of factorization.

$$5p^2 - 30 = -19\ p$$

A) $\left\{\dfrac{1}{6}, -5\right\}$ B) $\left\{\dfrac{5}{4}, -7\right\}$

C) $\left\{\dfrac{6}{5}, 1\right\}$ D) $\left\{\dfrac{6}{5}, -5\right\}$

14) Solve the below quadratic equation using the method of factorization.

$$8x^2 - 65\ x = -8$$

A) $\left\{\dfrac{1}{8}, 8\right\}$ B) $\left\{\dfrac{8}{5}, -8\right\}$

C) $\left\{-\dfrac{1}{8}, -8\right\}$ D) $\left\{\dfrac{7}{8}, -6\right\}$

15) Simplify the below and state the excluded values.

$$\dfrac{28n}{8n^2 + 24n}$$

16) If y varies inversely as x, and y = 9 when x = 14, find y when x = 8.

A) $\dfrac{57}{7}$ B) 63

C) $\dfrac{67}{2}$ D) $\dfrac{63}{4}$

17) If y varies directly as x, and y = 9 when x = 3, find y when x = 6.

A) 17   B) 15

C) 16   D) 18

18) Which expression is equivalent to the below.

$$\left(10n^5 - 9n^3 - n^2\right) - \left(14n^3 - 10n^5\right)$$

A) $20n^5 - 23n^3 - n^2$

B) $20n^5 - 24n^3 - n^2$

C) $8n^5 - 24n^3 - n^2$

D) $-12n^5 - 24n^3 - n^2$

19) Find the product of the below
$$(-12n + 9)(-12n - 9)$$

A) $144n^2 - 216n + 81$

B) $144n^2 - 81$

C) $144n^2 + 216n + 81$

D) $n^2 - 64$

20) If y varies directly as x, and y = 11 when x = 15, find y when x = 12.

A) $\dfrac{55}{4}$ B) 55

C) $\dfrac{55}{6}$ D) $\dfrac{25}{2}$

21) Does the below equation represent a direct or inverse variation ?

$$y = \frac{4}{x}$$

A) Direct   B) Inverse

22) Simplify the below.

$$\left(36v^3 + 54v^2 + 3v\right) \div 9v$$

A) $4v^2 + 6v + \dfrac{1}{3}$

B) $\dfrac{5v^5}{6} + \dfrac{v^4}{2} + \dfrac{7v^3}{6}$

C) $4v + 3 + \dfrac{1}{2v}$

D) $2 + \dfrac{1}{4v} + \dfrac{2}{v^2}$

23) Find the discriminant value for the below quadratic equation. Also, state the number and type of solutions.

$$-4x^2 = -8x + 8$$

24) Find the constant of variation ?

$$y = 9x$$

A) -9 B) 13

C) 9   D) -10

 **Algebra 1**

25) Find the product of the below

$$(x - 11) (x + 11)$$

A) $1 - 6x + 9x^2$

B) $1 - 9x^2$

C) $x^2 - 121$

D) $x^2 - 22x + 121$

26) Simplify the below.

$$\left(8r^3 - 160r^2 + 158r - 95\right) \div \left(r - 19\right)$$

A) $8r^2 - 8r + 10 + \dfrac{15}{r - 19}$

B) $8r^2 - 8r + 8 + \dfrac{16}{r - 19}$

C) $8r^2 - 8r + 7 + \dfrac{20}{r - 19}$

D) $8r^2 - 8r + 6 + \dfrac{19}{r - 19}$

27) Solve the below quadratic equation using the method of factorization.

$$5 v^2 - 9 = -12 v$$

A) $\left\{ -\dfrac{4}{5}, \dfrac{1}{2} \right\}$ B) $\left\{ -\dfrac{3}{5}, 3 \right\}$

C) $\left\{ -\dfrac{5}{7}, -3 \right\}$ D) $\left\{ \dfrac{3}{5}, -3 \right\}$

28) Solve the below quadratic equation using the method of factorization.

$$a^2 - 35a + 304 = 0$$

A) $\{ 19 , 16 \}$ B) $\{ -12 , 0 \}$

C) $\{ 12 , 15 \}$ D) $\{ -15 , 12 \}$

29) Simplify the below and state the excluded values.

$$\dfrac{k^2 - 6k - 40}{k^2 - k - 90}$$

30) Simplify the below expressions.

$$\left(10m^3 + 7m - 2m^2\right) - \left(11 + 5m + 9m^2\right)$$

31) Solve the below quadratic equation.

$$n^2 + 4n - 105 = -9$$

www.math-knots.com | www.a4ace.com

32) If y varies directly as x, and y = 15 when x = 5, find y when x = 13.

A) $\dfrac{58}{3}$ B) 65

C) $\dfrac{59}{6}$ D) 36

33) Simplify the below.

$$(2a^3 + 36a^2 + 76a - 204) \div (a + 15)$$

A) $2a^2 + 6a - 14 - \dfrac{4}{a+15}$

B) $2a^2 + 6a - 13 - \dfrac{2}{a+15}$

C) $2a^2 + 6a - 13 + \dfrac{1}{a+15}$

D) $2a^2 + 6a - 14 + \dfrac{6}{a+15}$

34) Find the product of the below

$$(k+8)^2$$

A) $k^2 - 64$ B) $k^2 + 16k + 64$

C) $k^2 + 64$ D) $k^2 + 30k + 225$

35) Simplify the below and state the excluded values.

$$\dfrac{3x^3 - 30x^2}{x^2 - 14x + 40}$$

36) Which expression is equivalent to the below.

$$(n^3 + 12n^2 + 7n^5) + (12n^3 - 3n^2)$$

A) $7n^5 + 17n^3 - 2n^2$

B) $7n^5 + 13n^3 + 9n^2$

C) $10n^5 + 17n^3 - 2n^2$

D) $7n^5 + 13n^3 - 2n^2$

37) Find the product of the below.

$$(7n - 10)(-n - 5)$$

A) $-7n^2 - 25n + 50$ B) $-24n^2 + 120n - 54$

C) $-24n^2 + 54$ D) $-24n^2 - 96n + 54$

38) Find the product of the below.

$$(9x^2 - 8x - 5)(6x^2 + x - 2)$$

39) Find the product of the below

$$(p + 12)(p - 12)$$

A) $p^2 + 24p + 144$

B) $p^2 - 144$

C) $p^2 + 14p + 49$

D) $p^2 - 49$

40) Identify the polynomial by degree and number of terms. Choose the right option.

$$2r^2$$

A) quadratic trinomial

B) linear trinomial

C) linear binomial

D) quadratic monomial

41) Simplify the below expressions.

$$(8a^2 + 6 + a) - (a^2 - 6a - 12a^3)$$

42) Simplify the below and state the excluded values.

$$\frac{10k - 5}{20}$$

43) If y varies inversely as x, and y = 6 when x = 11, find y when x = 10.

A) $\frac{28}{5}$ B) $\frac{33}{2}$

C) $\frac{33}{5}$ D) 6

44) Simplify the below.

$$(16n^3 - 307n^2 + 350n - 149) \div (n - 18)$$

A) $16n^2 - 19n + 5 - \dfrac{8}{n - 18}$

B) $16n^2 - 19n + 8 - \dfrac{5}{n - 18}$

C) $16n^2 - 19n + 2 - \dfrac{5}{n - 18}$

D) $16n^2 - 19n + 4 - \dfrac{2}{n - 18}$

45) Find the product of the below

$$(x + 4)(x - 4)$$

A) $x^2 - 6x + 9$ B) $x^2 - 16$

C) $x^2 + 8x + 16$ D) $x^2 - 9$

46) Find the product of the below

$$(-10 + 4a)^2$$

A) $-10 + 16a^2$

B) $100 + 16a^2$

C) $100 - 16a^2$

D) $100 - 80a + 16a^2$

www.math-knots.com | www.a4ace.com

47) Does the below equation represent a direct or inverse variation ?

$$y = \frac{15}{x}$$

A) Inverse   B) Direct

48) Find the constant of variation ?

$$y = \frac{3}{x}$$

A) 6   B) 3

C) 2   D) –3

49) Simplify the below.

$$\left(p^3 - 31p^2 + 237p + 17\right) \div \left(p - 17\right)$$

A) $p^2 - 14p - 3 + \dfrac{4}{p - 17}$

B) $p^2 - 14p - 8 + \dfrac{6}{p - 17}$

C) $p^2 - 14p - 6 + \dfrac{1}{p - 17}$

D) $p^2 - 14p - 1$

50) Solve the below quadratic equation using the method of factorization.

$$6x^2 - 35 = -37x$$

A) $\left\{-\dfrac{5}{3}, 4\right\}$ B) $\left\{\dfrac{5}{6}, -7\right\}$

C) $\left\{\dfrac{8}{7}, -\dfrac{8}{5}\right\}$ D) $\left\{-\dfrac{8}{5}, \dfrac{3}{2}\right\}$

51) Find the value of c for the below to complete the square.

$$x^2 - \frac{29}{17}x + c$$

52) Find the product of the below

$$\left(5k + 7\right)^2$$

A) $25k^2 + 70k + 49$

B) $5k + 49$

C) $25k^2 + 49$

D) $25k^2 - 49$

53) Solve the below quadratic equation.

$$n^2 + 6n + 2 = 9$$

Algebra 1

54) Which expression is equivalent to the below.

$$\left(9 - 4n + 11n^4\right) + \left(5n - 7n^4\right)$$

A) $-6n^4 + n + 20$

B) $4n^4 + n + 12$

C) $4n^4 + n + 9$

D) $4n^4 + n + 20$

55) Simplify the below.

$$\left(70v^3 + 14v^2 + 98v\right) \div 14v^2$$

A) $v + 1 + \dfrac{2}{v}$

B) $3v + 3 + \dfrac{4}{v}$

C) $5v + 1 + \dfrac{7}{v}$

D) $\dfrac{v}{4} + \dfrac{5}{12} + \dfrac{1}{12v}$

56) Which expression is equivalent to the below.

$$\left(6 + 7v^3 - 15v\right) - \left(20v^3 + 10\right)$$

A) $3v^3 - 15v - 7$

B) $-13v^3 - 15v - 7$

C) $-13v^3 - 15v - 4$

D) $3v^3 - 11v - 7$

57) Find the value of c for the below to complete the square.

$$x^2 + \frac{128}{13}x + c$$

58) Solve the below quadratic equation.

$$v^2 + 10v - 75 = 4$$

59) Which expression is equivalent to the below.

$$\left(12 - 4a^5 - a^2\right) + \left(11a^5 - 1\right)$$

A) $11 - 7a^2$

B) $7a^5 - a^2 + 11$

C) $11 - 10a^2$

D) $7a^5 - 10a^2 + 11$

60) Find the value of c for the below to complete the square.

$$p^2 + 42p + c$$

61) Which expression is equivalent to the below.

$$(8\,p^2 + 20 + 9\,p^4) - (20\,p^2 + 16)$$

A) $-8p^4 - 18\,p^2 + 4$

B) $-14p^4 - 18\,p^2 + 4$

C) $9p^4 - 12\,p^2 + 4$

D) $9p^4 - 18\,p^2 + 4$

62) Find the product of the below.

$$(5\,a^2 + 6\,a + 5)(10\,a^2 - a - 5)$$

63) If y varies directly as x, and y = 2 when x = 9, find y when x = 12.

A) $\dfrac{8}{3}$ B) 2

C) 4   D) $\dfrac{5}{3}$

64) Find the value of c for the below to complete the square.

$$y^2 - 6\,y + c$$

65) Simplify the below.

$$(a^3 + 2a^2 - 254a + 2) \div (a - 15)$$

A) $a^2 + 17a + 5 + \dfrac{18}{a - 15}$

B) $a^2 + 17a + 1 + \dfrac{17}{a - 15}$

C) $a^2 + 17a + 4 + \dfrac{19}{a - 15}$

D) $a^2 + 17a + 3 + \dfrac{17}{a - 15}$

66) Find the value of c for the below to complete the square.

$$x^2 - 11x + c$$

67) Find the product of the below

$$(6 - 9\,b)^2$$

A) $36 + 81b^2$

B) $36 - 108b + 81b^2$

C) $36 - 81b^2$

D) $6 + 81b^2$

68) Find the constant of variation ?

$$y = 4\,x$$

A) 11   B) 5

C) 12   D) 4

 **Algebra 1**

69) Solve the below quadratic equation using the method of factorization.

$$x^2 - 36 = 0$$

A) { 6 , − 6 }   B) { − 6 , − 15 }

C) { 15 , 12 }   D) { 9 , − 12 }

70) Simplify the below and state the excluded values.

$$\frac{121x^2 + 22x}{143x^2 + 11x}$$

71) Simplify the below expressions.

$$(12x + 5 + 3x^3) - (2 - 10x^3 + 8x)$$

72) If y varies directly as x, and y = 10 when x = 6, find y when x = 2.

A) $\dfrac{10}{3}$ B) 10

C) $\dfrac{14}{3}$ D) 3

73) Simplify the below.

$$(3b^3 + 5b^2 + 5b) \div 10b$$

A) $\dfrac{3b^2}{10} + \dfrac{b}{2} + \dfrac{1}{2}$

B) $1 + \dfrac{1}{3b} + \dfrac{1}{2b^2}$

C) $7b^2 + b + \dfrac{1}{3}$

D) $5b^2 + b + 1$

74) Find the product of the below.

$$8n^3(3n - 18)$$

A) − 40 n + 20   B) $24n^4 - 144n^3$

C) − 180 n − 48   D) −26 n + 234

75) Find the constant of variation ?

$$y = \frac{13}{x}$$

A) 2   B) 12

C) 13   D) 10

76) Find the product of the below

$$(n + 19)^2$$

A) $n^2 - 361$

B) $n^2 + 14n + 49$

C) $n^2 + 361$

D) $n^2 + 38n + 361$

77) If y varies directly as x, and y = 10 when x = 14, find y when x = 12.

A) $\dfrac{60}{7}$ B) 30

C) 12   D) $\dfrac{58}{9}$

78) Simplify the below expressions.

$$\left(11 + 12n + 12n^4\right) - \left(12n + 12n^4 - 8\right)$$

79) Find the product of the below.

$$12v^2\left(16v - 4\right)$$

A) $-152v^2 + 285\,v$ B) $-119\,v - 14$

C) $132\,v - 12$ D) $192v^3 - 48v^2$

80) Identify the polynomial by degree and number of terms. Choose the right option.

$$2r - r^4 + 6r^5 - 7r^2$$

A) quintic trinomial

B) constant trinomial

C) quintic polynomial with four terms

D) quartic binomial

81) Solve the below quadratic equation.

$$n^2 - 6n + 9 = 0$$

82) Solve the below quadratic equation using the method of factorization.

$$m^2 + 14\,m - 120 = 0$$

A) { 6 , − 20 }   B) { − 14 , 20 }

C) { 7 , 14 }   D) { 1 , 0 }

83) Simplify the below and state the excluded values.

$$\dfrac{26n + 18}{4n - 22}$$

84) Find the product of the below

$$\left(n + 14 \right)^2$$

A) $n^2 + 196$

B) $n^2 + 28n + 196$

C) $n^2 - 6n + 9$

D) $n^2 - 196$

www.math-knots.com | www.a4ace.com

85) Find the product of the below.

$$(-6p-2)(-5p+7)$$

A) $20p^2 + 11$ 　　　　 B) $20p^2 + 12p - 11$

C) $30p^2 - 32p - 14$ 　 D) $20p^2 + 32p + 11$

86) Find the product of the below.

$$(3b^2 + 2b + 9)(2b^2 + 6b - 8)$$

87) Identify the polynomial by degree and number of terms. Choose the right option.

$$-1 + 10v$$

A) constant binomial

B) quadratic monomial

C) linear binomial

D) linear monomial

88) Find the product of the below.

$$(5r + 2)(5r - 3)$$

A) $4r^2 + 2r - 2$ 　　　 B) $25r^2 - 5r - 6$

C) $4r^2 - 6r + 2$ 　　　 D) $25r^2 - 25r + 6$

89) Simplify the below.

$$(18v^3 + 3v^2 + 12v) \div 6v^2$$

A) $\dfrac{v^2}{2} + \dfrac{v}{14} + \dfrac{5}{14}$

B) $4 + \dfrac{5}{v} + \dfrac{1}{2v^2}$

C) $3v + \dfrac{1}{2} + \dfrac{2}{v}$

D) $7v^3 + 4v^2 + 2v$

90) Find the discriminant value for the below quadratic equation. Also, state the number and type of solutions.

$$15p^2 = 16p$$

91) Find the value of c for the below to complete the square.

$$a^2 + 8a + c$$

92) Does the below equation represent a direct or inverse variation ?

$$y = 7x$$

A) Direct 　　　 B) Inverse

93) Find the product of the below.

$$-13\,b\,(\,15\,b + 14\,)$$

A) $-195\,b^2 - 182b$ B) $-234b^2 - 126b$

C) $-168b^3 - 72b^2$ D) $-102b - 96$

94) Simplify the below expressions.

$$\left(4x + 4x^4 + 3x^2\right) - \left(4x + 3x^3 + 11\,x^2\right)$$

95) Find the discriminant value for the below quadratic equation. Also, state the number and type of solutions.

$$x^2 - 2\,x = 15$$

96) Find the product of the below.

$$(\,10\,b^2 - 7\,b + 9\,)\,(\,3\,b^2 + 8\,b - 6\,)$$

97) Simplify the below and state the excluded values.

$$\frac{3p + 3}{p^2 + 2p + 1}$$

98) Solve the below quadratic equation.

$$r^2 - 16r + 7 = -8$$

99) Find the product of the below.

$$(\,-6\,n - 13\,)\,(\,n + 4\,)$$

A) $-6\,n^2 - 37n - 52$

B) $-6\,n^2 - 52$

C) $121\,n^2 + 32$

D) $121\,n^2 + 132n + 32$

100) Find the product of the below.

$$(\,x^2 - 9\,x - 9\,)\,(\,2\,x^2 + 2\,x - 3\,)$$

101) Find the product of the below.

$$18 x (- 6 x + 7)$$

A) $- 260 x + 180$ B) $- 187 x - 306$

C) $-108x^2 + 126x$ D) $228 x + 76$

102) Find the discriminant value for the below quadratic equation. Also, state the number and type of solutions.

$$2 n^2 + 4 n = - 2$$

103) Find the value of c for the below to complete the square.

$$n^2 + 12 n + c$$

www.math-knots.com | www.a4ace.com

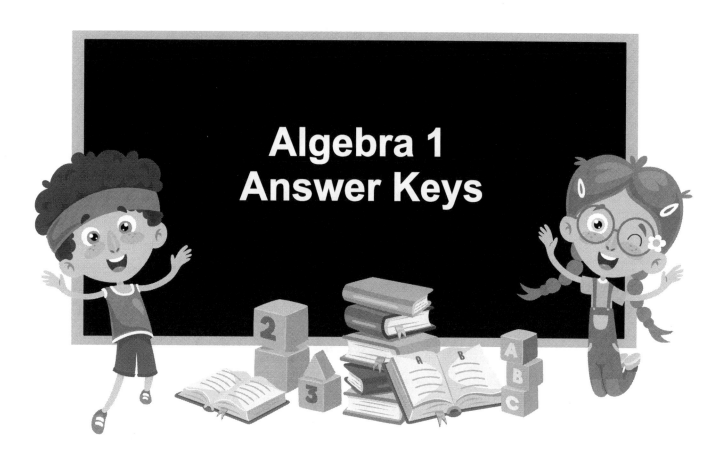

Algebra 1
Answer Keys

www.math-knots.com | www.a4ace.com

408   www.math-knots.com | www.a4ace.com

Week 19		Week 19		Week 19		Week 19	
1.	D	30.	B	59.	A	88.	D
2.	D	31.	A	60.	C	89.	C
3.	B	32.	D	61.	C	90.	B
4.	D	33.	B	62.	A	91.	A
5.	A	34.	D	63.	B	92.	A
6.	A	35.	D	64.	C	93.	C
7.	A	36.	A	65.	B	94.	D
8.	C	37.	C	66.	C	95.	D
9.	A	38.	D	67.	D	96.	C
10.	B	39.	D	68.	A	97.	B
11.	C	40.	C	69.	A	98.	C
12.	A	41.	A	70.	B	99.	D
13.	A	42.	B	71.	D	100.	D
14.	C	43.	C	72.	C	101.	A
15.	B	44.	D	73.	D	102.	A
16.	B	45.	C	74.	B	103.	B
17.	D	46.	D	75.	B	104.	D
18.	C	47.	A	76.	B	105.	C
19.	B	48.	D	77.	D	106.	D
20.	D	49.	C	78.	D	107.	B
21.	C	50.	C	79.	D	108.	B
22.	D	51.	D	80.	C	109.	D
23.	A	52.	B	81.	D	110.	B
24.	B	53.	C	82.	D	111.	C
25.	B	54.	D	83.	B	112.	C
26.	A	55.	C	84.	A	113.	C
27.	A	56.	D	85.	D	114.	A
28.	C	57.	C	86.	C	115.	A
29.	A	58.	D	87.	B	116.	B

www.math-knots.com | www.a4ace.com

Week 19		Week 19		Week 20		Week 20	
117.	B	146.	A	1.	C	30.	C
118.	A	147.	B	2.	C	31.	A
119.	B	148.	D	3.	A	32.	D
120.	B	149.	C	4.	B	33.	B
121.	A	150.	B	5.	A	34.	B
122.	B	151.	B	6.	D	35.	D
123.	A	152.	C	7.	D	36.	C
124.	A	153.	D	8.	A	37.	B
125.	B	154.	D	9.	C	38.	B
126.	C	155.	B	10.	B	39.	D
127.	A	156.	C	11.	D	40.	D
128.	A	157.	D	12.	C	41.	A
129.	B	158.	C	13.	B	42.	C
130.	B	159.	B	14.	B	43.	D
131.	D	160.	A	15.	D	44.	B
132.	A	161.	C	16.	C	45.	C
133.	D	162.	A	17.	A	46.	D
134.	B	163.	C	18.	C	47.	A
135.	C	164.	A	19.	D	48.	B
136.	D			20.	B	49.	A
137.	D			21.	C	50.	C
138.	C			22.	A	51.	C
139.	C			23.	C	52.	C
140.	D			24.	D	53.	D
141.	D			25.	B	54.	C
142.	B			26.	A	55.	B
143.	C			27.	D	56.	A
144.	B			28.	A		
145.	A			29.	C		

     www.math-knots.com | www.a4ace.com

57)

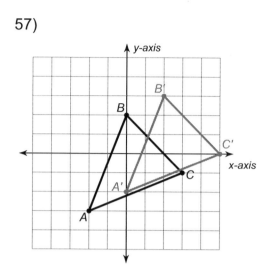

60)

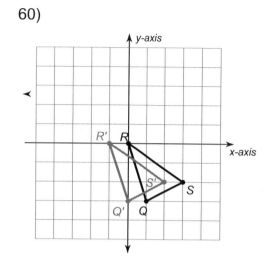

58)

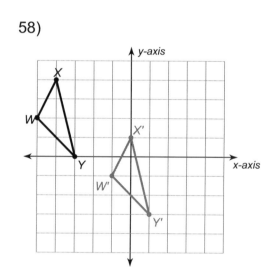

61)

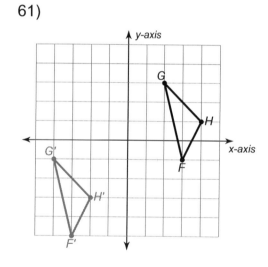

59)

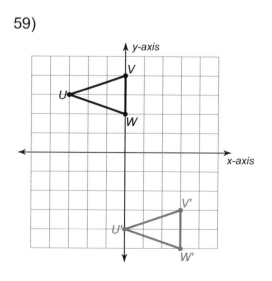

62)

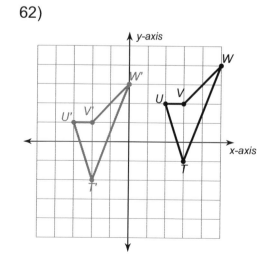

411

63)

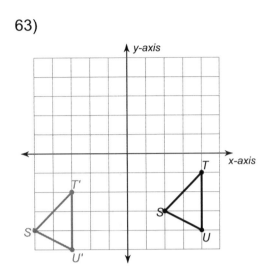

66)

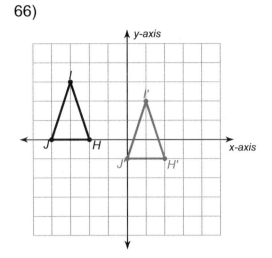

64)

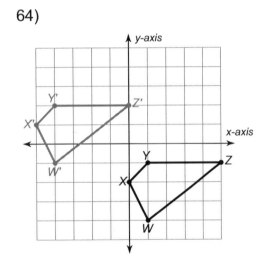

67)

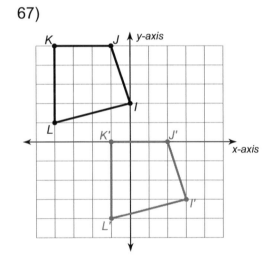

65)

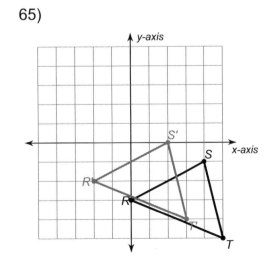

68)

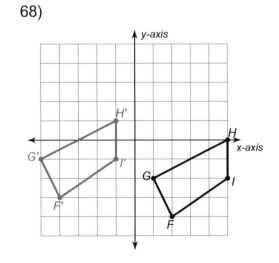

69)

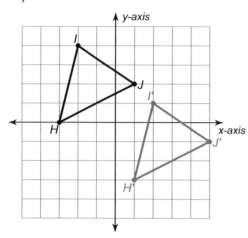

72)

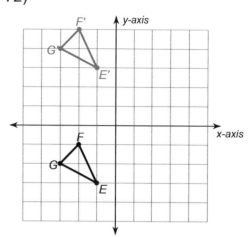

70)

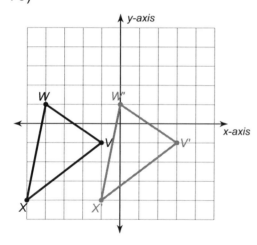

73)

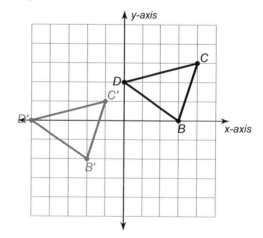

71)

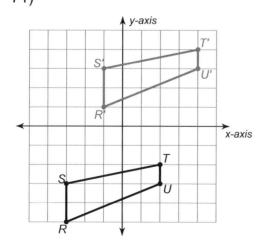

74)

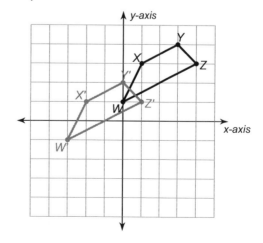

 www.math-knots.com | www.a4ace.com

75)

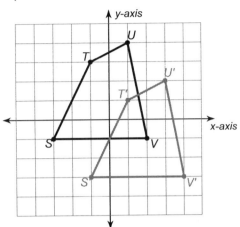

78)

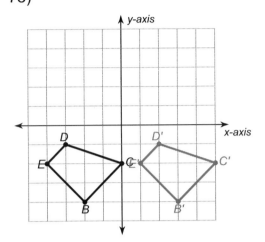

76)

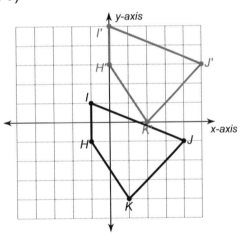

79)

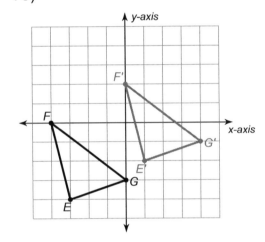

77)

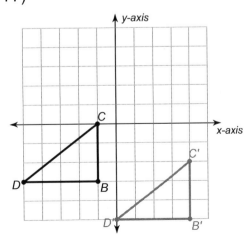

80)

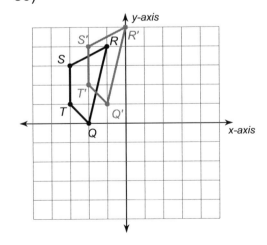

81)

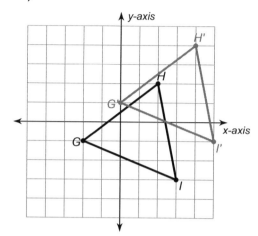

84)

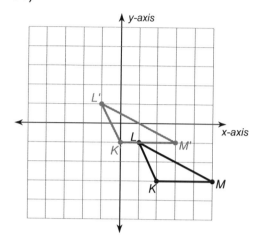

82)

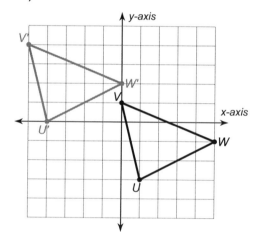

83)

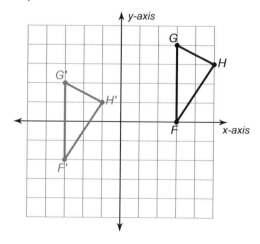

www.math-knots.com | www.a4ace.com

Week 20		Week 20	
85.	A	113.	A
86.	B	114.	D
87.	C	115.	A
88.	B	116.	A
89.	B	117.	C
90.	B	118.	C
91.	C	119.	D
92.	D	120.	A
93.	A	121.	A
94.	B	122.	A
95.	B	123.	B
96.	D	124.	D
97.	B	125.	D
98.	D	126.	C
99.	D	127.	A
100.	D	128.	A
101.	A	129.	B
102.	D	130.	C
103.	B	131.	A
104.	B	132.	B
105.	D	133.	C
106.	D	134.	D
107.	B	135.	B
108.	A	136.	A
109.	B		
110.	D		
111.	D		
112.	D		

137)

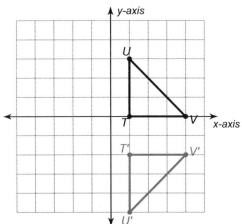

138)

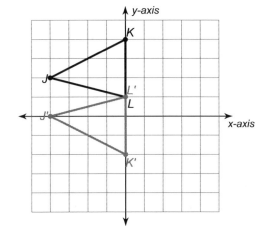

139)

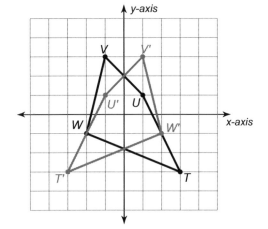

140)

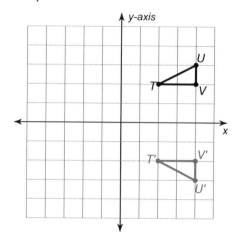

143)

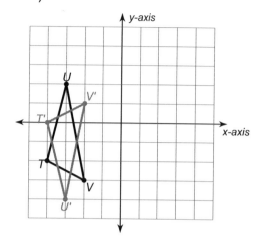

141)

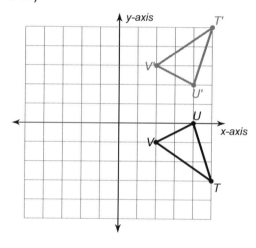

144)

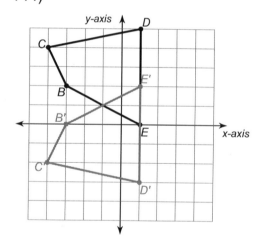

142)

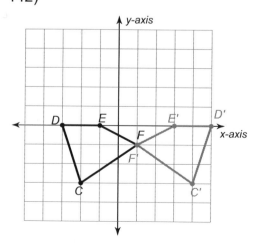

145)

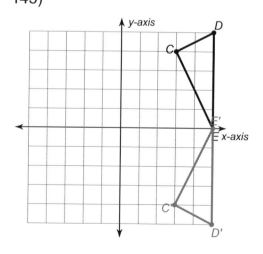

www.math-knots.com | www.a4ace.com

146)

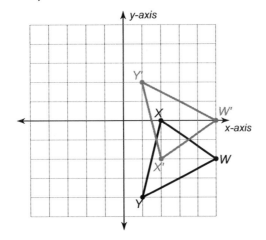

149)

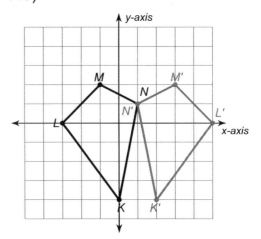

147)

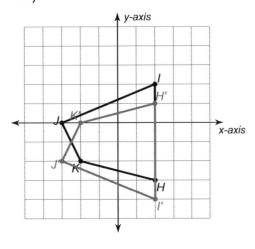

150)

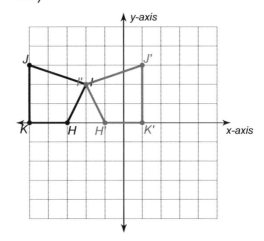

148)

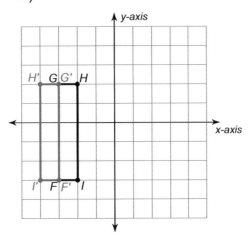

151)

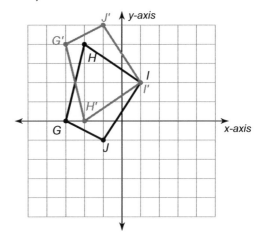

152)

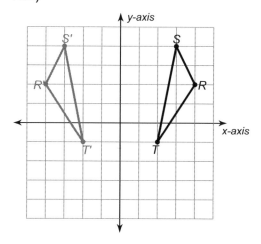

155)

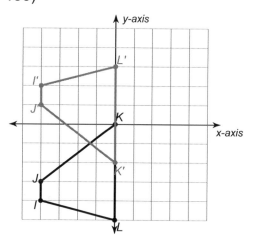

153)

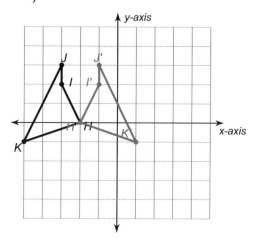

156)

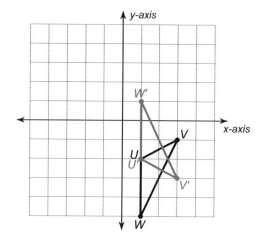

154)

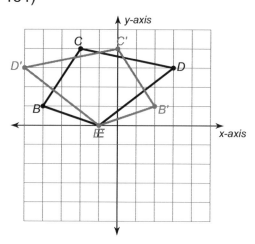

157)

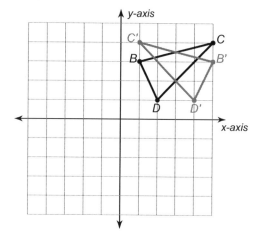

www.math-knots.com | www.a4ace.com

158)

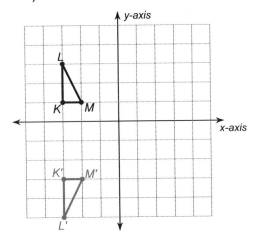

161)

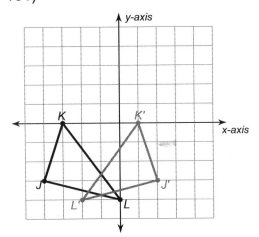

159)

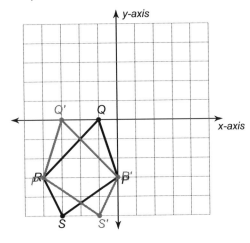

162)

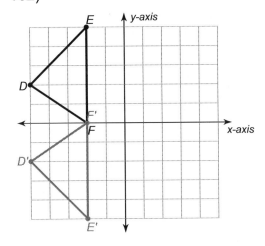

160)

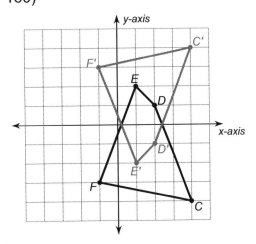

163)

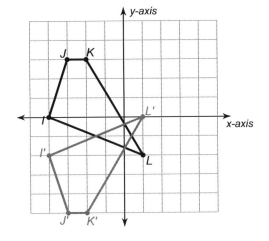

164)

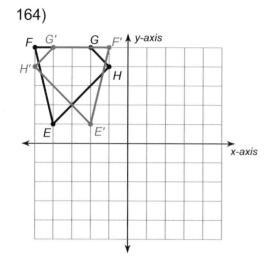

Week 21

#	Answer
1.	C
2.	B
3.	A
4.	A
5.	C
6.	D
7.	D
8.	D
9.	C
10.	C
11.	A
12.	D
13.	C
14.	A
15.	D
16.	B
17.	A
18.	D
19.	B
20.	A
21.	A
22.	B
23.	D
24.	B
25.	A
26.	D
27.	A
28.	D
29.	C

Week 21

#	Answer
30.	A
31.	C
32.	D
33.	A
34.	D
35.	B
36.	A
37.	D
38.	B
39.	D
40.	D
41.	B
42.	B
43.	A
44.	C
45.	C
46.	B
47.	D
48.	A
49.	A
50.	D
51.	B
52.	D
53.	C
54.	B
55.	A

421   www.math-knots.com | www.a4ace.com

56)

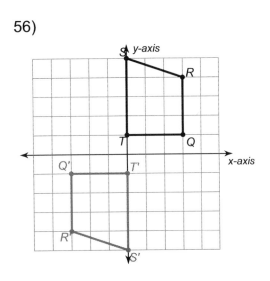

59)

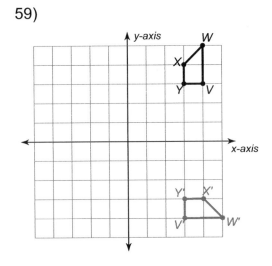

57)

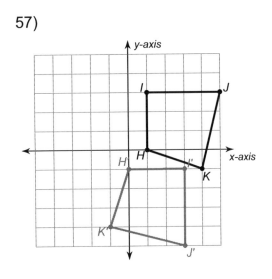

60)

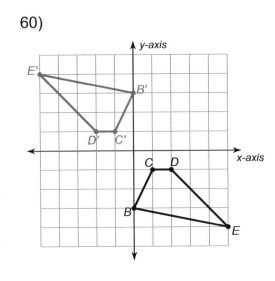

58)

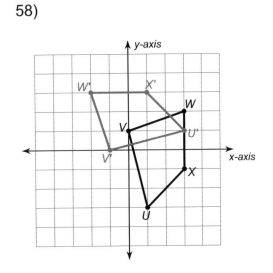

61)

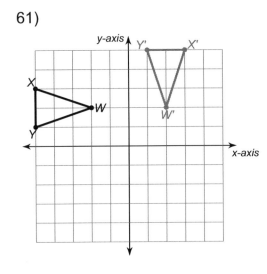

www.math-knots.com | www.a4ace.com

62)

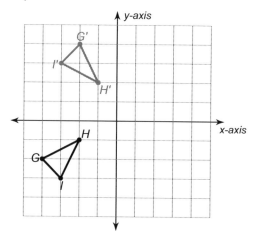

65)

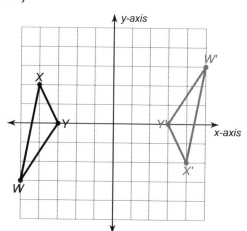

63)

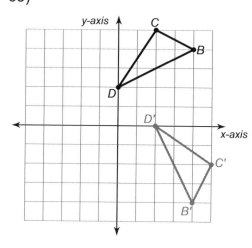

66)

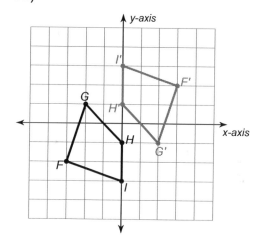

64)

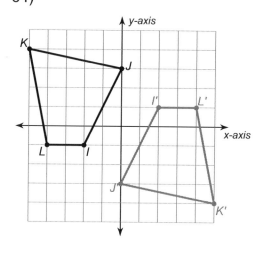

67)

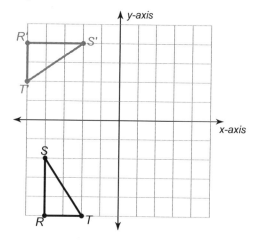

68)
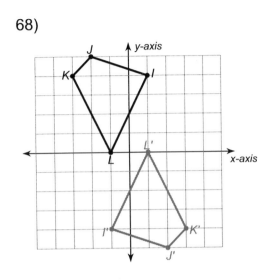

71)
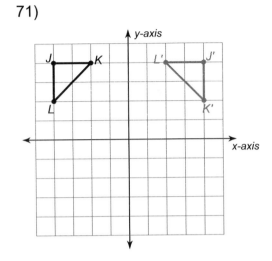

69)
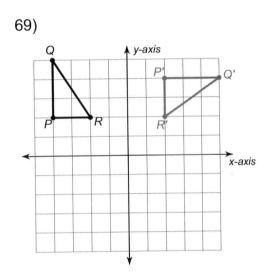

72)
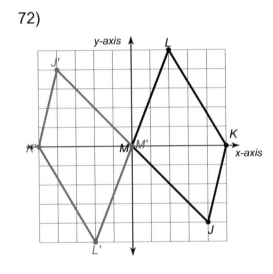

70)
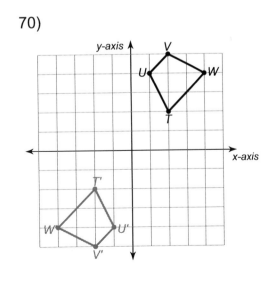

73)
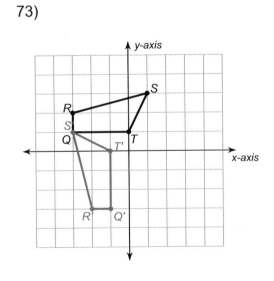

74)

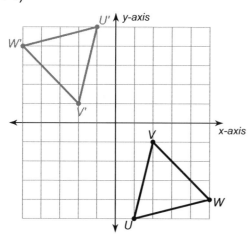

77)

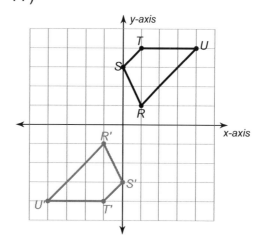

75)

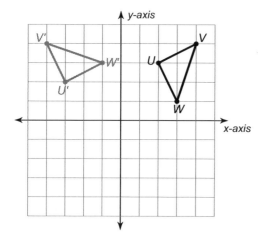

78)

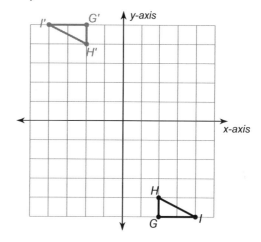

76)

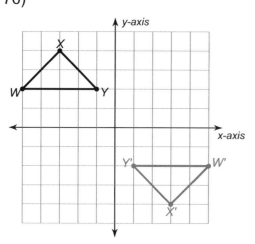

79)

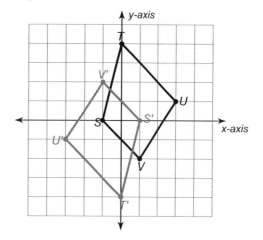

www.math-knots.com | www.a4ace.com

80)

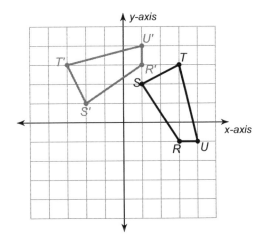

83)

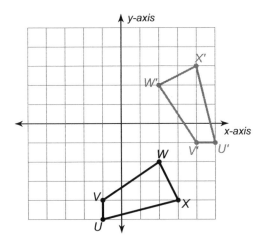

81)

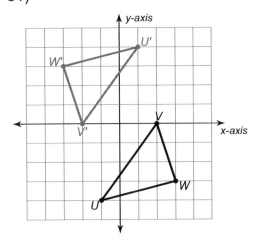

82)

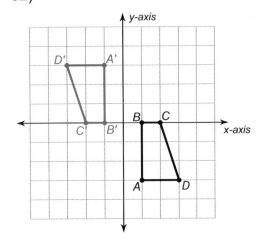

     www.math-knots.com | www.a4ace.com

Week 21

84. The number of pounds heavier a team member of 1 ft taller would weigh

85. The weight of a member with a height of 0 ft

86. 135.3 lbs

87. 163.2 lbs

88. 5.15 ft

89. The change in the wind speed for every 1 kpa of air pressure increase

90. The wind speed of a cyclone at 0 kpa of air pressure

91. 86 knots

92. 167 knots

93. 972 kpa

94. The change in wind speed when the air pressure increase by 1 kpa

95. The wind speed of a cyclone with 0 kpa air pressure

96. 99 knots

97. 168 knots

98. 979 kpa

99. For every additional Fahrenheit the consumption of electricity

Week 21

100. The average electricity consumption when the temperature equals to 0^0 F

101. 36 kWh

102. 54 kWh

103. 76^0 F

104. The additional electricity consumption for a change of temperature in Fahrenheit

105. The average electricity consumption at 0^0 F daily temperature

106. 37 kWh

107. 49 kWh

108. 69^0 F

109. The change in the wind speed with the increase in the air pressure of 1 kPa

110. The wind speed of a cyclone when the air pressure is 0 kPa

111. 91 knots

112. 141 knots

113. 960 kPa

114. The additional cost of the flight for every additional mile

115. The cost of a flight that travels 0 miles

116. $170

117. $372

Week 21

118.	1013 miles
119.	The number of pounds heavier than a flying club member of 1 ft taller would weigh
120.	The wind speed of a cyclone with 0 kPa air pressure
121.	160.5 lbs
122.	213.5 lbs
123.	5.15 ft
124.	C
125.	A
126.	B
127.	B
128.	A
129.	B
130.	B
131.	B
132.	B
133.	C
134.	C
135.	D
136.	C
137.	D
138.	B
139.	C
140.	C

Week 21

141.	B
142.	D
143.	A
144.	C
145.	D
146.	D
147.	D
148.	C
149.	C
150.	D
151.	C
152.	C
153.	C
154.	D
155.	D
156.	C
157.	A
158.	B
159.	B
160.	C
161.	B
162.	C
163.	A

www.math-knots.com | www.a4ace.com

Week 22

1.	B
2.	A
3.	B
4.	B
5.	The change in the wind speed for every 1 kPa increase in air pressure
6.	The cyclone wind speed at an air pressure of 0 kPa
7.	74 knots
8.	148 knots
9.	935 knots
10.	A
11.	D
12.	C
13.	D
14.	A
15.	B
16.	A
17.	D
18.	D
19.	D
20.	D
21.	A
22.	C
23.	A
24.	C

Week 22

25.	C
26.	D
27.	A
28.	C
29.	B
30.	A
31.	B
32.	B
33.	C
34.	B
35.	A
36.	D
37.	B
38.	A
39.	C
40.	A
41.	D
42.	A
43.	B
44.	A
45.	B
46.	D
47.	A
48.	D
49.	D
50.	B
51.	C
52.	C
53.	D

Week 22

54.	A
55.	A
56.	D
57.	A
58.	D
59.	D
60.	D
61.	C
62.	B
63.	B
64.	A

65.

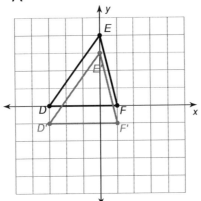

66.

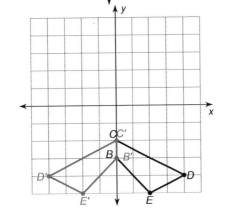

67.	A
68.	B
69.	A

Week 22

70.

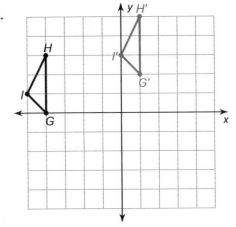

71. $-\dfrac{35}{8}$

72. y-intercept : -1

x-intercept : 0.2 , -4.2

73. C

74. -4

75. C

76.

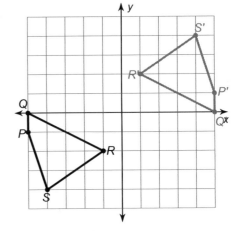

77. -5

78. B

Week 22

79.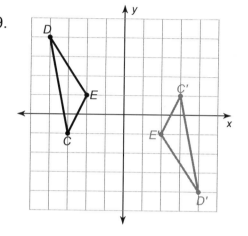

80. B

81. A

82. y-intercept : -3

   x-intercept : 3

83.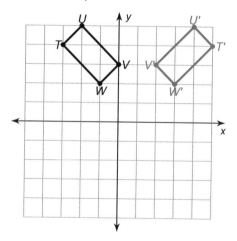

84. C

85. C

86. D

87. $-\dfrac{5}{6}$

Week 22

88. C

89. A

90. D

91. y-intercept : 14

   x-intercept : -2.6 , -5.5

92. D

93. A

94. C

95. C

96. C

97. y-intercept : -2

   x-intercept : 1.2

www.math-knots.com | www.a4ace.com

Week 23		Week 23		Week 23		Week 23	
1.	A	30.	A	59.	C	88.	13 sheep
2.	B	31.	D	60.	D	89.	9 chickens
3.	A	32.	B	61.	A	90.	19 horses
4.	D	33.	D	62.	A	91.	3 geese
5.	A	34.	A	63.	A	92.	11 goats
6.	B	35.	A	64.	C	93.	5 ducks
7.	B	36.	A	65.	D	94.	9 sheep
8.	C	37.	C	66.	D	95.	15 buffalo
9.	A	38.	D	67.	B	96.	10 chickens
10.	D	39.	C	68.	C	97.	10 ducks
11.	C	40.	B	69.	B	98.	10 cows
12.	C	41.	D	70.	D	99.	7 chickens
13.	C	42.	D	71.	8 chickens	100.	10 goats
14.	D	43.	A	72.	10 goats	101.	6 chickens
15.	C	44.	A	73.	9 chickens	102.	8 buffalo
16.	D	45.	B	74.	11 oxen	103.	7 ducks
17.	D	46.	B	75.	9 horses	104.	9 sheep
18.	B	47.	D	76.	9 geese	105.	6 ducks
19.	D	48.	D	77.	9 chickens	106.	4 horses
20.	B	49.	B	78.	3 sheeps	107.	8 ducks
21.	D	50.	B	79.	7 ducks	108.	15 goats
22.	D	51.	D	80.	18 sheeps	109.	3 ducks
23.	C	52.	C	81.	3 ducks	110.	12 horses
24.	D	53.	D	82.	5 buffalos	111.	6 oranges
25.	D	54.	C	83.	2 chickens	112.	2 tangerines
26.	D	55.	D	84.	3 buffalo	113.	3 dress pants
27.	B	56.	D	85.	9 geese	114.	2 jeans
28.	D	57.	C	86.	18 goats	115.	6 dress pants
29.	A	58.	A	87.	8 geese	116.	4 jeans

Week 23

117.	4 graph paper
118.	4 lined paper
119.	4 fancy shirts
120.	5 plain shirts
121.	6 corn chips
122.	4 potato chips
123.	2 spoons
124.	5 forks
125.	4 apples
126.	7 pears
127.	4 spoons
128.	2 forks
129.	4 spoons
130.	5 forks
131.	6 pens
132.	4 pencils
133.	5 pens
134.	5 pencils
135.	2 apples
136.	7 pears
137.	8 correction ink
138.	2 correction tape
139.	4 graph paper
140.	4 lined paper
141.	2 pens
142.	2 pencils
143.	6 fancy shirts
144.	2 plain shirts
145.	6 corn chips

Week 23

146.	10 potato chips
147.	4 tee shirts
148.	5 long sleeve shirts
149.	9 corn chips
150.	7 potato chips
151.	16 forks
152.	9 spoons
153.	14 forks
154.	13 spoons
155.	8 forks
156.	8 spoons
157.	10 plates
158.	13 cups
159.	10 forks
160.	16 spoons
161.	7 cups
162.	11 spoons
163.	16 plates
164.	8 spoons
165.	11 forks
166.	6 spoons
167.	18 cups
168.	12 bowls
169.	5 bowls
170.	6 spoons
171.	3 of Model A
172.	10 of Model B
173.	7 of Model A
174.	4 of Model B

www.math-knots.com | www.a4ace.com

Week 23		Week 23		Week 24		Week 24	
175.	7 of Model A	204.	5 buses	1.	D	30.	D
176.	2 of Model B	205.	8 cars	2.	A	31.	A
177.	4 of Model A	206.	5 buses	3.	B	32.	A
178.	9 of Model B	207.	8 cars	4.	C	33.	D
179.	10 of Model A	208.	3 vans	5.	B	34.	C
180.	7 of Model B	209.	6 cars	6.	D	35.	C
181.	6 of Model A	210.	3 buses	7.	A	36.	C
182.	8 of Model B	211.	5 cars	8.	A	37.	C
183.	6 of Model A	212.	7 buses	9.	D	38.	D
184.	2 of Model B	213.	4 cars	10.	B	39.	C
185.	5 of Model A	214.	7 buses	11.	D	40.	C
186.	6 of Model B			12.	D	41.	B
187.	10 of Model A			13.	A	42.	A
188.	6 of Model B			14.	B	43.	A
189.	2 of Model A			15.	D	44.	D
190.	5 of Model B			16.	A	45.	A
191.	4 vans			17.	C	46.	D
192.	8 buses			18.	D	47.	B
193.	6 cars			19.	C	48.	D
194.	8 vans			20.	D	49.	B
195.	2 cars			21.	D	50.	A
196.	4 vans			22.	B	51.	B
197.	8 cars			23.	C	52.	B
198.	8 buses			24.	C	53.	C
199.	5 cars			25.	C	54.	D
200.	5 vans			26.	C	55.	B
201.	5 cars			27.	B	56.	C
202.	5 vans			28.	C	57.	B
203.	5 vans			29.	A	58.	B

Week 24	
59.	D
60.	B
61.	C
62.	B
63.	D
64.	D
65.	C
66.	D
67.	A
68.	D
69.	D
70.	C
71.	D
72.	A
73.	A
74.	C
75.	D
76.	A
77.	B
78.	B
79.	A
80.	D
81.	D
82.	C
83.	B
84.	B
85.	A
86.	B
87.	B

Week 24	
88.	D
89.	D
90.	D
91.	A
92.	A
93.	D
94.	D
95.	Senior ticket = $15
96.	Child ticket = $6
97.	Senior ticket = $6
98.	Child ticket = $14
99.	Senior ticket = $10
100.	Child ticket = $5
101.	Adult ticket = $5
102.	Child ticket = $11
103.	Senior ticket = $13
104.	Child ticket = $9
105.	40
106.	52
107.	29
108.	15
109.	83
110.	78
111.	80
112.	67
113.	31
114.	97
115.	12

Week 24	
116.	Tulips = $13
117.	Daffodil = $15
118.	Senior ticket = $4
119.	Child ticket = $11
120.	Plain paper = $10
121.	Shiny paper = $16
122.	Small box = $5
123.	large box = $12
124.	A
125.	C
126.	D
127.	C
128.	D
129.	D
130.	B
131.	A
132.	C
133.	A
134.	B
135.	D
136.	C
137.	A
138.	A
139.	D
140.	A
141.	D
142.	A
143.	B
144.	D

 **Algebra 1**

Week 24		Week 25		Week 25		Week 25		Week 25	
145.	D	1.	C	30.	A	59.	A	88.	C
146.	C	2.	C	31.	A	60.	B	89.	B
147.	C	3.	C	32.	A	61.	D	90.	B
148.	A	4.	B	33.	B	62.	C	91.	A
149.	D	5.	A	34.	D	63.	B	92.	D
150.	B	6.	C	35.	D	64.	C	93.	A
151.	D	7.	D	36.	C	65.	C	94.	B
152.	A	8.	B	37.	B	66.	A	95.	A
153.	B	9.	D	38.	C	67.	A	96.	D
154.	A	10.	B	39.	A	68.	A	97.	C
155.	C	11.	C	40.	C	69.	A	98.	D
156.	D	12.	C	41.	C	70.	B	99.	A
157.	D	13.	B	42.	A	71.	B	100.	B
158.	B	14.	C	43.	D	72.	A	101.	C
159.	C	15.	B	44.	D	73.	C	102.	C
160.	A	16.	C	45.	D	74.	C	103.	A
161.	B	17.	A	46.	B	75.	C	104.	C
162.	A	18.	A	47.	D	76.	D	105.	B
163.	C	19.	A	48.	A	77.	A	106.	C
164.	A	20.	A	49.	D	78.	C	107.	D
165.	D	21.	A	50.	A	79.	A	108.	C
166.	D	22.	B	51.	D	80.	A	109.	B
167.	D	23.	C	52.	C	81.	B	110.	A
		24.	C	53.	D	82.	B	111.	C
		25.	C	54.	D	83.	B	112.	A
		26.	B	55.	A	84.	B	113.	C
		27.	D	56.	A	85.	D	114.	A
		28.	D	57.	D	86.	B	115.	A
		29.	C	58.	A	87.	A	116.	C

 www.math-knots.com | www.a4ace.com

Algebra 1

Week 25		Week 25	
117.	D	133.	$\dfrac{36x^3}{y^6z}$
118.	D		
119.	$30x^7y^8z^8$	134.	$49x^{14}y^{13}z^5$
120.	$\dfrac{250r}{q^2p^6}$	135.	$36x^{12}y^{16}z^{10}$
121.	$35c^5a^4b^2$	136.	$180c^{16}a^5b^9$
122.	$\dfrac{320z^5x^5}{y^5}$	137.	$60y^{11}x^6z^5$
123.	$\dfrac{9m^{10}q^{12}}{p^7}$	138.	$\dfrac{21m^3}{q^{11}p^2}$
124.	$60m^{11}p^{14}q^8$	139.	$30zx^6y^{10}$
125.	$\dfrac{12p^7}{m^3n}$	140.	$20p^{13}m^2q^4$
126.	$\dfrac{90}{n^9p^{10}}$		
127.	$\dfrac{7}{b^{10}c^9}$		
128.	$10x^{10}y^5z^2$		
129.	$\dfrac{49mp^{16}}{n^8}$		
130.	$\dfrac{6z^2}{x^6y^8}$		
131.	$\dfrac{80x^{10}y^{16}}{z^9}$		
132.	$\dfrac{24k^{19}}{j^2}$		

Week 26		Week 26	
1.	A	30.	A
2.	B	31.	D
3.	D	32.	D
4.	B	33.	D
5.	D	34.	C
6.	D	35.	B
7.	B	36.	C
8.	D	37.	B
9.	A	38.	B
10.	C	39.	D
11.	C	40.	B
12.	D	41.	A
13.	B	42.	A
14.	C	43.	D
15.	A	44.	A
16.	D	45.	A
17.	A	46.	A
18.	D	47.	D
19.	B	48.	B
20.	B	49.	B
21.	A	50.	A
22.	C	51.	D
23.	B	52.	B
24.	D	53.	B
25.	D	54.	B
26.	B	55.	D
27.	B	56.	B
28.	D	57.	D
29.	A	58.	B

Week 26		Week 26		Week 26		Week 26		Week 26	
59.	B	88.	A	117.	B	146.	B	175.	A
60.	C	89.	A	118.	B	147.	D	176.	D
61.	B	90.	B	119.	A	148.	C	177.	B
62.	A	91.	A	120.	D	149.	C	178.	B
63.	D	92.	D	121.	D	150.	C	179.	A
64.	A	93.	D	122.	C	151.	C	180.	B
65.	A	94.	A	123.	A	152.	C	181.	C
66.	B	95.	D	124.	D	153.	C	182.	A
67.	D	96.	B	125.	D	154.	A	183.	C
68.	D	97.	B	126.	D	155.	A	184.	D
69.	B	98.	B	127.	B	156.	A	185.	B
70.	C	99.	A	128.	D	157.	B	186.	B
71.	B	100.	A	129.	B	158.	B	187.	A
72.	C	101.	D	130.	B	159.	B	188.	B
73.	B	102.	B	131.	D	160.	D	189.	D
74.	A	103.	C	132.	B	161.	B	190.	D
75.	A	104.	D	133.	A	162.	A	191.	D
76.	A	105.	C	134.	D	163.	A	192.	B
77.	B	106.	D	135.	C	164.	D	193.	A
78.	D	107.	A	136.	D	165.	D	194.	D
79.	C	108.	B	137.	B	166.	B	195.	B
80.	C	109.	B	138.	B	167.	A	196.	A
81.	C	110.	A	139.	B	168.	D	197.	D
82.	D	111.	B	140.	A	169.	C	198.	C
83.	D	112.	B	141.	D	170.	D	199.	A
84.	C	113.	C	142.	B	171.	B	200.	D
85.	B	114.	D	143.	B	172.	B	201.	C
86.	D	115.	B	144.	D	173.	B	202.	B
87.	B	116.	D	145.	B	174.	A	203.	A

Week 26		Week 27		Week 27		Week 27		Week 27	
204.	B	1.	C	30.	B	59.	B	88.	A
205.	A	2.	B	31.	A	60.	B	89.	A
206.	C	3.	C	32.	A	61.	D	90.	C
		4.	C	33.	A	62.	A	91.	C
		5.	D	34.	A	63.	A	92.	C
		6.	D	35.	B	64.	C	93.	A
		7.	C	36.	A	65.	C	94.	C
		8.	C	37.	C	66.	C	95.	A
		9.	C	38.	A	67.	D	96.	C
		10.	C	39.	B	68.	B	97.	C
		11.	B	40.	B	69.	B	98.	C
		12.	C	41.	C	70.	B	99.	D
		13.	B	42.	B	71.	B	100.	B
		14.	D	43.	C	72.	A	101.	B
		15.	C	44.	A	73.	B	102.	B
		16.	D	45.	A	74.	B	103.	C
		17.	A	46.	A	75.	D	104.	A
		18.	C	47.	C	76.	B	105.	A
		19.	C	48.	D	77.	C	106.	C
		20.	C	49.	A	78.	B	107.	A
		21.	B	50.	A	79.	C	108.	A
		22.	B	51.	C	80.	D	109.	A
		23.	A	52.	D	81.	D	110.	A
		24.	D	53.	A	82.	B	111.	A
		25.	A	54.	D	83.	C	112.	A
		26.	A	55.	B	84.	C	113.	D
		27.	C	56.	A	85.	B	114.	D
		28.	A	57.	C	86.	C	115.	D
		29.	D	58.	C	87.	C	116.	C

Week 27

117.	A
118.	B
119.	B
120.	B
121.	A
122.	B
123.	B
124.	B
125.	C
126.	B
127.	C
128.	C
129.	B
130.	C
131.	A
132.	A
133.	D
134.	D
135.	A
136.	C
137.	A

Week 28

1.	B
2.	A
3.	D
4.	C
5.	B
6.	D
7.	C
8.	Adult ticket = $12
	Student ticket = $9
9.	A
10.	C
11.	D
12.	C
13.	Senior ticket = $5
	Student ticket = $6
14.	86
15.	D
16.	Adult ticket = $8
	Student ticket = $14
17.	C
18.	B
19.	Adult ticket = $11
	Student ticket = $3
20.	D
21.	C
22.	C
23.	C
24.	B
25.	A

Week 28

26.	Adult ticket = $14
	Student ticket = $9
27.	C
28.	C
29.	C
30.	C
31.	A
32.	B
33.	B
34.	D
35.	D
36.	Small box = $13
	Large box = $15
37.	A
38.	32
39.	C
40.	B
41.	B
42.	D
43.	D
44.	D
45.	B
46.	C
47.	A
48.	A
49.	D
50.	C
51.	A

Algebra 1

Week 28

52.	C
53.	A
54.	Van : 11,Bus : 29
55.	18
56.	A
57.	D
58.	B
59.	D
60.	D
61.	A
62.	D
63.	C
64.	A
65.	D
66.	C
67.	A
68.	43
69.	A
70.	D
71.	C
72.	C
73.	D
74.	C
75.	C
76.	Chocolate cake = $15
	French cake = $14
77.	B
78.	A
79.	A

Week 28

80.	C
81.	A
82.	A
83.	A
84.	C
85.	D
86.	A
87.	C
88.	A
89.	D
90.	B
91.	B
92.	B
93.	C
94.	D
95.	$336p^7r^{11}q$
96.	D
97.	D
98.	D
99.	D
100.	A
101.	C
102.	C
103.	C
104.	$4x^{14}y^8z^{17}$
105.	B
106.	B
107.	$25y^2z^3$
108.	C
109.	B

Week 29

1.	A
2.	C
3.	C
4.	C
5.	D
6.	C
7.	D
8.	D
9.	B
10.	D
11.	C
12.	C
13.	A
14.	C
15.	C
16.	C
17.	D
18.	D
19.	A
20.	B
21.	B
22.	B
23.	B
24.	B
25.	D
26.	C
27.	B
28.	A
29.	C

www.math-knots.com | www.a4ace.com

Week 29

30.	B
31.	C
32.	C
33.	A
34.	C
35.	B
36.	D
37.	A
38.	C
39.	B
40.	D
41.	A
42.	C
43.	C
44.	D
45.	C
46.	C
47.	C
48.	A
49.	C
50.	C
51.	C
52.	D
53.	B
54.	B
55.	B
56.	C
57.	C
58.	D

Week 29

59.	A
60.	A
61.	A
62.	A
63.	A
64.	B
65.	B
66.	A
67.	A
68.	D
69.	$-3r^4 - 4r^2 - 4r$
70.	$2r^3 + r^2 + 22$
71.	$3n^4 - 17n^3 + 2$
72.	$-5b^4 + 2b^2 + 8$
73.	$-9m^4 - 13m^2 + 23$
74.	$x^4 + 2x^3 + 3$
75.	$2x^4 - 14x^2 + 3$
76.	$12x^3 - 22$
77.	$-12a^3 + 23a + 10$
78.	$-6m^3 + 9m - 6$
79.	$-4x^4 - 16x^2 + 9$
80.	$4n^4 - 18n^3 + 7n + 15$
81.	$-2k^4 + 8k^2 + 7k + 5$
82.	$-x^4 + 2x^3 + 2x^2$
83.	$4x^4 + 11x^3 + 7x^2 + 40x + 31$
84.	$6x^4 - 3x^3 + 3x^2 - 7x - 24$
85.	$11m^4 + 24m^2 - 9m + 9$
86.	$20x^4 + 27x^3 - 13x^2 + 31x$
87.	$5x^4 - 7x^3 + 17x^2 - 25x - 13$

Week 29

88.	$4n^4 + 15n^2 + 21n + 3$
89.	$13r^4 + 17r^3 - 22r^2 - 7r + 13$
90.	$-11n^4 - 5n^3 + 3n^2 + 25n - 16$
91.	$-2b^4 - 10b^2 + 20b + 15$
92.	$-6x^4 + 15x^2 + 4x - 17$
93.	A
94.	B
95.	A
96.	D
97.	B
98.	C
99.	C
100.	D
101.	B
102.	D
103.	A
104.	C
105.	B
106.	B
107.	D
108.	A
109.	D
110.	C
111.	C
112.	A
113.	B
114.	C
115.	C
116.	A

Week 29

117.	B
118.	C
119.	B
120.	C
121.	A
122.	A
123.	D
124.	C
125.	$42x^3 - 66x^2 - 18x + 24$
126.	$8x^3 + 12x^2 + 14x + 5$
127.	$32m^3 + 28m^2 + 27m + 3$
128.	$15n^3 + 13n^2 + 11n + 6$
129.	$28b^3 - 44b^2 + 65b - 28$
130.	$18x^3 + 27x^2 + x - 6$
131.	$12a^3 - 18a^2 - 12a + 18$
132.	$49n^3 + 98n^2 + 76n + 32$
133.	$8n^3 - 6n^2 + 11n + 8$
134.	$30a^3 - 23a^2 - 34a + 5$
135.	$15k^3 - 2k^2 - 30k + 8$
136.	$25n^3 + 20n^2 - 52n - 48$
137.	$12a^3 + 50a^2 + 34a - 24$
138.	$7x^3 + 13x^2 - 30x - 18$
139.	$49m^3 - 21m^2 - 19m + 3$
140.	$63n^4 + 36n^3 + 24n^2 + 81n - 24$
141.	$70n^4 - 66n^3 + 127n^2 - 56n + 72$
142.	$42v^4 + 15v^3 - 76v^2 + 68v - 16$
143.	$36a^4 - 7a^3 + 44a^2 - 13a + 12$
144.	$16r^4 + 90r^3 + 113r^2 - 18r - 48$

Week 30	Week 30	Week 30	Week 30	Week 30
1. D	30. A	59. D	88. C	117. D
2. C	31. B	60. A	89. B	118. B
3. A	32. A	61. B	90. C	119. B
4. C	33. A	62. B	91. A	120. D
5. D	34. A	63. B	92. C	121. D
6. C	35. D	64. D	93. D	122. C
7. A	36. B	65. D	94. C	123. C
8. D	37. D	66. D	95. B	124. C
9. D	38. D	67. C	96. D	125. D
10. D	39. B	68. B	97. C	126. A
11. C	40. D	69. C	98. D	127. B
12. D	41. D	70. B	99. A	128. D
13. C	42. C	71. D	100. C	129. D
14. D	43. C	72. B	101. C	130. B
15. B	44. D	73. D	102. B	131. C
16. C	45. A	74. C	103. A	132. A
17. B	46. D	75. D	104. A	133. D
18. B	47. D	76. C	105. D	134. D
19. A	48. A	77. C	106. C	135. B
20. B	49. C	78. C	107. C	136. B
21. C	50. A	79. A	108. D	137. D
22. B	51. D	80. D	109. A	138. B
23. C	52. D	81. D	110. A	139. B
24. B	53. D	82. C	111. A	140. D
25. B	54. A	83. D	112. D	141. B
26. D	55. C	84. B	113. B	142. C
27. D	56. C	85. C	114. A	143. C
28. A	57. A	86. B	115. A	144. D
29. D	58. D	87. B	116. A	145. A

Week 30		Week 31		Week 31		Week 31	
146.	A	1.	B	30.	D	59.	B
147.	A	2.	A	31.	A	60.	B
148.	A	3.	A	32.	B	61.	D
149.	D	4.	D	33.	B	62.	D
		5.	A	34.	B	63.	A
		6.	C	35.	A	64.	D
		7.	C	36.	A	65.	D
		8.	A	37.	D	66.	B
		9.	B	38.	A	67.	B
		10.	B	39.	A	68.	C
		11.	A	40.	D	69.	D
		12.	C	41.	D	70.	B
		13.	A	42.	B	71.	B
		14.	D	43.	D	72.	D
		15.	C	44.	D	73.	D
		16.	D	45.	C	74.	B
		17.	C	46.	A	75.	D
		18.	A	47.	B	76.	B
		19.	A	48.	A	77.	A
		20.	B	49.	A	78.	D
		21.	C	50.	D	79.	A
		22.	A	51.	B	80.	C
		23.	C	52.	D	81.	A
		24.	B	53.	B	82.	D
		25.	A	54.	A	83.	A
		26.	C	55.	B	84.	D
		27.	A	56.	B	85.	D
		28.	D	57.	C	86.	A
		29.	D	58.	B	87.	B

 www.math-knots.com | www.a4ace.com

Week 31	
88.	A
89.	D
90.	A
91.	D
92.	D
93.	A
94.	B
95.	D
96.	C
97.	C
98.	C
99.	A
100.	C
101.	D
102.	A
103.	D
104.	C
105.	B
106.	D
107.	A
108.	D
109.	D
110.	D

111. $\dfrac{6}{5x-7}$; $\left\{\dfrac{7}{5}\right\}$

112. $\dfrac{1}{9}$; $\{-10\}$

113. $2n$; $\{-8\}$

Week 31

114. 6 ; $\{-2\}$

115. $\dfrac{1}{7x}$; $\{0, -4\}$

116. $\dfrac{5r}{3r+8}$; $\left\{-\dfrac{8}{3}\right\}$

117. $\dfrac{7k-2}{9k^2}$; $\{0\}$

118. $\dfrac{6n^2}{2n+7}$; $\left\{-\dfrac{7}{2}\right\}$

119. $\dfrac{1}{m+1}$; $\{6, -1\}$

120. $x-9$; $\{-3\}$

121. $\dfrac{3(x-3)}{7}$; $\{0\}$

122. $\dfrac{1}{r+9}$; $\{-9, -1\}$

123. $\dfrac{1}{m+10}$; $\{-10, -3\}$

124. $-\dfrac{1}{2}$; $\{1\}$

125. $\dfrac{4}{3(n+2)}$; $\{-2\}$

126. $\dfrac{8k}{3k+5}$; $\left\{-\dfrac{5}{3}\right\}$

127. $a-1$; $\{-2\}$

Week 31

128. $-\dfrac{1}{2x}$; $\{0, 3\}$

129. $\dfrac{1}{n+9}$; $\{9, -9\}$

130. $\dfrac{1}{7x}$; $\{0, 2\}$

131. $\dfrac{3x-10}{10}$; $\{0\}$

132. $4n$; $\{2\}$

133. $\dfrac{3x+4}{10}$; No excluded values.

134. $\dfrac{v+11}{v+9}$; $\{-13, -9\}$

135. $\dfrac{11n-10}{11n-1}$; $\left\{\dfrac{1}{11}\right\}$

136. $\dfrac{n-10}{n+1}$; $\{-12, -1\}$

137. $\dfrac{n-6}{n+4}$; $\{4, -4\}$

138. $\dfrac{11n+13}{3(n-3)}$; $\{3\}$

139. $\dfrac{11}{p-12}$; $\{12, -2\}$

140. $\dfrac{5v+12}{3(v-3)}$; $\{0, 3\}$

141. $\dfrac{r-10}{r-7}$; $\{11, 7\}$

Week 31

142. $-\dfrac{5}{2}$; $\{1\}$

143. $\dfrac{3x-8}{7x+3}$; $\left\{0, -\dfrac{3}{7}\right\}$

144. $\dfrac{x-4}{7x^2}$; $\{0, -4\}$

145. $\dfrac{5(a+2)}{7(a-2)}$; $\{2\}$

146. $\dfrac{x+9}{x-6}$; $\{14, 6\}$

147. $\dfrac{10}{p+11}$; $\{5, -11\}$

148. $\dfrac{5x-1}{2(x+6)}$; $\{0, -6\}$

149. $\dfrac{x+13}{x+10}$; $\{-10, -7\}$

150. $\dfrac{n+14}{11}$; $\{8\}$

151. $\dfrac{11}{b-10}$; $\{10, -10\}$

152. $\dfrac{x-14}{x+10}$; $\{-10, -2\}$

153. $\dfrac{x-3}{14}$; $\{7\}$

www.math-knots.com | www.a4ace.com

Week 32

1.	B
2.	B
3.	D
4.	B
5.	A
6.	A
7.	C
8.	C
9.	A
10.	C
11.	B
12.	C
13.	C
14.	A
15.	A
16.	D
17.	B
18.	C
19.	B
20.	A
21.	C
22.	D
23.	D
24.	C
25.	C
26.	C
27.	D
28.	B
29.	B

Week 32

30.	B
31.	A
32.	D
33.	C
34.	C
35.	D
36.	A
37.	A
38.	D
39.	D
40.	B
41.	B
42.	A
43.	B
44.	A
45.	D
46.	A
47.	C
48.	D
49.	D
50.	D
51.	C
52.	D
53.	A
54.	A
55.	A
56.	C
57.	D
58.	D

Week 32

59.	C
60.	B
61.	B
62.	A
63.	B
64.	C
65.	B
66.	B
67.	A
68.	C
69.	A
70.	A
71.	D
72.	B
73.	B
74.	B
75.	C
76.	D
77.	A
78.	B
79.	A
80.	B
81.	B
82.	C

Week 33

1.	−15 ; two imaginary solutions
2.	−48 ; two imaginary solutions
3.	0 ; one real solution
4.	0 ; one real solution
5.	16 ; two real solutions
6.	81 ; two real solutions
7.	−60 ; two imaginary solutions
8.	−20 ; two imaginary solutions
9.	0 ; one real solution
10.	−24 ; two imaginary solutions
11.	81 ; two real solutions
12.	289 ; two real solutions
13.	−79 ; two imaginary solutions
14.	36 ; two real solutions
15.	0 ; one real solution
16.	−144 ; two imaginary solutions
17.	−95 ; two imaginary solutions
18.	9 ; two real solutions
19.	−119 ; two imaginary solutions
20.	0 ; one real solution
21.	−44 ; two imaginary solutions
22.	−23 ; two imaginary solutions
23.	−143 ; two imaginary solutions
24.	−64 ; two imaginary solutions
25.	−44 ; two imaginary solutions
26.	−815 ; two imaginary solutions
27.	−39 ; two imaginary solutions
28.	−95 ; two imaginary solutions
29.	0 ; one real solution

Week 33

30.	−56 ; two imaginary solutions
31.	−164 ; two imaginary solutions
32.	−119 ; two imaginary solutions
33.	−420 ; two imaginary solutions
34.	−359 ; two imaginary solutions
35.	121 ; two real solutions
36.	49 ; two real solutions
37.	0 ; one real solution
38.	4 ; two real solutions
39.	−76 ; two imaginary solutions
40.	529 ; two real solutions
41.	−248 ; two imaginary solutions
42.	169 ; two real solutions
43.	−463 ; two imaginary solutions
44.	0 ; one real solution
45.	0 ; one real solution
46.	0 ; one real solution
47.	169
48.	289
49.	256
50.	25
51.	256
52.	196
53.	1
54.	196
55.	225
56.	64
57.	49
58.	400

www.math-knots.com | www.a4ace.com

Week 33

59.	144
60.	100
61.	324
62.	81
63.	4
64.	4
65.	144
66.	441
67.	324
68.	$\dfrac{121}{4}$
69.	$\dfrac{4}{9}$
70.	$\dfrac{289}{4}$
71.	$\dfrac{9}{4}$
72.	$\dfrac{25}{4}$
73.	$\dfrac{1}{4}$
74.	$\dfrac{169}{256}$
75.	$\dfrac{16}{25}$
76.	$\dfrac{4}{9}$
77.	$\dfrac{49}{4}$
78.	$\dfrac{225}{4}$
79.	$\dfrac{81}{4}$
80.	$\dfrac{1849}{676}$

Week 33

81.	$\dfrac{9}{4}$
82.	$\dfrac{225}{4}$
83.	$\dfrac{81}{4}$
84.	$\dfrac{361}{4}$
85.	$\dfrac{9}{64}$
86.	$\dfrac{169}{4}$
87.	$\dfrac{49}{4}$
88.	$\dfrac{361}{4}$
89.	$\{3, -5\}$
90.	$\{16, -2\}$
91.	$\{5 + 3\sqrt{13}, 5 - 3\sqrt{13}\}$
92.	$\{15, 3\}$
93.	$\{5, -15\}$
94.	$\{4 + \sqrt{29}, 4 - \sqrt{29}\}$
95.	$\{6, -16\}$
96.	$\{15, -1\}$
97.	$\{12, -6\}$
98.	$\{-7 + \sqrt{146}, -7 - \sqrt{146}\}$
99.	$\{14, -4\}$
100.	$\{12, -8\}$
101.	$\{4, -20\}$
102.	$\{3 + \sqrt{10}, 3 - \sqrt{10}\}$
103.	$\{11, 7\}$
104.	$\{4 + \sqrt{57}, 4 - \sqrt{57}\}$

Week 33

105.	$\{12, -4\}$
106.	$\{6, -2\}$
107.	$\{3, -19\}$

www.math-knots.com | www.a4ace.com

Algebra 1

Week 34		Week 34		Week 34		Week 34	
1.	A	30.	B	59.	A	88.	C
2.	A	31.	A	60.	D	89.	C
3.	A	32.	C	61.	D	90.	A
4.	B	33.	A	62.	B	91.	A
5.	A	34.	C	63.	A	92.	D
6.	A	35.	C	64.	D	93.	D
7.	B	36.	D	65.	B	94.	B
8.	A	37.	C	66.	C	95.	B
9.	A	38.	A	67.	A	96.	A
10.	B	39.	B	68.	C	97.	A
11.	B	40.	D	69.	D	98.	B
12.	B	41.	D	70.	C	99.	A
13.	A	42.	D	71.	C	100.	D
14.	A	43.	B	72.	B	101.	A
15.	A	44.	B	73.	A	102.	B
16.	B	45.	C	74.	B	103.	C
17.	A	46.	A	75.	A	104.	B
18.	A	47.	B	76.	B	105.	B
19.	B	48.	B	77.	A	106.	A
20.	A	49.	B	78.	B	107.	C
21.	B	50.	A	79.	A		
22.	D	51.	B	80.	C		
23.	A	52.	B	81.	B		
24.	A	53.	B	82.	B		
25.	C	54.	D	83.	D		
26.	A	55.	C	84.	C		
27.	C	56.	D	85.	B		
28.	B	57.	A	86.	C		
29.	D	58.	D	87.	B		

www.math-knots.com | www.a4ace.com

Algebra 1

Week 35

1.   A

2.   B

3.   D

4. $-4m^3 +10m^2 -18m + 5$

5.   A

6.   C

7. $\dfrac{25}{4}$

8.   A

9.   C

10. B

11. A

12. $\{-8 + 2\sqrt{2}, -8 - 2\sqrt{2}\}$

13. D

14. A

15. $\dfrac{7}{2(n + 3)}$; $\{0, -3\}$

16. D

17. D

18. A

19. B

20. A

21. B

22. A

23. $- 64$; two imaginary solutions

24. C

25. C

26. D

Week 35

27. D

28. A

29. $\dfrac{k + 4}{k + 9}$; $\{10, -9\}$

30. $10m^3 - 11m^2 + 2m - 11$

31. $8 , -12$

32. B

33. D

34. B

35. $\dfrac{3x^2}{x - 4}$; $\{10 , 4\}$

36. B

37. A

38. $54x^4 - 39x^3 - 56x^2 + 11x + 10$

39. B

40. D

41. $12a^3 + 7a^2 + 7a + 6$

42. $\dfrac{2k - 1}{4}$; No excluded values.

43. C

44. B

45. B

46. D

47. A

48. B

49. D

50. B

51. $\dfrac{841}{1156}$

www.math-knots.com | www.a4ace.com

Week 35

52.	A
53.	1 , -7
54.	C
55.	C
56.	C
57.	$\dfrac{4096}{169}$
58.	$\left\{-5 + 2\sqrt{26},\ -5 - 2\sqrt{26}\right\}$
59.	B
60.	441
61.	C
62.	$50a^4 + 55a^3 + 19a^2 - 35a - 25$
63.	A
64.	9
65.	B
66.	$\dfrac{121}{4}$
67.	B
68.	D
69.	A
70.	$\dfrac{11x + 2}{13x + 1}$; $\left\{0,\ -\dfrac{1}{13}\right\}$
71.	$13x^3 + 4x + 3$
72.	A
73.	A
74.	B
75.	C
76.	D
77.	A

Week 35

78.	19
79.	D
80.	C
81.	n = 3
82.	A
83.	$\dfrac{13n + 9}{2n - 11}$; $\left\{\dfrac{11}{2}\right\}$
84.	B
85.	C
86.	$6b^4 + 22b^3 + 6b^2 + 38b - 72$
87.	C
88.	B
89.	C
90.	256 , two real solutions
91.	16
92.	A
93.	A
94.	$4x^4 - 3x^3 - 8x^2$
95.	64 , two real solutions
96.	$30b^4 + 59b^3 - 89b^2 + 114b - 54$
97.	$\dfrac{3}{p + 1}$; $\{-1\}$
98.	15 , 1
99.	A
100.	$2x^4 - 16x^3 - 39x^2 + 9x + 27$
101.	C
102.	0 , one real solution
103.	36

www.math-knots.com | www.a4ace.com

Made in the USA
Middletown, DE
22 October 2024

63103921R00252